Controversial Herbal Drugs of Ayurveda

Controversial Herbal Drugs of Ayurveda

Amritpal Singh Saroya
Herbal Consultant

SCIENTIFIC
PUBLISHERS (INDIA)

Published by:

Scientific Publishers (India)
5 A, New Pali Road, P.O. Box 91
Jodhpur 342 001 (India)

E-mail: info@scientificpub.com
Website: www.scientificpub.com

Branch Office
Scientific Publishers (India)
4806/24, Ansari Road, Daryaganj
New Delhi - 110 002 (India)

Print: 2021

ISBN: 978-81-7233-821-3

Typeset by Rajesh Ojha
Printed in India

PREFACE

We feel glad for presenting the work 'Controversial Herbal Drugs of Ayurveda' before you. Many of plant drugs documented in Ayurvedic textbooks have a controversy on their accurate botanical linkages. Proper identification of medicinal plants has significant impact on finished product and therapeutics. In Ayurveda, identification of medicinal plants is major hurdle in laying pharmacopeial standards for formulations.

Since plants and plant drugs in Ayurveda were designated Sanskrit names, often based on the "doctrine of signature", morphological appearance, properties and action, the interpretation of these names during the later period of time led to acceptance of more than one botanical species for one plant drug. Due to lack of correct identification, similar looking plants are collected from the field site along with the genuine medicinal plant by mistake. But many times similar looking (inferior) cheap alternatives are intentionally mixed along with some quantity of genuine plant.

The adulterators not only mix similar species but also cheap and inferior quality produces such as rotten or substandard products which can be procured at very cheap rates. *Astavarga* (group of eight medicinal plants) as an instance is integral part of Ayurvedic formulations like Chyvanprasha and four plants viz, *Riddhi, Vriddhi, Jivaka and Rishbhaka* have been discussed as possible members of family Orchidaceae. In Ayurveda *Pasanbheda* is another drug of controversial origin. Further work on proper botanical identification of *Pasanbheda* is warranted. In Indian market several commercial lithontriptic preparations are sold for dissolving kidney stones. Some formulations are utilizing *Bergenia ligulata* and other one *Bryophyllum calycinum.*

The drugs in the book have been arranged in alphabetic form and details have been mentioned according to Ayurvedic and Herbal Materia Medica. Information on chemical composition, wherever possible, has been included. The work cited in the text is thoroughly referenced throughout the book. Hope the book will be useful for the Ayurvedic drug industry and practitioners. Critical comments from the readers are welcome.

Dr. Amritpal Singh Saroya

CONTENT

CONTROVERSIAL HERBAL DRUGS OF AYURVEDA

1. ADHOGUDA

Current representative: *Euphrobia nivulia*

Possible representatives: *Euphorbia acaulis* and *Ipomoea petaloides.*

Euphorbia nivulia Buch.-Ham.

Syn: *E. neriifolia* Auct. non Linn, *E. helicothele* (Lemaire) Boiss.

Family: Euphorbiaceae.

Habitat: India.

General Description: An open-crowned succulent-branched spiny, glabrous, deciduous tree up to c. 9 m tall and 1 m in girth. Bark thick, rough. Branches ± terete, 2.5 cm thick, with paws of sharp stipular spines arising from spirally-arranged tubercles. Leaves alternate, sessile or subsessile. Leaf-blades oboyate-spathulate to oblanceolate, 10-25 x 3-8 cm, rounded or retuse at the apex, tapered to the base, entire, lateral nerves 6-8 pairs, obscure, thick, fleshy, with the midrib prominent beneath. Stipules spiny, 5 mm long, paired on each spine-shield. Cyathia in axillary pedunculate 'cymes' of 3 each, towards the tips of the branches, reddish; peduncles c. 1 cm long; cyathophylls 1-2 mm long, scarious. Glands transversely oblong, yellowish. Fruits triradiate, the lobes laterally compressed, c. 6 x 13 mm, smooth, glabrous, exsert on a pedicel, c. 1 cm long. Styles stout, united to above the middle, c. 2 mm long. Seeds ovoid-quadrangular, 4 mm long, smooth (Joseph & Rangari, 2002).

Chemical constituents: Diterpenoids and triterpenoids from latex (Ravikanth *et al.,* 2002; Ravikanth *et al.,* 2003). A cycloartane epoxy triterpenoid, cyclonivuliaol has been reported (Satyanaryana *et al.,* 1991).

Bioactivity: Cytotoxic (Ravikanth *et al.,* 2003) and antimicrobial (Annapurna *et al.,* 2004).

Further reading

Annapurna J, Chowdhary IP, Lalitha G, Ramakrishna SV, Iyengar DS. Antimicrobial Activity of *Euphorbia nivulia* Leaf Extract. *Pharm Biol* 2004; 42: 91-93.

Joseph GVR, Rangari VD. Standardisation and quality evaluation of *Euphorbia nivulia* Buch.-Ham. *Ancient Sci Life* 2002; 21:160-166.

Ravikanth V, Lakshmi Niranjan Reddy V, Vijender Reddy A, Ravinder K, Prabhakar Rao T, Siva Ram T, Anand Kumar K, Prakesh Vamanarao D, Venkateswarlu Y. Three new ingol diterpenes from *Euphorbia nivulia*: evaluation of cytotoxic activity. *Chem Pharm Bull* (Tokyo) 2003; 51: 431-434.

Ravikanth V, Niranjan Reddy VL, Prabhakar Rao T, Diwan PV, Ramakrishna S, Venkateswarlu Y. *Phytochemistry* 2002; 59:331-335.

Satyanaryana V, Krupadanam GLD, Srimannarayana G. Cyclonivuliaol, a New Cycloartane Epoxy Triterpenoid from *Euphorbia nivulia* Buch-Ham. *ChemInform* 1991; 22:34.

Euphorbia acaulis Roxb.

Syn: *Euphorbia fusiformis* Buch.-Ham. ex D.Don

Family: Euphorbiaceae

Habitat: Tropical Himalayas.

General Description:

Chemical constituents: A diterpene lactone, caudicifolin has been isolated from the cold petrol extract of rhizomes (Satti *et al.,* 1986). A diterpenoid, 3-Oxoatisane-16α, 17-diol has been reported from hexane extract (Satti *et al.,* 1987).

Actions: Alterative and irritant (Chopra & Badhwar, 1940, Behl *et al.,* 1966).

Therapeutics: In Ayurvedic system of medicine, roots of *E. acaulis* are used in the treatment of chronic eczema and worm infestation. The tubers are traditionally used by the Malayali tribes of Chitteri hills, Eastern Ghats, Tamil Nadu to treat liver disorders.

Parts used: Whole plant.

Bioactivity: Antibacterial (Natarajan *et al.,* 2005), antinociceptive (Ashok *et al.,* 2011), antifungal (Natarajan *et al.,* 2007), anti-inflammatory and diuretic (Singh *et al.,* 2004) and hepatoprotective (Anusuya, Raju & Manian, 2010).

Clinical studies: 18 out of 23 patients reported beneficial effect in eczema with 50 mg of powdered drug (Agarwal, Chandra & Raju, 1971).

Further reading

Agarwal DK, Chandra RJ, Raju, TV. Clinical studies of *Euphorbia acaulis* Roxb. in cases of eczema-a preliminary report. *Indian J Dermatol* 1971; 16:57-59.

Anusuya N, Raju K, Manian S. Hepatoprotective and toxicological assessment of an ethnomedicinal plant *Euphorbia fusiformis* Buch.-Ham.ex D.Don. *J Ethnopharmacol* 2010; 127:463-467.

Ashok BK, Bhat SD, Rao SR, Ravishankar B. Screnning of antinociceptive activity of *Euphorbia fusiformis*. *Int J Res Ayur Pharm*.2011; 2:27-29.

Natarajan D, John Britto S, Srinivasan K, Nagamurugan N, Mohanasundari C, Perumal G. Anti-bacterial activity of *Euphorbia fusiformis* — A rare medicinal herb. *J Ethnopharmacol* 2005; 102:123-126.

Natarajan D, Nagamurugan N, Ramchandran A, Mohanasundari C, Srinivasan K. Anticandidial and anticryptococcal activity of *Euphorbia fusiformis*, a rare medicinal plant. *World J Microbiol Biotech* 2007; 23:719-721.

Satti NK, Suri OP, Dhar KL, Atal CK, Kawasaki T, Miyahara K, Kawano S. High resolution *nmr* and X-ray crystallography data of caudicifolin from *Euphorbia acaulis*. *Phytochemistry* 1986; 25:1411-1413.

Satti NK, Suri OP, Dhar KL, Atal CK, Kawasaki T, Miyahara K, Kawano S. *J Nat Prod.* 1987; 50: 790-793.

Singh GB, Kaur S, Satti NK, Atal CK, Mageswaru JK. Anti-inflamatory activity of *Euphorbia acaulis* Roxb. *J Ethnopharmacol* 1984; 10:225-233.

Ipomoea petaloides Choisy.

Syn: *Merremia peltata* (L.) Merr., *Operculina petaloides* Choisy.

Ayurvedic names: *Shyama, Vriddhdaaruka.*

Family: Convolvulaceae.

Habitat: Throughout India; ascending to 300m.

Substitute: *Ipomoea turpethum* L.

Note: *Argreyia speciosa* Sweet is equated with *Vriddhdaruka* (Prasad & Chauhan, 1974; 1975), while *I.petaloidea* and *I. pes-carpae* are also known by identical synonyms (Prasad & Chauhan, 1975; Nair *et al.*, 1986).

Futher reading

Prasad S, Chauhan RBPS. Pharmacognostical studies on `vidhara', part I - *Ipomoea petaloidea* Chois. *J Res Indian Med* 1974; 9:50-56.

Prasad S, Chauhan RBPS. Pharmacognostical studies on `vidhara', part II - *Ipomoea petaloidea* Chois. *J Res Indian Med* 1975; 10:60-70.

Prasad S, Chauhan RBPS. Pharmacognostical studies on `vidhara', part III - *Argyreia speciosa* Sw. *J Res Indian Med* 1975; 10:23-31.

Singh RS, Gode KD. *Ipomoea petaloidea* Chois - its medicinal and economic importance. *J Res Indian Med* 1966; 1:144-148.

Vasudevan Nair K, Gopakumar K, Yoganarasimhan SN, Shantha TR, Keshavamurthy KR. Medico-botany of Andaman and Nicobar Islands -- IV (Ayurvedic drugs- 2) *Ancient Sci Life* 1986; 5:191-196.

Note: Vajravalli has been used for *E. nivulia*. *Guda* is one of the synonyms for *snuhi* (*Euphorbia nerifolia* L. N.O.: Euphorbiaceae). *Adhoguda* seems to be closely allied to *Guda*. *Charaka* has described *Adhoguda* under *mulini* (roots or rhizome of some plants are used in medicine) drugs. Some researchers identify *Saptala* as *E. nivulia*.

2. AGNIGURBHA

Current representative: *Solanum trilobatum*

Possible representatives: *Solanum xanthocarpum* and *Artemisia vulgaris*

Solanum trilobatum Linn.

Common name: Purple-fruited pea egg plant.

Ayurvedic names: *Achuda, agnidamani, alarka* and *vallikantakarika*.

Family: Solanaceae.

Habitat: Deccan Peninsular region in India.

General Description: A spiny shrub of climbing habit having irregularly lobed or sinuate leaves. The plant bears purple-blue or violet-purple coloured, attractive flowers. The fruit is a red globose berry (Ramana, Steve Thomas & Ganapathy, 2002-2003).

Chemical constituents: Calcium, iron, phosphorus, carbohydrates, fat, crude fibre and minerals, tannins and saponins in the leaves (Purushothaman, Saradambal & Narayanaswamy, 1969;

Jawahar, Rabert & Jeyaseelan, 2004; Doss & Rangasamy, 2008; Kannabiran, 2009). β-sitosterol has been reported from sobatum (Mohanan & Devi, 1998). *In vitro* production of solasodine has been reported (Anirudh & Nayar, 2009). A crude glycoalkaloid mixture, isolated from the plant material, contained about 20% beta-solamarine.

Actions: Bronchodilator (Madhvan & Balu, 1999).

Therapeutics: *S. trilobatum* is widely used to treat respiratory diseases in Siddha system of medicine (Madhvan & Balu, 1999). It is used to cure asthma, arrest blood vomiting, to reduce blood glucose level and bilious matter phlegmatic rheumatism and several kinds of leprosy.

Parts used: Whole plant.

Dose: Powder (1-3G).

Preparations:

Bioactivity: Cytotoxic (Mohanan & Devi, 1996; Ramachandran *et al.*, 2007; Venkatesan *et al.*, 2008; hepatoprotective (Shahjahan *et al.*, 2004), antioxidant (Shahjahan & Vani, 2005), anti-inflammatory and analgesic (Emmanuel *et al.*, 2006; Pandurangan *et al.*, 2011), and anti-bacterial (Doss, Mubarack & Dhanabalan, 2009).

Clinical study: A pilot study confirmed the effectiveness of *S. trilobatum* in the treatment of mild to moderate bronchial asthma (Govindan *et al.*, 1999).

Further reading

Anirudh K, Nayar AS. *In vitro* production of solasodine from *Solanum trilobatum*. *Biol Plant* 2009; 53(4):719-722.

Doss A, and Rangasamy D. Preliminary Phytochemical Screening and Antibacterial Studies of Leaf Extract of *Solanum trilobatum* Linn. *Ethnobot Leaflets* 2008; 12:638-642.

Doss A, Mubarack M, Dhanabalan R. Antibacterial activity of tannins from the leaves of *Solanum trilobatum* Linn. *Indian J Sci Technol* 2009; 2:41-44.

Emmanuel S, Ignacimuthu S, Perumalsamy R, Amalraj T. Antiinflammatory activity of *Solanum trilobatum*. *Fitoterapia* 2006; 77: 611-612.

Govindan S, Viswanathan S, Vijayasekaran V, Alagappan R. A pilot study on the clinical efficacy of *Solanum xanthocarpum* and *Solanum trilobatum* in bronchial asthma. *J Ethnopharmacol* 1999; 66: 205-210.

Jawhar M, Amalan Rabert G and Jeyaseelan M. Rapid proliferation of multiple shoots in *Solanum trilobatum* L. *Plant Tissue Culture* 2004; 14:107-112.

Madhvan S, Balu S. Ethnobotanical studies on *Solanum trilobatum* Linn.: An Indian drug plant *J Econ Taxon Bot* 1999; 23:453-456.

Kannabiran K. Antibacterial activity of saponin isolated from the leaves of *Solanum trilobatum* Linn. *J Pharm Res* 2009; 2:22.

Mohanan PV, Devi KS. Cytotoxic potential of the preparations from *Solanum trilobatum* and the effect of sobatum on tumour reduction in mice. *Cancer Lett* 1996; 110:71-76.

Pandurangan A, Khosa RL, Hemalatha S. Anti-inflammatory activity of an alkaloid from *Solanum trilobatum* on acute and chronic inflammation models. *Nat Prod Res* 2011; 25:1132-41.

Pandurangan A, Khosa RL, Hemalatha S. Evaluation of Anti-inflammatory and Analgesic Activity of Root Extract of *Solanum Trilobatum* Linn. *Iranian J Pharma Res* 2008; 7:217-221.

Purushothaman KK, Saradambal S, Narayanaswamy V. Chemical estimation of *Solanum trilobatum*.L. *Aust J Chem* 1969; 22: 1569-1570.

Ramachandran V, Palaniyandi S, Magesh V, Balasubramanian RK, Dhanapal Sakthisekaran. Chemotherapeutic Efficacy of *Solanum trilobatum* along with Paclitaxel in Lung Cancer Bearing Swiss Albino Mice. *Int J Can Res* 2007; 3:119-126.

Ramana V, Steve Thomas KP, Ganapathy S. Pharmacognostical studies on *Solanum trilobatum* Linn. *Aryavaidyan* 2002-3; 16: 81-85.

Shahjahan M, Sabitha KE, Jainu M, Shyamala Devi CS. Effect of *Solanum trilobatum* against carbon tetrachloride induced hepatic damage in albino rats *Indian J Med Res* 2004;120:194-198.

Shahjahan M, Vani G, Shyamaladevi CS. Effect of *Solanum trilobatum* on the antioxidant status during diethyl nitrosamine induced and phenobarbital promoted hepatocarcinogenesis in rat.*Chem-Biol Inter* 2005; 156:113-123.

Venkatesan PN, Rajendran P, Ekambaram G, Sakthisekaran D. The combination of cisplatin with *S. trilobatum* could effectively treat the B[a] P-induced lung cancer in mice by offering protection from reactive oxygen species damage and also by suppressing cell proliferation. *Nat Prod Res* 2008; 22:1094-106.

Note: Sobatum is partially purified portion of *Solanum trilobatum*.

Solanum xanthocarpum **Schrad et Wendl.**

Syn: *Solanum surattense* Burm.f.

Common name: Yellow-berried-night shade.

Ayurvedic names: *Dhavani, dusparsha, duspradarisini, kantarika, kantkini, ksudra, nidigdhika* and *vyaghri*.

Family: Solanaceae.

Habitat: India.

Description

(a) *Macroscopic*: Root - 10-45 cm long, few mm to two cm in diameter, almost cylindrical and tapering, bearing a number of fine longitudinal and few transverse wrinkles with occasional scars or a few lenticels and small rootlets, transversely smoothened surface shows a thin bark and wide compact cylinder of wood, fracture, short, taste, bitter.

Stem-herbaceous, prickly with prominent nodes and internodes, green when fresh, young branches, covered with numerous hairs, mature ones glabrous, furrows more prominent in young stem appearing almost circular towards basal region, stem pieces 8-10 mm thick of variable length, external surface light green, when dry, surface yellowish green and smooth, transversely smoothened surface shows a very thin bark and prominent wood, centre shows a large and distinct, pith, mr ture and dry stem often with hollow pith, fracture short to slightly fibrous.

Leaves-petiolate, exstipulate, ovate--oblong or elliptic, sinuate or sub-pinnatifid, subacute hairy, 4-12.5 cm long and 2-7.5 cm wide, green, veins and midrib full with sharp prickles, odour and taste not distinct.

Flower- ebracteate, pedicellate, bisexual, pentamerous, regular, complete, bright blue or bluish purple, calyx-persistent, gamosepalous, tube short, globose, linear-lanceolate, acute, hairy, 0.5-1 .3 cm long and densely prickly, corollagamopetalous, lobes deltoid, acute, hairy, 1-2 cm long and purple in colour, stamens 5, epipetalous, basifixed, filament short 1-1.5 mm long, anther, oblong lanceolate, 0.7-0.8 cm long, ovary superior, ovoid, glabrous, bilocular with axile placentation having numerous ovules.

Fruit - Berry globular, measuring 0.8-1 cm in diameter, surrounded by persistent calyx at base unripe fruits variegated with green and white strips, ripe fruit shows different yellow and white shades.

Seeds-circular, flat, numerous, embedded in a fleshy mesocarp about 0. 2 cm in diameter, glabrous taste, bitter and acrid.

(b) *Microscopic*

Root- transverse section of mature root shows cork composing of 3-6 layers of thinwalled, rectangular and tangentially elongated cells, cork cambium single layered followed by 6-15 layers of thin-walled, tangentially elongated to oval or circular parenchymatous cells, stone cells either single or in groups of 2-20 or even more present in this region, secondary phloem composed of sieve elements and phloem parenchyma traversed by medullary rays, stone cells present in singles or in groups of 2-20 or more in outer, and middle phloem regions, phloem rays 1-4 cells wide and 2-22 cells high, cambium 3-5 layered of thin-walled rectangular cells, xylem composed of vessels, tracheids, fibre trachieds, parenchyma and transversed by medullary rays, all elements being lignified, vessels and tracheids with bordered pits, fibres with a few simple pits, xylem parenchyama rectangular or lightly elongated with simple pits and rarely with reticulate thickening, xylem rays 1-3 cells wide and 1-20 cells high, microsphenoidal crystals of calcium oxalate as sandy masses and simple starch grains present in secondary cortex, phloem and medullary rays.

Stem-transverse section of mature stem, 1.5-2 cm thick consists of 6-12 layers of cork of thin- walled somewhat rectangular cells, epidermis remains intact for a long time, secondary cortex consists of 7-11 layers of parenchymatous cells, some cells thickened and lignified forming stone cells primary cortex remains intact even in quite mature stage but later gets crushed, pericyclic fibre, occur singly or in small groups of 2-3, secondary phloem consists of sieve elements, parenchyama, a few fibres, stone cells and traversed by phloem rays, fibres found scattered in singles or in small groups in outer and middle phloem region, inner phloem devoid of fibres, stone cells present in singles or in small groups of 2-4, phloem rays, 1-2 or rarely 3 cells wide, cambium composed of 2-3 layers, xylem consists of vessels, tracheids, parenchyma, fibres and traversed by xylem rays, vessels vary grea tly in shape and size and show bordered pits, tracheids elongated with irregular walls and bordered pits, fibres much elongated, thick-walled and lignified with tapering and pointed ends, some having truncated ends or bifurcated at one or both ends with a few simple pits, trancheids fibres smaller than fibres, with both ends tapering and have reticulate thickening, xylem parenchyma cubical to rectangular with simple or bordered pits or reticulate thickening, xylem rays conspicuous by their pitted thickenings, longer size and radial elongation of cells, 1-2 or rarely 3 cells wide and 2-25 cells high, internal phloem composed of sieve elements and parenchyma, forming more or less continuous band and embedded in perimedullary zone, a few phloem fibres similar to those of outer phloem region also present, central region occupied by a large pith, microsphenoidal crystals of calcium oxalate as sandy massesand simple starch grains present in cortex, secondary cortex, phloem, medullary rays and pith cells.

Leaves - Petiole-transverse section of petiole shows circular to wavy outlines, epidermis single layered, covered externally by a thick cuticle, hypodermis consists of 3-4 layers of collenchymatous, cells, one large-crescent-shaped, bicollateral, central vascular bundle and two small lateral bundles present, rest of tissue of petiole composed of polygonal, angular, thin-walled, parenchymatous cells, epidermis shows mostly stellate and rarely urn to tricellular hairs.

Midrib-transverse section of midrib shows a biconvex structure, epidermis on either side covered externally by a thick cuticle, below epidermis 3-4 layers of collenchyma present, stele composed of crescent-shaped, bicollateral, central vacscular bundle and two small lateral vascular bundles, rest of tissue composed of thin-walled, parenchyma, some stellate hair present on epidermis.

Lamina-transverse section shows dorsiventral structure, epidermis on either side, wavy in outline, covered externally by a thick cuticle, on upper side mesophyll composed of a single layered palisade and 4-6 layers of loosely arranged spongy parenchyma, some stellate hairs (4-8 armed) present on both sides of epidermis, anisocytic stomata present on both surfaces, vein-islet number 46-80 on lower epidermis (mean 63), 61-80 on upper epidermis (mean 70), stomatal index 20-25 (mean 22.5) on lower epidermis, 14-24 (mean 19) on upper epidermis, palisade ratio 1.7-4 (mean 2 .85).

Fruit-transverse section of mature fruit shows single layered epidermis, covered externally by a thin cuticle, 1-2 layers of collanchyma present below epidermis, esocarp composed of thin-walled, oval to polygonal cells, some fibre., vascular undles present scattered, seed consists of thick-walled radially elongated testa, narrowendosperm with embryo, some cells of endosperm contain oil globules.

Powder - Greenish, under microscope shows single or groups of stone cells, groups of aseptate fibre with tapering ends, pitted vessels, groups of spongy parenchyma, fragments of palisade tissue, anisocytic stomata, stellate hairs and simple, rounded to oval starch grains measuring 2.75-11 μ in dia.

Chemical constituents: Alkaloids: solasonine, solamargine and khasianine (Gupta *et al.,* 1967, Gupta & Santra 1967, Chand *et al.,* 1995, Shanker *et al.,* 2011) and steroid saponins (Singh *et al.,* 2007).

Action: Antitussive.

Therapeutics: Stems possess carminative, anthelmintic and bitter properties, root is expectorant and used in treatment of toothache, chest pain due to cough, asthma and bronchitis. The leaves are applied externally as a pain relieving agent (Kumar, Dwivedi & Chaturvedi, 1981).

Parts used: Whole plant.

Preparations: *Kantkaryavleha* and *vyaghriharitaki*.

Bioactivity: Antispermatogenic (Mali, Chaturvedi & Dixit, 1966; Rao, 1988), cardiac stimulating (Gupta *et al.*, 1966), antifungal (Singh *et al.*, 2007), hypoglycaemic (Kar *et al.*, 2006) and adaptogenic (Gupta, 2009).

Clinical study: Clinical efficacy of *S. xanthocarpum* has been documented in several clinical trials in bronchial asthma (Bector & Puri 1971, Jain, 1980, Govindan *et al.*, 1999; Govindan *et al.*, 2004). A study reported efficacy in epilepsy (Dharmadhikari, 1976).

Further reading

Bector NP, Puri AS. *Solanum xanthocarpum* (kantakari) in chronic bronchitis, bronchial asthma and non-specific unproductive cough (An experimental and clinical correlation). *J Assoc Physicians India* 1971; 19:741-744.

Chand R, Kumar S, Sharma AK, Srivastava L. Variation of solasodine in *Solanum sisymbrifolium* and *Solanum xanthocarpum* with plant growth and development. *Indian Drugs* 1995; 32:362-365.

Dharmadhikari SV. Antiepileptic action of *Solanum xanthocarpum*. *Nagarjun* 1976; 20:4-5.

Govindan S, Viswanathan S, Vijayasekaran V, Alagappan R. A pilot study on the clinical efficacy of *Solanum xanthocarpum* and *Solanum trilobatum* in bronchial asthma. *J Ethnopharmacol* 1999; 66: 205-210.

Govindan S, Viswanathan S, Vijayasekaran V, Alagappan R. Further studies on the clinical efficacy of *Solanum xanthocarpum* and *Solanum trilobatum* in bronchial asthma. *Phytother Res* 2004; 18:805-809.

Gupta AK. Adaptogenic effects of total extracts and steroidal saponins of *Solanum xanthocarpum* and *Solanum nigrum*. *J Pharm Res.* 2009; 2:16.

Gupta MP, Santra DK. Isolation of solasodine from *Solanum xanthocarpum* Schrad and Wendl. *J Res Indian Med* 1967; 1:239-241.

Gupta SS, Verma SC, Singh C, Khandelwal P, Gupta NK. Chemical and pharmacological studies on *Solanum xanthocarpum* (Kantakari). *Indian J Med Res* 1967; 55: 723-732.

Gupta SS, Verma SC, Singh C, Khandelwal P. Cardiac stimulating effects of kantakari (*Solanum xanthocarpum*). *Indian J Med Sci* 1966; 20: 554-559.

Jain, J.P. A clinical trial of kantakari (*Solanum xanthocarpum*) in cases of kasa. *J Res Ayur Siddha* 1980; 1:35-51.

Kar DM, Maharana L, Pattnaik S, Dash GK. Studies on hypoglycaemic activity of *Solanum xanthocarpum* Schrad. & Wendl. fruit extract in rats. *J Ethnopharmacol* 2006 24; 108:251-256.

Kumar S, Dwivedi RN, Chaturvedi. Scientific appraisal of *Solanum xanthocarpum* (kantakari). *J Nat Integ Med Assoc* 1981; 33: 282-285.

Mali PC, Chaturvedi M, Dixit VP. Antispermatogenic activity of *Solanum xanthocarpum* S and W root (50 per cent EtOH extract) in rats. *J Phytol Res* 1966; 9:13-17.

Rahman MT, Ahmed M, Alimuzzaman M, Shilpi JA. Antinociceptive activity of the aerial parts of *Solanum xanthocarpum*. *Fitoterapia* 2003; 74:119-21.

Rao MV. Effects of alcoholic extract of *Solanum xanthocarpum* seeds in adult male rats. *Indian J Exp Biol* 1988; 26: 95-98.

Shanker K, Gupta S, Srivastava P, Srivastava SK, Singh SC, Gupta MM. Simultaneous determination of three steroidal glycoalkaloids in *Solanum xanthocarpum* by high performance thin layer chromatography. *Pharm Biomed Anal* 2011; 54:497-502.

Singh OM, Subharani K, Singh NI, Devi NB, Nevidita L. Isolation of steroidal glycosides from *Solanum xanthocarpum* and studies on their antifungal activities. *Nat Prod Res* 2007; 21:585-590.

Artemisia vulgaris Willd.

Common name: Indian Wormwood or mugwort.

Ayurvedic names: *Bhujagaparini, brahamhajata, damani, gandhotakata, nagagandha, nagahava, pushachamara* and *tapodhana.*

Family: Asteraceae.

Habitat: It is native to temperate Europe, Asia, northern Africa and Alaska and is naturalized in North America.

General Description: *A. vulgaris* is a tall herbaceous perennial plant growing 1–2 m (rarely 2.5 m) tall, with a woody root. The leaves are 5–20 cm long, dark green, pinnate, with dense white tomentose hairs on the underside. The erect stem often has a red-purplish tinge. The rather small flowers (5 mm long) are radially symmetrical with many yellow or dark red petals. The narrow and numerous capitula (flower heads) spread out in racemose panicles. It flowers from July to September (Issar, 1975; Ramay, 1987).

Chemical constituents: The extracted oil contains camphor, alpha-thujone, germacrene D, camphene, 1,8-cineole and beta-caryophyllene (Mishra *et al.,* 1968; Govindaraj *et al.,* 2008). Phytochemical studies of the hexane and methanolic extracts showed the presence of carbohydrate, saponins, phytosterol, proteins, amino acids, tannins, phenols and flavonoids (Ashok & Upadhyaya, 2011).

Action as per Ayurveda: *Vatanulomana* (carminative).

Therapeutics: Skin disorders, asthma, liver ailments, loss of libido, amenorrhea, dysmenorrhoea and general debility. With *Lauha bhasama*, it is given in anemia.

Parts used: Whole plant.

Bioactivity: Antifertility (Narwaria, Khosa & Dhar, 1994).

Further reading

Ashok PK, Upadhyaya K. Preliminary phytochemical screening and physico-chemical paramaterts of aerial parts of *Artemisia vulgaris*. *Int J Res Ayur Pharm* 2010; 1:206-211.

Govindaraj S, Kumari BD, Cioni PL, Flamini G. Mass propagation and essential oil analysis of *Artemisia vulgaris*. *J Biosci Bioeng* 2008; 105:176-83.

Issar RK. A note on the botanical sources of a snake-bite antidote drug "nagdaun". *J Res Indian Med* 1975; 10:171-172.

Mishra MB, Singh SW, Misra RK, Mishra SS. Preliminary pharmacological screening of *Artemisia vulgaris* Linn. *Indian J Med Sci* 1968; 22:141-143

Narwaria A, Khosa RL, Dhar SK. Experimental studies on *Artemisia vulgaris* L.-A possible antifertility drug. *Ancient Sci Life* 1994; 14:10-15.

Ramay B. Botany of Artemisia. *Allerg Immunol* (Paris) 1987; 19:250-252.

3. AGNIMANTHA

Current representative: *Premna integrifolia*

Possible representative: *Clerodendron phlomidis*

Premna integrifolia Linn.

Syn: *Premna mucronata* Roxb.

Common name: Headache tree.

Ayurvedic names: *Agnimantha, agni, arni, ganikarika, harimantha, ketu, mantha, sriparna, vahnimantha, vaijayantika and vataghani.*

Family: Verbenaceae.

Habitat: India and Sri Lanka.

General Description: Medium-sixed tree, 5-7m high. Young twigs quadrangular, large branches sometimes horny. Leaves opposite, entire or minutely dentate in the upper part, shining glabrous above, pale beneath. Infloresence in terminal corymb; flowers greenish-white. Drupe globose or ovoid, black (Gokani, Kapadia & Shah, 2008).

Chemical composition: Tannin, alkaloids: premnine, graniarine and (Basu & Dandiya, 1947; Barik *et al.,* 1992; Singh, *et al.,* 1994), flavonoids (Dasgupta *et al.,* 1984), resin, phenolic diterpenoids (Rao, Aruna, *et al.,* 1987), isoprenoids (Rao, Krishna, *et al.,* 1985), flavone: premnoside A (Rao *et al.,* 1986), and antibiotic substance (Kurup & Kurup, 1964). Three novel diterpenoids have been reported from the root bark namely1β,3α,8β-trihydroxy-pimara-15-ene,6α,11,12,16-tetrahydroxy-7-oxo-abieta-8,11,13-triene and 2α, 19-dihydroxy-pimara-7,15-diene (Yadava, Tiwaria & Gupta, 2010). The roots yield an essential oil.

Specific action as per Ayurveda: Anti-inflammatory.

Therapeutics: Loss of appetite, general debility and rheumatism. Decoction of roots is given in urinary tact infection.

Parts used: Roots and leaves.

Bioactivity: Antibiotic (Kurup & Kurup, 1964), antifungal (Singh, Singh *et al.,* 1994), anti-inflammatory and antioxidant (Barik *et al.,* 1992; Gokani *et al.,* 2011), hypoglycemic (Alamgir *et al.,* 2001).

Clinical trial: A study reported efficacy of *P. integrifolia* in the treatment of diabetes mellitus (Shankaran *et al.,* 1963).

Further reading

Alamgir M, Rokeya B, Hannan JM, Choudhuri MS. The effect of *Premna integrifolia* Linn. (Verbenaceae) on blood glucose in streptozotocin induced type 1 and type 2 diabetic rats. *Pharmazie* 2001; 56:903-4.

Barik BR, Bhowmik T, Dey AK, Patra A, Chatterjee A, Joy S, Susan T, Alam M, Kundu AB. Premnazole, an isoxazole alkaloid of *Premna integrifolia* and *Gmelina arborea* with anti-inflammatory activity. *Fitoterapia* 1992; 63: 295-299.

Basu NK, Dandiya PC. Chemical investigation of *Premna integrifolia* Linn. *J Am Pharm Assoc Am Pharm Assoc* 1947; 36:389-91.

Dasgupta B, Sinha NK, Pandey VB, Ray AB. Major Alkaloid and Flavonoid of *Premna integrifolia. Planta Med* 1984; 50:281.

Gokani RH, Kapadia NS, Shah MB. Comparative pharmacognostic study of *Clerodendrum phlomidis* and *Premna integrifolia. J Nat Rem* 2008; 8:222-231.

Gokani RH, Lahiri SK, Santani DD, Shah MB. Evaluation of Anti-inflammatory and Antioxidant Activity of *Premna integrifolia* Root. *J Comp Integr Med* 2011; 8: Article 25. DOI: 10.2202/1553-3840.1216.

Kurup KK, Kurup PA. Antibiotic substance from the root bark of *Premna integrifolia*. *Naturwissensachaften* 1964; 51: 484.

Rao, C.B., Krishna, PG *et al.,* Chemical examination of *Premna* spp.: 10. New isoprenoids from the root barks of *Premna integrifolia* and *Premna latifolia* var. mollissima. *Indian J Chem* 1985; 24: 403-407.

Rao, C.B., Krishna, PG *et al.,* Chemical examination of *Premna* species: Part XI. Structure of premnoside A from *Premna latifolia* var. *cuneata*. *Indian J Chem* 1986; 25:100-101.

Shankaran PS, Pathak SP, Shukla KN, Sen SP. The treatment of diabetes mellitus with the indigenous herb *Premna integrifolia*. *Curr Med Prac* 1963; 7: 697-703.

Singh, UP, Singh KP *et al.,* Antifungal activity of some naturally occurring plant alkaloids. *Int J Trop Plant Dis* 1994; 12: 209-212.

Yadava D, Tiwaria N, Gupta MM. Diterpenoids from *Premna integrifolia*. *Phytochem Lett* 2010; 3:143-147.

Clerodendron phlomidis Linn.

Syn: *C.multiflorum* (Burm. f.) O. Kuntze

Ayurvedic name: *Ganikarika.*

Family: Verbenaceae.

Habitat: It is distributed more or less throughout India, Ceylon, Malay and Peninsula.

General Description (Gokani, Kapadia & Shah, 2008)

(a) *Macroscopic*: *C. phlomidis* is a small herb. The leaves are opposite, deltoid, ovate, hairy from below and wavy. The flowers bloom in August to February. The flowers are in small rounded terminal panicle. The flowers are white or pinkish and are very fragrant. Drug pieces 7-15 cm long, 0.2 -3.0 cm thick, occasionally branched, cylindrical, tough, yellowish-brown externally, bark thin, occasionally easily peeled, outer surface rough due to exfoliation, wood light yellow, fracture hard; taste, slightly astringent.

(b) *Microscopic*: Root shows exfoliating cork, consisting of 10-15, occasionally more, rows of tangentially elongated, thin-walled cells; secondary cortex consists of round to oval parenchymatous cells, a few containing rhomboidal crystals of calcium oxalate; secondary phloem consists of isodiametric, thin-walled, parenchymatous cells, a few of them containing rhomboidal crystals of calcium oxalate; phloem rays distinct, consisting of radially elongated cells; secondary xylem shows a wide zone, consisting of usual elements, all being lignified; vessels found in single as well as in groups of 2-3, scattered throughout xylem region; xylem parenchyma simple pitted, squarish wide lumen; xylem rays 1-5 seriate, consisting of radially elongated cells; rhomboidal crystal of calcium oxalate packed in xylem parenchyma and xylem rays; abundant simple, round starch grains measuring 6-17 μ in dia., found scattered throughout.

Powder - Dull yellow; shows fragments of cork cells, small, pointed, aseptate, lignified fibres, simple, pitted vessels, lignified cells packed with rhomboidal crystals of calcium oxalate and numerous simple, round to oval starch grains having narrow hilum, measuring 6-11 μ in diameter

Chemical constituents: Chalcone glycosides (Anam, 1999) and lupeol ester (Pandey *et al.*, 2008) have been reported. Pectolinaringenin, scutellarein, clerodin, clerodendrin, clerosterol, 24beta-ethylcholesta-5,22E,25-triene-3beta-ol, lup-20(29)-en-3-triacontanoate, 4,2',4'-trihydroxy-6'-methoxychalcone-4,4'alpha-D-diglucoside, 7-hydroxyflavone, 7-hydroxyflavanone-7-O-glucoside and alpha-L-rhamnopyranosyl-(1-2)alpha-D-glucopyranosyl-7-O-naringin-4'-O-alpha-D-glucopyranoside-5-methyl ether had been isolated (Seth, Pandey & Dasgupta, 1982; Roy & Pandey, 1995; Raja & Mishra, 2010).

Actions: Antitussive, alterative, laxative, anti-inflammatory and cardiac tonic.

Therapeutics: Common cold, cough, constipation and indigestion. The sandals rub the plant over their bodies in dropsy and also give it to their cattle to cure them of diarrhea and worms, or when the stomach swells (Kirtikar & Basu, 1947). The decoction of roots is used as a demulcent in gonorrhea. The juice of leaves is used as bitter tonic (Nadkarni, 2001) and also given in neglected syphilitic complaints (Chopra, Nayar & Chopra, 1996).

Parts used: Bark.

Bioactivity: Hypoglycemic (Bhattacharya & Bajpai, 1975), anti-diarrhoeal (Rani *et al.*, 1999), psychopharmacological effects (Murugesan *et al.*, 2001), antifungal (Rajasekaran & Ponnusamy, 2006), immunomodulatory (Gokani *et al.*, 2007), antiamnesic (Joshi & Megeri, 2008), anti-hepatotoxic (Verma & Bahar, 2009), Antarthritic (Kilimozhi, Parthasarathy & Amuthavalli, 2009) and analgesic (Vijayamirtharaj, Vincent & Senthilkumar, 2011).

Clinical trial: A study reported efficacy of *C. phlomidis* in the treatment of diabetes mellitus (Chaturvedi *et al.*, 1984).

Further reading

Anam EM. Novel Flavone and chalcone glycosides from *Clerodendrum phlomidis* (Verbenaceae). *Indian J Chem* 1997; 36B: 897-900.

Anam EM. Novel Flavone and chalcone glycosides from *Clerodendrum phlomidis* (Verbenaceae). *Indian J Chem* 1999; 36B: 1307-1310.

Bhattacharya SK, Bajpai HS. Hypoglycaemic effect of *Clerodendron phlomidis* (arani). *J Res Indian Med* 1975; 10:1-6.

Chaturvedi GN, Subramaniyam PR, Tiwari SK, Singh KP. Experimental and clinical studies on diabetes mellitus evaluating the efficacy of an indigenous hypoglycaemic drug-arani (*Clerodendron phlomidis*). *Ancient Sci Life* 1984; 3:216-224.

Chopra RN, Nayar SL and Chopra IC. Glossary of Indian Medicinal Plants 1996; 71.

Gokani RH, Kapadia NS, Shah MB. Comparative pharmacognostic study of *Clerodendrum phlomidis* and *Premna integrifolia*. *J Nat Rem* 2008; 8:222-231.

Gokani RH, Lahiri SK, Santani DD and. Shah MB. Evaluation of Immunomodulatory Activity of *Clerodendrum phlomidis* and *Premna integrifolia* Root. *Indian J Pharmacol* 2007;3: 352-356.

Joshi A, Megeri K. Antiamnesic evaluation of *C. phlomidis* Linn. bark extract in mice. *Brazilian J Pharm Sci* 2008; 44:717-725.

Kilimozhi D, Parthasarathy V, Amuthavalli N. Effect of *Clerodendrum phlomidis* on adjuvant induced arthritis in rats-A radiographic densitometric analysis. *Int J of Pharm Tech Res* 2009; 1:1434-1441.

Kirtikar KR and Basu BD. Indian Medicinal Plants, International Book Distributors, Dehradun (1999): 1947.

Mohan Maruga Raja MK, Mishra SH. Comprehensive review of *Clerodendrum phlomidis*: a traditionally used bitter. *Zhong Xi Yi Jie He Xue Bao* 2010; 8:510-24.

Murugesan T, Saravanan KS, Lakshmi S, Ramya G, Thenmozhi K. Evaluation of psychopharmacological effects of *Clerodendrum phlomidis* Linn. extract. *Phytomedicine* 2001; 8:472-476.

Nadkarni AK. Indian Materia Medica, Popular Prakashan, Bombay, 2002: 353.

Pandey R, Kaur R, Malasoni R, Gupta MM. Lupeol Ester from *Clerodendrum phlomidis* L. *ChemInform* 2008; 39: 23.

Rajasekaran A, Ponnusamy K. Antifungal activity of *Clerodendron inerme* (L) and *Clerodendron phlomidis* (L). *Turk J Biol* 2006; 30: 139-142.

Rani S, Ahamed N, Raja ram S and Saluja R, Thenmozhi S, Murugesan T., Antidiarrhoeal evaluation of *Clerodendrum phlomidis* Linn. Leaf extract in rats. *J.Ethanopharmacol* 1999: 68; 315.

Roy R, Pandey VB. Flavonoids of *Clerodendron phlomidis*. *Indian J Nat Prod* 1995; 11: 13-14.

Seth KK, Pandey VB, Dasgupta B. Flavonoids of *Clerodendrum phlomidis* flowers. *Pharmazie* 1982; 37: 74-75.

Verma A, Bahar A. Anti-hepatotoxic activity of *Clerodendrum phlomidis*. *Int J Pharm Tech Res* 2009; 1:1028-1031.

Vijayamirtharaj R, Vincent S, Senthilkumar N. Analgesic activity of *Clerodendrum phlomidis* Linn. (Aerial parts). *Int J Res Pharm Biomed Sci* 2011; 2:120-122.

4. AJAGANDHA

Current representative: *Gynandropsis gynandra*

Possible representative: *Clemone viscosa*

Gynandropsis gynandra L. Briq.

Syn: *Gynandropsis pentaphylla* D.C., *Cleome gynandra* L., *Pedicellaria pentaphylla* Shrank

Common name: Dog mustard.

Ayurvedic names: *Babraka, putighanda, tilaparni* and *ugragandha.*

Family: Capparidaceae.

Habitat: India.

General Description: This plant is an erect, branched, somewhat hairy herb, growing from 0.4 to 1 meter in height, and with usually purplish stems. The leaf has a very long petiole and has five leaflets. The leaflets are obovate or oblong obovate 2 to 5 centimeters in length, pointed at the base, and abruptly pointed at the tip. The flowers are borne in racemes which terminate the branches and are showery and somewhat spidery in appearance. The sepals are lanceolate, green, and about 4 millimeters long. The petals are white or tinged with purple and 1 to 1.2 centimeters long, with the elliptic to obovate blades. The fruit is a cylindrical capsule, 4 to 10 centimeters in length, and about 5 millimeters in diameter, containing numerous, small, dark-brown seeds (Puri, 1971).

Chemical constituents: Fixed oil (containing cleomine) and flavonoids: apigenin and penduletin (Petit *et al.,* 2005) and phenolics (Jain & Gupta, 1985).

Specific action as per Ayurevda: Vidahi (irritant).

Therapeutics: The plant is considered to be antidote to snake and scorpion poisoning. The juice of the leaves is used as application in filariasis. The drug is useful in infantile convulsions. The seeds are antispasmodic, sudorific, anthelmintic and carminative. A decoction of the seeds is employed as a wash for piles and for rheumatism and malarial disorders. The seeds are febrifuge. The powdered seeds mixed with sucrose and given in ascariasis and this are followed by castor oil intake.

Parts used: Leaves, seeds and root.

Bioactivity: Antibacterial (Rajeswara Rao *et al.,* 1998), anticancer (Petit, *et al.,* 2005; Sivanesan & Begum, 2007), antimicrobial (Borgio *et al.,* 2008), antinociceptive (Ghogarea *et al.,* 2009), antipyretic, analgesic and hypoglycaemic (Rajeswara Rao *et al.,* 1997).

Further reading

Borgio JF, Thorat PK, Lonkar AD. Toxicity of *Gynandropsis pentaphylla* DC Extracts against Microbials and Its Phytochemical Profile. *Ethnobot Leaflets* 2008; 12: 320-336.

Ghogarea UR, Nirmala SA, Patila RY, Kharya MD. Antinociceptive activity of *Gynandropsis gynandra* leaves. *Nat Prod Res* 2009; 23: 327-333.

Jain AC, Gupta SM. Minor phenolic components of the seeds of *Gynandropsis gynandra. J Nat Proceed* 1985;48:332-333.

Pettit GR, Meng Y, Herald DL, Stevens AM, Pettit RK, Doubek DL. Antineoplastic agents 540. The Indian *Gynandropsis gynandra* (Capparidaceae). *Oncol Res* 2005; 15:59-68.

Puri HS. Macro- and micromorphology of the leaf and seed of *Gynandropsis pentaphylla* L. *Quarterly J Crude Drug Res* 1971; 11:1805-1811.

Rajeswara Rao P, Babu MR, Krishna Rao RV, Mohan Babu GV. Studies on antipyretic, analgesic and hypoglycaemic activities of root of *Gynandropsis gynandra* Linn. *Indian Drugs* 1997; 34:690-694.

Rajeswara Rao P, Murali Mohan Babu GV. Antibacterial activity of root of *Gynandropsis gynandra* Linn. *Indian Drugs* 1998; 35:721-722.

Sivanesan D, Begum VH (2007) Preventive role of *Gynandropsis gynandra* L., against aflatoxin B-1 induced lipid peroxidation and antioxidant defense mechanism in rat. *Indian J Exp Biol* 45:299-303.

Sivanesan D, Begum VH. Modulatory effect of *Gynandropsis gynandra* L. on glucose metabolizing enzymes in aflatoxin B1-induced hepatocellular carcinoma in rats. *Indian J Biochem Biophys* 2007; 44:477-480.

***Clemone viscosa* L.**

Syn: *Clemone isocandra* L.

Common name: Asian spider flower.

Ayurvedic name: *Adityabhakta.*

Family: Capparidaceae.

Habitat: Throughout India growing wild in planes as a weed.

General Description: *C. viscosa* is a widely distributed herb. Leaves compound, 3-5 foliate, palmate, yellow flowers and long slender pods containing seeds. The flowers are hermaphrodite. Fruits capsules, hairy and covered with sticky material. Seeds kidney shaped and turns dark on ripe. The whole plant is sticky in nature and has a strong odour resembling asafoetida.

Chemical constituents: Fixed oil containing viscosin and lignan, clemiscosin-D (Saxena, Koli & Saxena, 2000). Lactam nonanoic acid has been isolated and purified from the root exudates of *C. viscosa* (Anirban & Biswas, 2011).

Specific action as per *Ayurveda*: *Vidahi* (irritant).

Therapeutics: Traditionally, this plant is used in various disorders such as diarrhea, fever, inflammation, liver diseases, bronchitis, skin diseases and malarial fever. The juice is useful in piles, lumbago and earache (Kirtikar & Basu, 1975).

Parts used: Leaves, seeds, and root.

Bioactivity: Anti-diarrheal (Devi, Boominathan & Mandal, 2002), antipyretic (Devi, Boominathan & Mandal, 2003a), analgesic (Devi, Boominathan & Mandal, 2003b), immunomodulatory (Tiwari *et al.*, 2004) and hepatoprotective (Mobiya *et al.*, 2010).

Further reading

Anirban J, Biswas SM. Lactam nonanic acid, a new substance from *Cleome viscosa* with allelopathic and antimicrobial properties. *J. Biosci* 2011; 36 http://www.ias.ac.in/jbiosci/jana_187.pdf.

Devi BP, Boominathan R and Mandal SC. Evaluation of antipyretic potential of *Cleome viscosa* Linn. (Capparidaceae) extract in rats. *J. Ethnopharmacol* 2003a; 87:11–13.

Devi BP, Boominathan R, Mandal SC. Evaluation of anti-diarrheal activity of *Cleome viscosa* L. extract in rats. *Phytomedicine* 2002; 9:739-742

Devi BP, Boominathan R, Mandal SC. Studies on analgesic activity of *Cleome viscosa* in mice. *Fitoterapia* 2003b; 74: 262–266.

Kirtikar KR, Basu BD. Indian Medicinal Plants. 2nd ed. Vol 1. New Delhi: Periodical Experts; 1975: 183-185.

Mobiya AK, Patidar AK, Selvam G, Jeyakandan M. Hepatoprotective Effect of *Cleome viscosa* L. Seeds in Paracetamol Induced Hepatotoxic Rats. *Int J Pharm Biol Arch* 2010; 1: 399-403.

Saxena, BR, Koli MC, Saxena RC. Preliminary ethnomedical and phytochemical study of *Cleome viscosa* L. *Ethnobotany* 2000; 12:47–50.

Tiwari U, Rastogi B, Thakur S, Jain S, Saraf DK. Studies in the immunomodulatory effects of *Cleome viscosa*. *Indian J Pharm Sci* 2004; 66:171–176.

5. ALAMBUSHA

Current representative: *Sphaeranthus indicus*

Possible representative: *Mimosa pudica* and *Biophytum senstivitum*

Sphaeranthus indicus L.

English name: East Indian Globle-thistle.

Ayurvedic name: Gorkhmundi.

Family: Asteraceae.

Habitat: India.

General Description: A much branched, strongly-scented annual with winged stem and the wings toothed. Leaves obovate-oblong, narrowed at the base, dentate and serrate. Flowers compound heads, globose avoid, Flowering time November to January in Indian conditions; glandular hairy. Achene staled.

Microscopy

Leaf: The leaf is dorsiventral and shows abundant trichomes of varying types on both the epidermis. Simple trichomes are three to four celled, thick walled and measure 130.8–145.2 µm in length and 29.0–43.5 µm in width. Trichomes are straight/knee shaped, with a swollen base and with collapsed cell at the middle or at the apex. Midrib shows three to four collateral vascular bundles associated with a group of sclerenchymatous cells on either side (Vikani *et al.,* 2008).

Stem: The stem shows cork with two to three layers of parenchymatous cells covered with papillose cuticle having trichomes and can be distinguished by the presence of a discontinuous ring of lignified pericyclic fibers and a well-developed ring of bicollateral vascular bundle surrounding the pith. Medullary rays are pitted, lignified and about unitetraseriate.

Root: The root shows on its outer side metaderm, a typical brown colored tissue. It consists of suberized cells, arranged irregularly and forms a protective layer. Radial groups of pericyclic fibers and few stone cells are seen alternating with radially arranged secretory canals in the secondary cortex. Phloem is parenchymatous and radially arranged. Medullary rays are pitted, lignified and about two to five seriate.

Chemical constituents: Stems, leaves and flowers have been found to contain fatty acids, phytosterols, a glycoside composed of the alkaloid sphaeranthine and a polysaccharide, β-sitosterol, n-pentacosane, n-triacontanol, stigmasterol, hentri-acontane and β-D-glucoside of β-sitosterol. Flowers also contain a number of eudesmanolides and sesquiterpene lactones, viz., hydroxy-, dihydroxy- and methoxy frullanolides. Stems, leaves and flowers contain essential oil composed of methyl chavicol, α-ionone, d-cadinene, p-methoxycinnamaldehyde as major constituents and α-terpinene, citral, geraniol, geranyl acetate, β-ionone, sphaerene, indicusene and sphaeranthol α-cadenine, ocimene as minor constituents. β-Sitosterol, n-triacontanol, phenylurethan, n-pentacosane have also been isolated from oil (Baslas,1959; Sohoni *et al.,* 1988; Singh, Tripathi & Singh,1989; Rojatkar & Nagasampagi,1992; Yadava & Kumar, 1998; Yadava & Kumar, 1999; Pujar *et al.,* 2000; Mishra *et al.,* 2007).

Specific action as per *Ayurveda*: *Rakat prasadana* (alterative).

Therapeutics: *S. indicus* has long been used in the treatment of skin infection, bronchitis, jaundice and nervous depression (Nadkarni 1976). The roots and seeds are considered as anthelmintic. The herb is reported to be useful as a tonic to treat indigestion, asthma, leucoderma and dysentery (Chopra *et al.,* 1956). It is also used in acne vulgaris and gout.

Parts used: Whole plant.

Bioactivity: Haemostatic (Srivastav, Khan & Vohra, 1971), antifungal (Garg & Kasera, 1982), antimicrobial (Dubey, Ansari & Harduha, 2000), immunomodulatory (Bafna & Mishra, 2004), neuroleptic (Mhetre, Ambavade & Bodhankar, 2006), antioxidant (Shirwaikar, Prabhu & Punitha, 2006), anxiolytic (Ambavade *et al.,* 2006), hepatoprotective (Nayak, Maity & Maiti, 2007; Tiwari & Khosa, 2009), antidiabetic (Prabhu, Lobo & Shirwaikar, 2008), renoprotective (Srinivasan, Jessy & Anand, 2009), antihyperlipidemic (Pande & Dubey, 2009), analgesic and antipyretic (Nanda *et al.,* 2009), bronchodilator (Sarpate, Deore & Tupkari,

2009), mast cell stabilizing (Mathew *et al.,* 2009), psychotropic (Galani & Patel, 2009), wound healing (Jha, Garud & Nema, 2009) antidepressant and anticonvulsant (Galani & Patel, 2010)

Further reading

Ambavade SD, Mhetre NA, Tate VD, Bodhankar SL. Pharmacological evaluation of the extracts of *Sphaeranthus indicus* flowers on anxiolytic activity in mice. *Indian J Pharmacol* 2006; 38:254–259.

Bafna AR, Mishra SH. Immunomodulatory activity of methanol extracts of flower heads of *Sphaeranthus indicus* Linn. *Ars Pharmaceutica* 2004; 45:281–291.

Baslas KK. Essential oil from *Spharanthus indicus*. *Perf Essent Oil Rec* 1959; 50:765-768.

Chopra RN et al., Glossary of Indian Medicinal Plants, Publication and Information Directorate, New Delhi, 1956; pp. 232.

Chopra RN, Chopra IC, Honda KL, Kapur LD. Indigenous drugs of India, 2nd edn., UN Dhur and Sons (P) Ltd, Calcutta.

Dubey KS, Ansari AH, Harduha M. Antimicrobial activity of the extract of *Sphaeranthus indicus*. *Asian J Chem* 2000; 12:577-578.

Duraipandiyan V, Kannan P, Ignacimuthu S. Antimicrobial Activity of *Sphaeranthus indicus* L. *Ethnobot Leaflets* 2009; 13: 320-25.

Galani VJ, Patel BG. Effect of hydro-alcoholic extract of *Sphaeranthus indicus* against experimentally induced anxiety, depression and convulsions in rodents. *Int J Ayu Res* 2010; 2 In press.

Galani VJ, Patel BG. Psychotropic activity of *Sphaeranthus indicus* Linn.in experimental animals. *Pharmacog Res* 2009; 1:307–313.

Garg SC, Kasera HL. Antifungal activity of the essential oil of *Sphaeranthus indicus* Linn. *Pafai J* 1982; 4:23–24.

Jha RK, Garud N, Nema RK. Excision and incision wound healing activity of flower head alcoholic extract of *Sphaeranthus indicus* Linn. in albino rats. *Global J Pharmacol* 2009; 3:32–37.

Mathew JE, Srinivasan KK, Dinakaran V, Joseph A. Mast cell stabilizing effects of *Sphaeranthus indicus*. *J Ethnopharmacol* 2009; 122:394–396.

Mhetre NA, Ambavade SD, Bodhankar SL. Neuroleptic activity of extract of *Sphaeranthus indicus* in mice. *Indian J Nat Prod* 2006; 22:24–27.

Mishra BB, Yadav SB, Singh RK, Tripathi V. Novel flavonoid C-glycoside from *Sphaeranthus indicus* L. (Family Compositae) *Molecules* 2007; 12: 2288-2291.

Nanda BK, Jena J, Rath B, Behera BR. Analgesic and Antipyretic activity of whole parts of *Sphaeranthus indicus* Linn. *J Chem Pharma Res* 2009; 1:207-212.

Nayak SS, Maity TK, Maiti BC. Hepatoprotective activity of *Sphaeranthus indicus* Linn. *Int J Green Pharm* 2007; 1:32-36.

Pande VV, Dubey S. Antihyperlipidemic activity of *Sphaeranthus indicus* on atherogenic diet induced hyperlipidemia in rats. *Int J Green Pharm* 2009; 3:159–161.

Prabhu KS, Lobo R, Shirwaikar A. Antidiabetic properties of the alcoholic extract of *Sphaeranthus indicus* in streptozotocin-nicotinamide diabetic rats. *J Pharm Pharmacol* 2008; 60:909-916.

Pujar PP, Sawaikar DD, Rojatkar SR, Nagasampagi BA. Eudesmanoids from *Sphaeranthus indicus*. *Fitoterapia* 2000; 71:264-268.

Rojatkar SR, Nagasampagi BA. 7-hydroxyeudesmanolides from *Spharanthus indicus*. *Phytochem*. 1992; 31:3270-3271.

Sarpate RV, Deore TK, Tupkari SV. Bronchodilatory effect of *Sphaeranthus indicus* Linn against allergen induced bronchospasm in guinea pigs. *Pharmacog Mag* 2009; 5:74–77.

Shekhani MS *et al.*, An immunostimulant sesquiterpene glycoside isolated from *Sphaeranthus indicus*. *Phytochem* 1990; 29:2573-2576.

Shirwaikar A, Prabhu KS, Punitha IS. *In vitro* antioxidant studies of *Sphaeranthus indicus* (Linn.) *Indian J Exp Biol* 2006; 44:993-996.

Singh SK, Saroj KM, Tripathi VJ, Singh AK, Singh RH. An Antimicrobial Principle from *Sphaeranthus indicus* L. (Family Compositae) *Int J Crude Drug Res.*1988; 26: 235-239.

Singh SK, Tripathi VJ, Singh RH. β-D-glucoside of (24S)-24-ethylcholesta-4, 22-dien-3- β-ol from *Sphaeranthus indicus* L. *Indian drugs* 1989; 26:317-18.

Sohoni JS, Rojatkar SR, Kulkarni MM, Dhaneshwar NN, Tavale SS, Gururow TN *et al.*, A new eudesmenolide and 2-hydroxycostic acid from *Spharanthus indicus* Linn. x-ray molecular structure of 4-alpha, 5-alpha-epoxy-7-alpha-hydroxyeudesmanolide. *J Chem Soc Perkin* 1 1988; 2:157-160.

Srinivasan VM, Jessy KK, Anand Alex EM. Effect of *Sphaeranthus indicus* Linn. on Gentamicin induced acute renal failure in rats. *Indian J Pharmacol* 2008; 40:71.

Srivastav SC, Khan MS, Vohra SB. Pharmacological and haemostatic investigation on *Sphaeranthus indicus* Linn. *Indian J Physiol Pharmacol* 1971; 15:27–33.

Tiwari Bk, Khosa RL. Hepatoprotective and antioxidant effects of *Sphaeranthus indicus* against acetaminophen induced hepatotoxicity in rats. *J Pharma Sci Res* 2009; 1:26-130.

Vikani KV, Dangar RD, Kapadia NS, Shah MB. A pharmacognostic study on *Sphaeranthus indicus*. *J Nat Rem* 2008; 8:61–67.

Yadava RN, Kumar S. A novel isoflavone glycoside from the leaves of *Sphaeranthus indicus*. *Fitoterapia* 1999; 70:127-129.

Yadava RN, Kumar S. Chemical examination of the leaves of *Spharanthus indicus*. *Asian J Chem* 1998; 10:764-766.

***Mimosa pudica* L.**

Common name: Sensitive Plant, Touch me not.

Ayurvedic names: *Gandhakarini, khadiri, lajjalu, mohini, namaskari, patri, raktapadika, samanga, sami* and *sprkka.*

Family: Fabaceae.

Habitat: *M. pudica* is native to South America and Central America. It has been introduced to many other regions and is regarded as an invasive species in Tanzania, South Asia and South East Asia.

General Description: *M. pudica* is a creeping annual or perennial herb. The stem is erect in young plants, but becomes creeping or trailing with age. The stem is slender, branching, and sparsely to densely prickly, growing to a length of 1.5 m (5 ft). The leaves are compound leaves. The leaves are bipinnately compound, with one or two pinnae pairs, and 10-26 leaflets per pinna. The petioles are also prickly. Pedunculate (stalked) pale pink or purple flower heads arise from the leaf axils. The globose to ovoid heads are 8–10 mm in diameter (excluding the stamens). The fruit consists of clusters of 2-8 pods from 1–2 cm long each, these prickly on the margins. The pods break into 2-5 segments and contain pale brown seeds some 2.5 mm long (Rajendran & Sundararajan, 2010).

Table 1: Botanical difference among the major species of *Mimosa* (*Source:* Society for Parthenium Management (SOPAM), Raipur.

Characters	*M. pudica*	*M. himalayana* syn. *M. rubicaulis*	*M. hamata*
Plant	Small woody herbs or low-spreading undershrub with hairy and prickly branches, hairs glandular	A large straggling shrub, studded with straw-coloured, hooked prickels	A much branched, armed shrub, branches downy, with numerous straw-coloured, curced or straight prickles
Leaves	Bipinnate, sensitive to touch, pinnae 1-2 pairs, leaflets 10-20 pairs, linear, glabrous	Bipinnate, main rachis with hooked prickles, pinnae 5-11 pairs, linear-oblong	2-pinante, main rachis pubescent, some timely prickly, leaflets 6-10 pairs

Flowers	Heads small, peduncled, globose, axilalry, pink-purple, Calyx campanulate, Petals crenate towards base	Numerous, in globose heads, peduncles crowded at the ends of branchlets	4-merous in globose heads, peduncles axillary, crowded at the end of branches
Pods	1.5-2.5 cm long, closely prickly on the sutures	7-10 cm long, falcate, glabrous, one seeded joints, persistant but not prickly	5-7 cm long, falcate, consisting 4-8 one seeded joints, pubescent

Chemical constituents: Tannin (10%), mucilage (composed of d-xylose and d-glucuronic acid), ash and alkaloid: mimosine, identical with leucenine from *Leucaena glauca*. An adrenaline-like substance has been identified in the extracts of leaves. The plant contains tubulin which shows the ability to bind colchicine with its sulfhydryl groups. A new class of phytohormones-turgorines is active in the plant. These periodic leaf movement factors are derivatives of 4-O-(β-D-glucopyranosyl-6-sulphate) gallic acid (Pande & Pathak, 2010; Rajendran & Sundararajan, 2010)

Specific action as per Ayurveda: Sandhaniya.

Therapeutics: According to Ayurveda, root is bitter, acrid, cooling, vulnerary, alexipharmic and used in treatment of biliousness, leprosy, dysentery, vaginal and uterine complaints, inflammations, burning sensation, fatigue, asthma, leucoderma, blood diseases etc. According to the Unani system of medicine, root is resolvent, alternative, useful in diseases arising from blood impurities and bile, bilious fevers, piles, jaundice, leprosy etc (Nayagam & Pushparaj, 1999).

Parts used: Whole plant and roots.

Bioactivity: Diuretic (Pillai *et al.,* 1978), hyperglycemic (Amalraj & Ignacimuthu, 2002), wound healing (Kokane *et al.,* 2009), antibacterial (Rajendran & Sundararajan, 2010), and antinociceptive (Karthikeyan, 2010).

Clinical trials: Menorrhagia (Vaidya & Sheth, 1986) and uterine prolapse (Shivanandaiah & Indudhar, 2010)

Further reading

Amalraj T, Ignacimuthu S. Hyperglycemic effect of leaves of *Mimosa pudica* Linn. *Fitoterapia* 2002; 73: 351-352.

Karthikeyan, M. Antinociceptive activity of *Mimosa pudica* Linn. *Iranian J Pharmacol Ther* 2010; 9:1.

Kokane DD, More RY, Kale MB, Nehete MN, Mehendale PC, Gadgoli CH. Evaluation of wound healing activity of root of *Mimosa pudica*. *J Ethnopharmacol* 2009; 124: 311-315.

Nayagam MC, Pushparaj MS. `Touch me not': A medicinal plant of the Nilgiri tribals - a study. *J Econ Taxon Bot* 1999; 23:417-420.

Pande M, Pathak A. Preliminary pharmacognostic evaluations and phytochemical studies on roots of *Mimosa pudica*. *Int J Pharm Sci Rev Res* 2010; 1: 50-52.

Pillai NR, Uma R, Thyagarajan R, Venugopal PM. Preliminary studies on the diuretic activity of *Mimosa pudica* Linn. in experimental animals. *J Res Indian Med* 1978; 13:114-117.

Rajendran R, Sundararajan R. Preliminary phytochemical analysis and anti-bacterial activity of *Mimosa pudica* Linn. leaves. *Int J Pharma Bio Sci* 2010; 1, 1 (article 29; 8 p.)

Shivanandaiah TM, Indudhar TM. Lajjalu treatment of uterine prolapse. *J Ayur Integ Med* 2010; 1:125-128.

Vaidya GH, Sheth UK. *Mimosa pudica* (Linn.): Its medicinal value and pilot clinical use in patients with menorrhagia. *Ancient Sci Life*.1986; 5:156-160.

Biophytum senstivitum R.Br.

Syn: *Biophytum petersianum* Klotz.; *Oxalis senstivita* L.

Common name: Life Plant.

Ayurvedic names: *Viparita lajjalu, jhulapushpa, peethapushpa.*

Family: Oxalidaceae.

Habitat: Tropical Africa and Asia.

General Description: A slender erect annual; leaves abruptly pinnate, leaflets opposite, 6-12 pairs, the terminal pair is the largest; flowers yellow with many peduncles; fruits ellipsoid capsules.

Chemical constituents: Amentoflavone.

Actions: Diuretic and styptic.

Therapeutics: Plant pacifies vitiated *kapha, pitta,* urinary calculi, wounds, abscesses, asthma and stomachalgia.

Parts used: Whole plant.

Bioactivity: Hypoglycemic (Puri and Baral, 1998), insulinotropic (Puri, 2001), hypocholesterolemic (Jachak, Bucar & Kartnig, 2003), immunomodulatory and antitumor Guruvayoorappan & Kuttan, 2007; Guruvayoorappan & Kuttan, 2008). Amentoflavone from the roots is selective inhibitor of cycloxgenase catalyzed prostaglandin biosynthesis.

Further reading

Guruvayoorappan C, Kuttan G. *Biophytum sensitivum* (L.) DC inhibits tumor cell invasion and metastasis through a mechanism involving regulation of MMPs, prolyl hydroxylase, lysyl oxidase, nm23, ERK-1, ERK-2, STAT-1, and proinflammatory cytokine gene expression in metastatic lung tissue. *Integ Can Ther* 2008; 7:42-50.

Guruvayoorappan C. Kuttan G. Immunomodulatory and antitumor activity of *Biophytum sensitivum. Asian Pac J Cancer Prev* 2007; 8: 27-32.

Jachak SM, Bucar F, Kartnig T. Anti-inflammatory activity of *Biophytum sensitivum* in carrageenin induced rat paw oedema. *Phytother Res* 1999; 13:73-74.

Puri D, Baral N. Hypoglycemic effect of *Biophytum sensitivum* in alloxan diabetic rabbits. *Indian J Physiol Pharmacol* 1998, 42:401-406.

Puri D. Hypocholesterolemic Effect of *Biophytum sensitivum* leaf water extracts. *Pharm Biol* 2003; 41: 253-258.

Puri D. The insulinotropic activity of a Nepalese medicinal plant *Biophytum sensitivum*: preliminary experimental study. *J Ethnopharmacol* 2001; 78: 89–93.

6. AMARVALLI

Current representative: *Cuscuta reflexa*

Possible representative: *Cassyatha filiformis*

Cuscuta reflexa Roxb.

Common name: Giant dodder.

Ayurvedic names: *Amaravallari, akasha vela* and *khavalli.*

Family: Convolvulaceae.

Habitat: Subtropical and tropical countries.

General Description: Dodder can be identified by its thin stems appearing leafless, with the leaves reduced to minute scales. From mid-summer to early autumn, the vines can produce small fruit that take the same color as the vine, and are approximately the size of a common pea. Flowers range in color from white to pink to yellow to cream (Issar, 1980).

Chemical constituents: Cuscutin, sterol, flavonoid: quercetrin, glucoside, alkaloid and oil. Lupeol has been reported (Joshi *et al.*, year not published). Reflexin (5-hydroxy-7-methoxy-6-(2, 3-epoxy-3-methylbutyl)-flavanone) has bene reported (Tripathi, Yadav & Upadhyay, 2005).

Specific action as per Ayurevda: *Stambhana* (astringent).

Therapeutics: Diarrhea, worm infestation, rheumatoid arthritis, skin diseases and fever. Expressed juice of the plant is used in conjunctivitis and pruritis.

Parts used: Whole plant.

Bioactivity: Antioxidant (Yadav *et al.,* 2000), muscle relaxant and analgesic (Pal *et al.,* 2003), hair-growth promotion (Pandit, Chauhan & Dixit, 2008) and diuretic (Sharma *et al.,* 2009).

Further reading

Issar RK. Pharmacognostic study of three parasitic herbs (*Cassytha filiformis* Linn., *Cuscuta reflexa* Roxb. and *Cuscuta chinensis* Lamk.) used differently in traditional systems of medicine. *Bull Med Ethnobot Res* 1980; 1:478-493.

Joshi H, Shailajan S, Menon SK, Tiwari B. Microwave assisted extraction of Lupeol from *Cuscuta reflexa* Roxb. growing on different hosts. mayagroup.org.in/confpaper/Harshvardhan%20Joshi.doc.

Pal DC, Panda SS, Dutta A, Bhattacharya S. Evaluation of psychopharmacological effects of petroleum ether extract of *Cuscuta reflexa* Roxb. stem in mice. *Acta Poloniae Pharm Drug Res* 2003; 60:481-486.

Pandit S, Chauhan NS, Dixit VK. Effect of *Cuscuta reflexa* Roxb on androgen-induced alopecia. *J Cosmet Dermatol* 2008; 7:199-204.

Sharma S, Hullatti SN, Prasanna SM, Kuppast IJ, Sharma P. Comparative study of *Cuscuta reflexa* and *Cassytha filiformis* for diuretic activity *Pharmacog Res* 2009; 1: 327-330.

Tripathi VJ, Yadav SB, Upadhyay AK. A new flavanone, reflexin, from *Cuscuta reflexa* and its selective sensing of nitric oxide. *Appl Biochem Biotechnol* 2005; 127:63-67.

Yadav SB, Tripathi V, Singh RK, Pandey HP. Antioxidant activity of *Cuscuta reflexa* stems. *Indian J Pharm Sci* 2000; 62: 477-478.

Cassyatha filiformis L.

Common name: Love vine.

Ayurvedic name: *Akashavela.*

Family: Lauraceae.

Habitat: Subtropical and tropical countries.

General Description: Stems twining, pale green to yellow-green to orange, filiform, glabrous or pubescent. Leaves alternate, ca. 1 mm. Inflorescences spikes, rarely reduced to single flower. Flowers bisexual, subtended by ciliate bract and bracteoles; outer 3 tepals 0.1-1 mm, similar to bracteoles, inner 3 tepals 1-1.8 mm, apex incurved; pistil 1.2 mm, glabrous. Drupe to 7 mm diameter.

Chemical constituents: Aporphine alkaloids: laurotetanine, cassythine and cassthyidine (Major *et al.,* 1979) and dulcitol.

Therapeutics: Bloody diarrhoea and urinary tract infections.

Parts used: Whole plant.

Further reading

Major P *et al.* Aporphine alkaloids *of Cassytha filiformis. BMEBR.* 1979; 1: 478-493.

Srivastava J. On *Cassytha filiformis* Linn. *Curr Sci.* 1945; 14: 242.

7. AMLA VETASA

Current representative: *Garcinia pedunculata*

Possible representative: *Rheum emodi*

Garcinia pedunculata Roxb. ex Buch.-Ham.

Ayurvedic names: *Amla*, *cukra*, *sarahedaka* and *vetasa.*

Family: Guttiferae.

Habitat: Found and cultivated in India.

General Description: Trees 20 m tall. Bark thick, corky; branchlets obtusely 4-angled or subterete, striate, lenticellate, glabrous. Petiole 2-2.5 cm; leaf blade oblong, obovate, or oblong-lanceolate, (12-)15-25(-28) × 7-12 cm, papery, midvein robust, raised abaxially, somewhat impressed adaxially; secondary veins regular, oblique, 9-14 pairs, near margin arching and joining together; tertiary veins nearly parallel, almost inconspicuous, base cuneate, margin conspicuously narrowly involute, apex usually rounded, rarely obtusely acuminate. Plant dioecious, flowers 4-merous. Male flowers 8-12 in an erect 8-15 cm paniculiform cyme; peduncle 3-6 cm; pedicels robust, 3-7 cm; sepals broadly ovate or suborbicular, thick, fleshy, margin membranous; petals yellow, oblong-lanceolate, 7-8 mm; stamen fascicles connate in capitate ring ca. 3 mm high, anthers sessile, or a few near pistillode with short filaments, anthers 2-celled, cells longitudinally dehiscent; pistillode columnar-cuneate, slightly angular; stigma peltate, inconspicuously tuberculate. Female flowers usually in pairs or solitary at apex of branchlet; pedicels robust, slightly tetragonous, with 2 suborbicular bracts at base; staminodes basally united, surrounding ovary, 80-100, upper parts free; ovary subglobose, 8-10-loculed; stigma radiate, 8-10-lobed, papillate. Fruit yellow, large, oblate, concave on both ends when mature 10-18 × 11-20 cm, smooth; fruiting pedicel 5-6 cm. Seeds 8-10, reniform (Togunashi & Venkataram, 1976).

Chemical constituents: Malic acid and polyisoprenylated benzophenones (Das & Chatterjee, 1989).

Specific action as per Ayurevda: *Triptighana*.

Therapeutics: Indigestion, constipation, dysuria and loss of appetite (Sharma, 2000).

Parts used: Fruit.

Further reading

Das SB, Chatterjee A. Polyisoprenylated benzophenones from *Garcinia pedunculata*. *Phytochemistry* 1989; 28:1233-1235.

Sharma PV. Amlavetas par punarvicar. *Sachitra Ayurved* 2000; 52:1007-1011; repr. in: S.D. Dubey and A.N. Singh (Eds.) (2005), 483-486.

Togunashi VS, Venkataram BS, Yoganarasimhan SN. Discussion and identification of amlavetasa. *Nagarjun* 1976; 20:15-17.

Rheum emodi Wall.

Common name: Indian rhubarb.

Ayurvedic names: *Amlaparni*, *revtika* and *pitamuli.*

Family: Polygonaceae.

Habitat: Himalayas and Kashmir.

General Description: *R. emodi* is a leafy perennial herb, streaked green. Radical leaves, orbicular or broadly ovate, cordate at base, often long, 5 to 7 nerved papilose beneath. Petiole long and stout. Inflorescence panicles, flower dark purple, sepals 5, stamens 6 to 9, ovary 2 to 4, fruit long, ovoid-oblong, purple.

Chemical constituents: Anthraquinones (emodin, rhein, rhaponticin, chrysophanol, sennoside A and B, resins (aporetin, erythroretin and phoeoretin), gallic acid, rheo-tannin acid, catechin, cinnamic acid, starch, calcium oxalate and essential oil (eugenol is main constituent). Two new oxanthrone esters, revandchinone-1, revandchinone-2, a new anthraquinone ether revandchinone-3 and a new oxanthrone ether, revandchinone-4 have been reported.

Specific action as per Ayurveda: *Pitta virecaka* (cholagouge purgative).

Therapeutics: Rhubarb is stomachic, bitter, tonic, cathartic. Useful for simple diarrhea, but not in constipation or any affection in which a continuous aperient action is necessary; its stimulating combined with its aperient properties renders it valuable in atonic dyspepsia. Best used for the ailments of children and aged persons. Combined with ginger, it may be given in the form of pill in cases where the bowels are sluggish. Ordinary dose of the powder is from 5 to 20 grains.

Parts used: Rhizome.

Bioactivity: Antifungal activity of rhein, physcion, aloe-emodin and chrysophanol (Agarwal, Singh, Verma and Kumar, 2000) and alpha-glucosidase inhibitory activity of rhapontigenin, desoxyrhapontigenin, chrysophanol-8-O-beta-d-glucopyranoside, torachrysone-8-O-beta-d-glucopyranoside isolated from methanolic extract of rhizome (Suresh Babu *et al.*, 2004).

Further reading

Agarwal SK, Singh SS, Verma S, Kumar S. Antifungal activity of anthraquinone derivatives from *Rheum emodi*. *J Ethnopharmacol* 2000; 72:43-46.

Azhar Alam MM, Javed K, Jafri MA. Effect of *Rheum emodi* (Revand Hindi) on renal functions in rats. *J Ethnopharmacol* 2005; 96: 121-125.

Babu KS, Srinivas PV, Praveen B, Kishore KH, Murty US, Rao JM. Antimicrobial constituents from the rhizomes of *Rheum emodi*. *Phytochemistry* 2003; 62:203–207.

Babu KS, Tiwarib AK, Srinivasa PV, Alib AZ, Rajua BC, Rao JM. Yeast and mammalian α-glucosidase inhibitory constituents from Himalayan rhubarb *Rheum emodi* Wall. ex Meisson. *Bioorg Med Chem Lett* 2004; 14:3841-3845.

Khorana ML, Kaji NN. Studies in *Rheum emodi*, Indian rhubarb. *Indian J Pharm* 1965; 27: 91.

Sharma AP, Sharma RK. Pharmacognostic studies on the rhizome of *Rheum emodi* Wall. *J Sci Res Plant Med* 1980; 1: 32-36.

Venkatadri RK, Guha G, Rangasamy AK. Antioxidant and Anti-Cancer Potentials of *Rheum emodi* Rhizome Extracts. Evidence-Based Comp Alter Med. 2011; Article ID 697986, 9 pages doi:10.1093/ecam/neq048.

8. ANSHUMATI

Current representative: *Desmodium gangeticum*

Possible representatives: *Uraria picta*, *Stereospermum tetragonum* and *Pseudoarthria viscida*

Desmodium gangeticum DC. var. *maculatum* (L.) DC. (L.) Baker

Syn: *Desmodium natalitium* Sond., *Hedysarum lanceolatum* Schum. & Thonn.

Other Ayurvedic names: *Shalparni.*

Family: Fabaceae.

Habitat: Africa and Asia.

General Description: An erect, branched shrub grows up to 1 meter in height. Stem is woody with numerous prostrate branches and covered with soft hairs. Leaves are single, ovate-lanceolate, and covered with numerous gray hairs. Flowers white, purple or lilac found in elongated terminal or axillary recemes. Fruits are moniliform 6-8 joined pods. Joints separate when ripe into one seeded segments. Seeds are compressed reniform (Mehra & Raina, 1969; Janardan, 1978).

Root morphology

(a) *Macroscopic*: Tap root, poorly developed, but lateral roots 15-30 cm long, and 0.1-0.8 cm thick, uniformly cylindrical with a number of branches; surface smooth bearing a number of transverse, light brown lenticels, bacterial nodules frequently present; light yellow; fracture fibrous; odour not characteristic; taste, sweetish and mucilaginous.

(b) *Microscopic*: Mature root shows cork, 3-7 layers of thin-walled, tangentially elongated cells, having a few prismatic crystals of calcium oxalate; cork cambium single layered; secondary cortex 4-10 layers of thin-walled, tangentially elongated cells having a few isolated cortical fibres; secondary phloem composed of parenchyma, sieve tubes, companion cells and fibres, traversed by phloem rays; sieve tubes collapsed in outer region, but intact in inner region; phloem fibres slightly elongated, lignified; phloem rays uni to multiseriate, 1-4 cells wide and 4-15 cells high; outer phloem region having occasionally prismatic crystals of calcium oxalate; cambium 2-3 layers; secondary xylem having 1-2 growth rings, consisting of vessels, tracheids, xylem parenchyma, and xylem fibres, traversed by xylem rays; vessels, lignified, large, narrow, with both reticulate thickening or bordered pits; xylem parenchyma with rectangular or slightly elongated cells, resem-bling those of phloem parenchyma in shape but larger in size and xylem fibres resemble those of phloem fibres in shape but larger in size; xylem rays thickwalled possessing simple pits, 1-5 cells wide and 4-12 cells high; simple, round to oval starch grains measuring 7-25 μ in dia. and prismatic crystals of calcium oxalate present in secondary phloem and secondary xylem.

Powder - Light brown; shows fragments of rectangular cork cells, vessels having reticulate thickening and bordered pits, xylem fibres, ray cells, prismatic crystals of calcium oxalate and simple round to oval starch grains, measuring 7-25 μ in dia.

Chemical constituents: Alkaloid, resin and pterocarpene: gangetin (Niranjan and Tewari, 2008).

Specific action as per Ayurveda: *Angamurda prashamana.*

Therapeutics: It is a general tonic and aphrodisiac, has a calming, sedative effect and is used to control inflammation, fever and neurological imbalances.

Parts used: Whole plant.

Bioactivity: Anti-inflammatory and analgesic (Ghosh & Anandakumar, 1983), antiamnesic (Hanumanthachar & Milind, 2007), antiinflammatory and antioxidant (Govindarajan *et al.,* 2007; Niranjan & Tewari, 2008) and hypoglycemic (Raghavan *et al.,* 2007) activites. Gangetin (pterocarpene) isolated from hexane extract has anti-inflammatory and anlagesic (Ghosh & Anandakumar, 1983) and antifertility effects (Latha & Govindasamy, 1997; Latha, Govindaswamy & Balakrishna, 1997).

Further reading

Ghosh D, Anandakumar A. Anti-inflammatory and analgesic activities of gangetin, a pterocarpenoid from *Desmodium gangeticum. Indian J Pharmacol* 1983; 15:391-402.

Govindarajan RM, Vijayakumar Ch. V. Rao, Shirwaikar A, Kumar S, Rawat AKS, Pushpangadan P. Antiinflammatory and antioxidant activities of *Desmodium gangeticum* fractions in carrageenan-induced inflamed rats. *Phytother Res* 2007; 21: 975-979.

Hanumanthachar J, Milind P. Pharmacological evidences for the antiamnesic effects of *Desmodium gangeticum* in mice. *Iranian J Pharma Res* 2007; 6:198-200.

Janardan S. Comparative pharmacognostical studies on Desmodium species. *Nagarjun* 1978; 21:18-20.

Latha P, Govindasamy S. Gangetin - a reproductive inhibitor in male rats. *Phytother Res* 1997; 11: 466-468.

Latha P, Govindaswamy S, Balakrishna K. Effect of gangetin on fertility of male rats. *Phytother Res* 1997; 11:372-375.

Meena AK, Rao MM, Ajit Kandale *et al.*, Standardisation of *Desmodium gangeticum* - A Tradition Ayurvedic Plant. *Drug Invention Today* 2010; 2:182-184.

Mehra PN, Raina MK. Pharmacognostical studies on the roots of *Desmodium gangeticum* {DC}. *Indian J Pharm* 1969; 31:171.

Niranjan, A. and S.K. Tewari. Phytochemical composition and anti-oxidant potential of *Desmodium gangeticum* (Linn.) DC. *Nat Prod Rad* 2008; 7:35-39.

Raghavan G, Anane HS, Persaud S, Jones P, Houghton PJ. Effect of *Desmodium gangeticum* extract on blood glucose in rats and on insulin secretion *in vitro*. *Planta Med* 2007; 73:427-432.

Uraria picta (Jacq.) L.

Syn: *Doodia picta* Roxb., *Hedysarum pictum* Jacq., *Uraria aphrodisiaca* Welw.

Ayurvedic names: *Prshniaparni.*

Family: Fabaceae.

Habitat: Throughout India.

General Description: An erect, under shrub upto 90 cm high.

(a) *Macroscopic*

Root - Occur in pieces of varying size, thickness of 1 to 2 cm, gradually tapering, tough,woody, cylindrical; externally light yellow to buff, internally pale yellow; surface bearing fine longitudinal striations; fracture, splintery or fibrous; taste, slightly acrid.

Stem - About 8.0 to 16.0 cm long, 0.2 to 0.4 cm in diameter, in cut pieces; cylindrical, branched, pubescent, external surface light yellow to brown; transversely cut and smoothened surface shows buff-white colour, mature stem longitudinally wrinkled, leaf scar present at nodes; fracture, fibrous.

Leaf - Very variable, imparipinnate, upto 20 cm or more long, upto 2 cm wide; leaflets on the upper part of the stem 5 to 7, rigidly sub-coriaceous, linear-oblong, acute, blotched with white; glabrous above, finely reticulately veined and minutely pubescent beneath, base rounded; leaflets on the lower part of the stem 1 to 3, sub-orbicular or oblong.

(b) *Microscopic*

Root - Shows 5 or 6 layers of thin-walled, tabular, regularly arranged cork cells; cork cambium single layered; secondary cortex composed of 4 to 6 layers of oval, tangentially arranged, thin-walled, parenchymatous cells, a few fibres present singly or in groups; secondary phloem composed of sieve elements, parenchyma and fibres traversed by phloem rays; sieve elements somewhat collapsed towards periphery but intact in inner phloem region; phloem parenchyma composed of rounded to somewhat oval cells, larger towards periphery; fibres thick-walled, lignified with narrow lumen and tapering ends;phloem rays 1 to 5 cells wide, their cells being oval or rectangular in the portion nearer the wood but broader towards their distal ends; secondary xylem composed of vessels, tracheids, fibres, crystal fibres and parenchyma traversed by xylem rays; vessel very few, mostly confined to inner and outer part of xylem; fibres similar to those of phloem fibres and arranged in close set concentric bands; in isolated preparation vessels are cylindrical, pitted with transverse to oblique perforation; tracheids possess bordered pits; xylem parenchyma mostly rectangular with simple pits; xylem ray cells isodiametric showing simple pits; starch grains simple, round to oval, measuring 6 to 17μ in dia., distributed throughout parenchymatous cells of secondary cortex, phloem and xylem; prismatic crystals of calcium oxalate present in crystal fibres, as well as in many parenchymatous cells of secondary cortex, phloem and ray cells.

Stern - Shows single layered epidermis covered with cuticle, a few epidermal cells elongate outwards forming papillae; cortex 8 to 10 cells wide, consisting of oval to circular, thin walled, parenchymatous cells; groups of pericyclic fibres present in the form of discontinuous ring; phloem consisting of usual elements except phloem fibres; phloem rays 2 to 4 cells wide; xylem consisting of usual elements; vessels mostly simple pitted; fibres simple with blunt tips; xylem rays 1 to 4 cells wide and 2 to 8 cells in height; pith wide, consisting of thin-walled, round to oval parenchymatous cells.

Leaf- Midrib - single layered epidermis on either surfaces covered with striated cuticle having a few unicellular or bicellular, hooked or straight and pointed tipped hairs present on both surfaces but more on lower surface; collenchyma 2 or 3 layered, followed by 2 layers of parenchyma cells; single row of pericyclic fibers present on both sides; vascular bundle located centrally.

Lamina - shows single layered epidermis on either surfaces, a few unicellular or bicellular, hooked or straight, pointed tipped hairs present on lower surface; mesophyll differentiated into single layered palisade and spongy parenchyma; spongy parenchyma cells oval to rounded having small intercellular spaces; numerous paracytic stomata present on lower surface; stomatal index 27 to 36 on lower surface; palisade ratio 4 or 5; vein-islet number 29 to 32 per sq. mm.; vascular bundle present centrally.

Powder - Greenish-yellow; shows simple pitted vessels; fragments of fibres, tracheids, parenchyma cells; pieces of hairs; palisade cells; a few prismatic crystals of calcium oxalate; epidermal cells wavy walled in surface view showing paracyic stomata and starch grains simple, round to oval, measuring 6 to 17 μ in dia. (Prasad & Ansari, 1967).

Chemical constituents: Isoflavanones.

Specific action as per *Ayurveda*: *Angamurda prashamana.*

Therapeutics: Loss of appetite, fever and general debility.

Parts used: Whole plant.

Bioactivity: It helps in fracture healing in animal models (Prasad, Sankaran & Deshande, 1964).

Further reading

Prasad S, Ansari MS. Pharmacognostical studies on the root of *Uraria picta* Desv (*prishniparni*). *J Res Indian Med* 1967; 2:16-24.

Prasad GC, Shankaran PS Deshpande PJ. Effect of *Uraria picta* on fracture healing. *J Exp Med Sci* 1964; 8:43-48.

Pseudarthria viscida (L.) Wight & Arn.

Family: Fabaceae.

Habitat: *P. viscida* grows throughout south India up to 900 m in the hills and in Gujrat.

General Description: A semi erect, pubescent, perennial shrub grows up to 50 cm to 1 meter in height. Branches are covered with whitish hairs. Leaves are hairy, compound; trifoliate, terminal leaflet rhomboid ovate, lateral ones are obliquely ovate or oblong. Flowers are purplish or pink, small numerous, found on terminal and axillary spikes. Fruits pods, flattened and covered with hairs, seeds 4-6.

Chemical constituents: Leucopelargonidin.

Therapeutics: Roots of this plant have been traditionally used for various medicinal properties such as anti-inflammatory, diuretic, cardiotonic, diabetes, diarrhoea, fever, asthma, worms and urinary diseases.

Parts used: Whole plant.

Bioactivity: Antifungal (Deepa, Narmatha and Basker, 2004), antidiabetic (Masirkar *et al.*, 2008), anti-diarrhoeal (Vijayabaskaran *et al.*, 2010), antioxidant (Vijaybaskaran *et al.*, 2010) and neuroprotective (Singh *et al.*, 2011).

Further reading

Deepa MA, Narmatha Bai V, Basker S. Antifungal properties of *Pseudarthria viscida Fitoterapia* 2004; 75:581-584.

Masirkar VJ, Deshmukh VN, Jadhav JK, Sakarkar DM. Anti Diabetic Activity of the Ethanolic Extract of *Pseudarthria Viscida* Root. against Alloxan Induced Diabetes in Albino Rats. *Res J Pharm Tech* 2008; 1: 541-542.

Shama N, Mohana Lakshmi S. Standardisation of *Pseudarthria viscida* (L.) Wight & Arn. (Leguminosae). *Int J Phytopharm Res* 2011; 2:58-61.

Singh JC, Alagarsamy V, Parthiban P, Selvakumar P, Reddy YN. Neuroprotective potential of ethanolic extract of *Pseudarthria viscida* (L) Wight and Arn against beta-amyloid (25-35)-induced amnesia in mice. *Indian J Biochem Biophys* 2011; 48:197-201.

Vijayabaskaran M, Venkateswaramurthy N, Babu G, Khatale PN. Antidiarrhoeal activity of *Pseudarthria viscida roots. Int J PharmTech* 2010; 2:307-313.

Vijaybaskaran M, Venkateswaramurthy N, Perumal N, Jayakar B. *In vitro* antioxidant evaluation of *Pseudarthria viscida Linn. Int J Curr Pharma Res* 2010; 2:20-22.

Prasad NBR, Krishnan PN. Leucopelargonidin from the roots of *Pseudarthria viscida* W & A. *J Res Indian Med* 1976; 11:104-105.

Note: See notes on *Stereospermum tetragonum.*

9. ARJUNA

Current representative: *Terminalia arjuna*

Possible representative (s): *Sterculia urens, Terminalia tomentosa and Lagerstromia flos-reigna*

Terminalia arjuna Roxb. W. & A.

Syn: *Terminalia berryi* Wight & Arn., *Terminalia crenulata* Roth, *Terminalia glabra* (Roxb.) Wight & Arn., *Terminalia ovalifolia*

Rottler, *Pentaptera angustifolia* Roxb., *Pentaptera arjuna* Roxb. ex DC., *Pentaptera glabra* Roxb.

Common name: Malabar almond.

Ayurvedic names: *Arjuna, dhavala, indradaru, kakubha, nadisraja and viravruksha.*

Family: Combretaceae.

Habitat: *T. arjuna* is a deciduous tree (about 60-70 feet height), is abundantly found in India and Ceylon, also in Myanmar and Sri Lanka.

General Description: Tree up to 25 meter hight; bark grey, smooth; leaves sub-opposite, 5–14 × 2–4.5 cm., oblong or elliptic oblong, glabrous, often inequilateral, margin often crenulate, apex obtuse or sub-acute, base rounded or sometimes cordate; petioles 0.5–1.2 cm; glands usually two. Flowers small, white. Fruit 2.3–3.5 cm long, fibrous woody, glabrous with 5 hard wings, striated with numerous curved veins. Flowering time April–July in Indian conditions.

Detailed Description of the Bark

(a) *Macroscopic*

Bark available in pieces, flat, curved, recurved, channelled to half quilled, 0.2-1.5 cm thick, market samples upto 10 cm in length and upto 7 cm in width, outer surface somewhat smooth and grey, inner surface somewhat fibrous and pinkish, transversely cut smoothened bark shows pinkish surface, fracture, short in inner and laminated in outer part; taste, bitter and astringent.

(b) *Microscopic*

Stem Bark -Mature bark shows cork consisting of 9-10 layers of tangentially elongated cells, a few outer layers filled with brown colouring matter; cork cambium and secondary cortex not distinct and medullary rays observed traversing almost upto outerbark; secondary phloem occupies a wide zone, consisting of sieve tubes, companion cells, phloem parenchyma and phloem fibres, traversed by phloem rays, usually uniseriate but biseriate rays also occasionally seen; in the middle and outer phloem region, sieve tubes get collapsed and form ceratenchyma; phloem fibres distributed in rows an present in groups of 2-10; rosette crystals of calcium oxalate measuring 80-180μ in dia., present in most of the phloem parenchyma, alternating with fibres; idioblasts consisting of large cells having aggregates of prismatic and rhomboidal crystals of calcium oxalate in row throughout the zone, measuring

260-600 μ in dia., starch grains, mostly simple, compound of 2-3 components, sometimes upto 5 components, round to oval, elliptical, measuring 5-13 μ in dia., distributed throughout the tissue (absent in *T. alata*); in a tangential section the uniseriate phloem rays 2-10 cells high and biseriate, 4-12 cells high; in longitudinal section rosette crystals of calcium oxalate found in the form of strands in phloem parenchyma.

Powder - Reddish-brown; shows fragments of cork cells, uniseriate phloem rays, fibres,a number of rosette crystals of calcium oxalate, a few rhomboidal crystals, starch grains simple and compound, round to oval, elliptic, having 2-3 components with concentric striations and small narrow hilum, measuring 5-13 μ in diameter (Gupta , Sen & Udupa,1976; Dwivedi & Udupa,1989).

Chemical composition: ß-sitosterol, tannic acid, ellagic acid, temic acid, triterpene glycosides: arjunetin, fridelin, arjunolic acid, arjunic acid, arjunin, arjunetoside, 3-O-beta-D-glucopyranosyl-(1--> 6)-O-beta-D-galactopyranoside, terminoside A, termiarjunoside I and termiarjunoside II, arjunetoside and naphthanol glycoside (Anjaneyulu & Prasad 1982a; Anjaneyulu & Prasad, 1982 b; Anjaneyulu & Prasad, 1983; Ali *et al.,* 2003a; Ali *et al.,* 2003 b; Pandey & Kori, 2009.

Action: Cardiac tonic.

Therapeutics: Angina pectoris, hyperlipidemia, spermatorrhoea, bloody diarrhea and pruritis. Locally the drug finds application in hemorrhage and fractures. According to Ayurvedic texts it also very useful in the treatment of any sort of pain due to falls, ecchumosis, spermatorrhoea and sexually transmitted diseases such as gonorrhoea. It is thought to be a useful astringent, cooling, aphrodisiac, cardiotonic, tonic and is used for ulcers, leucorrhoea, diabetes, cough, tumour, excessive perspiration, asthma, inflammation and skin disorders etc .

According to *Bhava Prakash*, It is cooling, *kapha* and *pitta* pacifying, cardiac restorative, helpful in wounds, tuberculosis and poisoning; good for obesity and urinary disorders. Chakradatta advised to take it by processing in milk for cardiac disorders alone or with *Panchamula* or with *Bala*. It would be helpful in cardiac disorders, indigestion, fever and bleeding disorders. He advises it in fractures with milk.

Parts used: Bark.

Formulations based on Arjuna: Arjunaristha and Arjunaghrita (Narayana & Kumaraswamy,1996).

Bioactivity: Antifertility (Lal & Udupa, 1993), antimutagenic (Kaur, Grover & Kumar, 1997), hypoglycemic and hypolipidemic (Ragavan & Krishnakumari, 2001), gastroprotective (Devi *et al.*, 2007), hypoglycemic and hypolipidemic (Ragavan & Krishnakumari, 2001). Arjunolic acid has platelet aggregation, coagulation and antioxidant acvitites (Sumitra *et al.*, 2001). Arjunaphthanoloside isolated from *T. arjuna* has antioxidant and nitric oxide inhibitory activities. Terminoside A has nitric oxide inhibitory activity (Ali *et al.*, 2003).

Clinical trial: A study reported cardio protective activity of bark extract of *T. arjuna* in 12 patients with refractory congestive cardiac failure (Bharani *et al.*, 2002). The extract demonstrated significant cardio protective activity as compared to placebo. A clinical reported beneficial effect of *T. arjuna* in patients with history of ventricular tachyarrhythmias and angina pectoris (Dwivedi, Avasthi & Mahajan, 1994; Bharani *et al.*, 2002).

Further reading

Ali A, Kaur G, Hamid H, Abdullah T, Ali M, Niwa M, Alam MS. Terminoside A, a new triterpene glycoside from the bark of *Terminalia arjuna* inhibits nitric oxide production in murine macrophages. *J Asian Nat Prod Res* 2003; 5:137-42.

Ali A, Kaur G, Hayat K, Ali M, Ather M. A novel naphthanol glycoside from *Terminalia arjuna* with antioxidant and nitric oxide inhibitory activities. *Pharmazie* 2003; 58:932-934.

Anjaneyulu ASR, Prasad AVR. Chemical examination of roots of *Terminalia arjuna* (Roxb.) Wight & Arnot. Part I. Characterisation of two new triterpenoid glycosides. *Indian J Chem* 1982a; 21B:530–533.

Anjaneyulu ASR, Prasad AVR. Chemical examination of roots of *Terminalia arjuna*-the structure of arjunoside III and arjunoside IV, two new triterpenoid glycosides. *Phytochemistry* 1982b; 21:2057-2060.

Anjaneyulu ASR, Prasad AVR. Structure of terminic acid, a dihydroxy triterpene carboxylic acid from *Terminalia arjuna*. *Phytochemistry* 1983; 22:993–998.

Bharani A, Ganguli A, Mathur LK, Jamra Y, Raman PG. Efficacy of *Terminalia arjuna* in chronic stable angina: a double-blind, placebo-controlled, crossover study comparing *Terminalia arjuna* with isosorbide mononitrate. *Indian Heart J* 2002; 54: 170-175.

Damodaran KS, Prabhakar YS. On the ethnomedical significance of the arjun tree, *Terminalia arjuna* (Roxb.) Wight et Arnot. *J Ethnopharmacol* 1987; 20: 173-190.

Devi RS, Narayan S, Vani G, Shyamala Devi CS. Gastroprotective effect of *Terminalia arjuna* bark on diclofenac sodium induced gastric ulcer. *Chem Biol Interact* 2007; 167:71–83.

Dwivedi S, Avasthi S, Mahajan S. Role of *Terminalia arjuna* in ventricular tachyarrhythmias. *Indian Pract* 1994; 57: 523-525.

Dwivedi S, Udupa N. *Terminalia arjuna*: pharmacognosy, phytochemistry, pharmacology and clinical use - A review. *Fitoterapia* 1989; 60: 413-420.

Gupta LP, Sen SP, Udupa KN. Pharmacognostical and pharmacological studies on *Terminalia arjuna. J Res Indian Med* 1976; 11:16-24.

Jain V, Poonia A, Agarwal RP, Panwar RB, Kochar DK, Mishra SN. Effect of *Terminalia arjuna* in patients of angina pectoris (a clinical trial) *Indian Med Gaz* (New Series) 1992; 36:56.

Lal B, Udupa KN. A preliminary study of antifertility effect of an indigenous drug arjuna (*Terminalia arjuna* W. et A.). *J Res Ayur Siddha* 1993; 14:165-169.

Maulik SK. Focused Conference Group: P16 - Natural products: Past and future? Role of *Terminalia arjuna* an Indian medicinal plant in cardiovascular diseases. *Basic Clin Pharmacol Toxicol* 2010; 107:445-446.

Narayana A, Kumaraswamy R. A medico-historical review of arjuna (*Terminalia arjuna* W. et A.). *BIIHM* 1996; 26:1-10.

Pandey AK, Kori DC. Variations in tannin and oxalic acid content in *Terminalia arjuna* (arjuna) bark. *Pharmacog Mag* 2009; 5:159-164.

Ragavan B, Krishnakumari S. Hypoglycemic and hypolipidemic activities of *Terminalia arjuna* stem bark in alloxan induced diabetic rat. *J Nat Rem* 6:124-130.

Satya P, Thakur AK. Evaluation of extract of arjuna (*Terminalia arjuna*) in stable angina. *Sach Ayur* 2001; 53: 625.

Shah CS, Bhavsar GC. Pharmacognosy of the bark of *Terminalia tomentosa* {W}. et {A}. and comparison with *Terminalia arjuna* {W}. et {A}. bark. *Indian J Pharm* 1956; 18:81-84.

Shaila HP, Udupa SL, Udupa AL. Hypocholesterolaemic activity in rats of different fractions from *Terminalia arjuna. Pharm Pharmacol Commun* 2000; 6:327-330.

Shukla SK, Dwivedi S, Singh SB, Sharma UR. *Terminalia arjuna* as a therapeutic and preventive modulator in experimentally induced myocardial infarction. *Diab Vas Dis Res* 2011; 1:80-88.

Singh DV, Gupta MM, Santha Kumar TR, Saikia D, Khanuja SPS. Antibacterial principles from the bark of *Terminalia arjuna. Curr Sci* 2008; 94: 27-29.

Singh G. Role of jaharamohara khatai and arjuna (*Terminalia arjuna*) in the management of ischaemic heart disease. *Ayu Intern* 2001; 1:1.

Sumitra M, Manikandan P, Kumar DA, Arutselvan N, Balakrishna K, Manohar BM, Puvanakrishnan R. Experimental myocardial necrosis in rats: role of arjunolic acid on platelet aggregation, coagulation and antioxidant status. *Mol Cel Biochem* 2001; 224:135–142.

Tiwari AK, Gode JD, Dubey GP. Effect of *Terminalia arjuna* on lipid profiles in rabbits fed hypercholesteromic diet. *Int J Crude Drug Res* 1990; 28:43-47.

Tiwari AK, Gode JD, Dubey GP. Effect of *Terminalia arjuna* on lipid profiles of rabbit fed hypercholesterolemic diet. *Int J Crude Drug Res* 1990; 28:43–47.

Yadav BL, Mathur R, Gupta RB. Antifertility activity of *Terminalia arjuna* W. et A. in female albino rats. *Probe* 1980; 19:196-198.

Sterculia urens Roxb.

Common name: Gum karaya, Indian Tragacanth, Thapsi tree.

Family: Sterculiaceae.

Habitat: Native to India, Myanmar and Sri Lanka.

General Description: *S. urens* is a medium-sized, deciduous tree to 15 m in height, usually with a clean, crooked, short bole up to 2 m DBH; branches large, spreading; bark thick, greyish-white or reddish, smooth, shining with a thin, white transparent outer coat, peeling off in papery flakes. Leaves on long petioles, crowded at the ends of branches, palmately 5-lobed, 20-30 cm diameter; tomentose beneath, glabrous above, entire, acuminate; stipules caducous. Flowers greenish yellow, small, in terminal panicles; follicles 4-6, ovoid-oblong, about 2.5 cm diameter, coriaceous, red, covered with stinging hairs. Fruit consists of 5 sessile, radiating, ovate-lanceolate hard, coriaceous carpels, 7.5 cm long, red when ripe, covered outside with many stiff bristles. Seeds 6 mm long, oblong, dark chestnut-brown, 3-6 per carpel.

Further reading

Khera IP, Kudalkar VG, Khorana ML. Study of gum tragacanth and gum karaya. *Indian J Pharm.* 1961; 23(4): 99.

Dikshith TSS, Raizada RB, Misra RB, Srivastava K. Toxicological evaluation of karaya gum; acute and subacute oral toxicity in rats. *J Biosci.* 1983; 5:25-30.

Kumar P, Ananthanarayana AK, Sharma, SN. Studies on physical and mechanical properties of *Sterculia urens. Indian Forester.* 1998; 114(4): 230-237.

Terminalia tomentosa W. & A.

Syn: *T. alata* Heyne ex Roth, *T. elliptica* Willd.

Habitat : Native to southern and southeast Asia in India, Bangladesh, Myanmar, Thailand, Laos, Cambodia, and Vietnam.

General Description: It is a tree growing to 30 m tall, with a trunk diameter of 1 m. The fruit is ovoid, 3 cm long, with five wings not extending beyond the fruit apex. The bark is fire-resistant. The wood is coarse, fairly straight grained; dull to somewhat lustrous and without any smell or taste. The heartwood varies from light brown with few markings to dark brown or brownish black and figured with darker streaks (Shah & Bhavsar, 1962; Sharma, 1997).

Chemical composition: Triterpene carboxylic acid: tomentosic acid (Row & Rao, 1962). The bark and especially the fruit yield pyrogallol and catechol. A new terpene glycoside, 2α, 3β, 19α-trihydroxy-olean-12-en-28-oic acid methylester 3β-*O*-rutinoside, a new flavanone, 8-methyl-5,7,2′, 4′-tetramethoxy-flavanone and a new chalchone glycoside, 2-*O*-β-glucosyloxy-4,6,2′,4′-tetramethoxy-chalchone have been reported (Srivastava, Srivastava 7 Chouksey, 1999).

Bioactivity: Antifungal (Srivastava, Srivastava & Chouksey, 2001).

Further reading

Hanumantha Rao K. A note on *Terminalia tomentosa. Indian J Pharm* 1959. 21:285.

Row LR, Rao GSR. Chemistry of *Terminalia* species—VI: The constitution of tomentosic acid, a new triterpene carboxylic acid from *Terminalia tomentosa* W. & A. *Tetrahedron* 1962; 18:827-838.

Shah CS, Bhavsar GC. Pharmacognosy of the bark of *Terminalia tomentosa* {W}. et {A}. and comparison with *Terminalia arjuna* {W}. et {A}. bark. *Indian J Pharm* 1956; 18; 81-84.

Sharma S, Kumar R. Controversy about *asvakarna*. *Sachitra Ayurved* 1997; 50: 298-303.

Sharma S. The identity of *asvakarna*. *Sachitra Ayurved* 1995; 28:18-122.

Srivastava SK, Srivastava SD, Chouksey BK. New antifungal constituents from *Terminalia alata*. *Fitoterapia* 2001; 72:106-112.

Srivastava SK, Srivastava SD, Chouksey BK. New constituents of *Terminalia alata*. *Fitoterapia* 1999; 70:390-394.

Lagerstroemia flos-regina Retz

Syn: *Lagerstroemia speciosa* (Linn.) Pers.

English name: Queens Flower Tree, Pride of India.

Family: Lythraceae.

Habitat: Asia.

General Description: A medium-sized to large much branched deciduous tree. Leaves elliptic or oblong-lanceolate, 10-20 cm long, acuminate. Flowers large, 5-7.5 cm across, showy, mauve purple, in large, terminal panicles, sometimes reaching 30 cm long. Fruits a subglobose capsule, 2-3.2 cm long.

Chemical constituents: Corosolic acid and maslinic acid.

Actions: Narcotic and purgative.

Parts used: Whole plant.

Bioactivity: Antidiabetic and anti-obesity.

Further reading

Web reference: *Lagerstroemia flos-regina*. Gautam Global, Dehra Dun, India

10. ARKA

Current representative: *Calotropis gigentia* and *Calotropis procera*

Calotropis gigentia (Linn) R. Br. ex Ait.

English name: Crown flower, milkweed or swallow-wort.

Ayurvedic names: *Alaraka*, *dirghapatraka*, *rajahava*, *sadapushpa* and *vasuka*.

Family: Asclepidaceae.

Habitat: Native to India (Lindley, 1985), *Calotropis* grows wild up to 900 meters throughout the country (Sastry & Kavathekar, 1990) on a variety of soils in different climates, sometimes where nothing else grows.

General Description: *C. gigentia* is a large shrub growing to 4 m tall. It has clusters of waxy flowers that are either white or lavender in colour. Each flower consists of five pointed petals and a small, elegant "crown" rising from the centre, which holds the stamens. The plant has oval, light green leaves and milky stem.

Chemical composition: The latex contains cardiac glycosides calotropin, calotoxin, uscharin, voruscharin, uschridin, uzarigenin, syriogenin, calotonic acid, proceroside, calotropin DI and DII (Sengupta *et al.*, 1984; Pal & Sinha, 1980) and flavonol glycosides (Sen, Sahu & Mahato, 1992).

Action: Drastic purgative.

Therapeutics: According to Ayurveda, dried whole plant is a good tonic, expectorant, depurative, and anthelmintic. The root bark is febrifuge, anthelmintic, depurative, expectorant, and laxative. The powdered root used in asthma, bronchitis, and dyspepsia. The leaves are useful in the treatment of paralysis, athralgia, swellings, and intermittent fevers. The flowers are bitter, digestive, astringent, stomachic, anthelmintic, and tonic (Joshi H, Gururaja MP, Suares, 2011). The milky juices of *C. gigantea* and *Euphorbia neriifolia* are made into tents with the powdered wood of *Berberis asiatica*, for introduction into sinuses and fistula in ano. The milky juice is applied to carious teeth for relief of pain.

Parts used: Root bark, flowers and leaves.

Bioactivity: Anti-diarrheal (Chitme, 2004), antipyretic (Chitme, Chandra & Kaushik, 2005), anticonvulsant, anxiolytic and sedative (Argal & Pathak, 2006), analgesic (Pathak & Argal, 2007), antibacterial (Subramanian & Saratha, 2010) and anti-inflammatory (Jagtap *et al.,* 2010).

Further reading

Ahmed M, Rana AC and Dixit VK. Free radical scavenging activity of Calotropis species. *Indian Drugs* 2003; 40: 654-655.

Argal A, Pathak AK. CNS activity of *Calotropis gigantean* roots. *J Ethnopharmacol* 2006; 106:142-145.

Chitme HR, Chandra R, Kaushik S. Evaluation of antipyretic activity of *Calotropis gigantea* (Asclepiadaceae) in experimental animals. *Phytother Res* 2005; 19:454-456.

Chitme HR. Studies on anti-diarrhoeal activity of *Calotropis gigantea* R.Br. in experimental animals. *J Pharm Pharmaceut Sci* 2004; 7:70-75.

Jagtap VA, Md Rageeb, Md Usman, Salunkhe P S, Gagrani M B. Anti-inflammatory Activity of *Calotropis gigantea* Linn. Leaves Extract on *In-vitro* Models. *Int J Curr Pharm Rev Res* 2010; 1:1-5.

Joshi H, Gururaja MP, Suares D. *Calotropis gigantea* R.Br. (Asclepiadaceae): A Review. *Int J Pharma Res* 2011; 3: 10-14.

Lindley, J. 1985. Flora medica. Ajay Book Services, New Delhi.

Pal G, Sinha NK. Isolation, Crystallization, and Properties of Calotropins DI and DII from *Calotropis gigantea. Arch Biochem Biophys* 1980; 202: 321-329.

Pathak AK, Argal A. Analgesic activity of *Calotropis gigantea* flower. *Fitoterapia* 2007; 78: 40-42.

Sastry, C.S.T. and K.Y. Kavathekar. 1990. In: Plants for reclamation of wasteland. Publication and Information Directorate, CSIR, New Delhi. p. 175-179.

Sen S, Sahu NP, Mahato SB. Flavonol glycosides from *Calotropis gigantea*. *Phytochemistry* 1992; 31: 2919-2921.

Sengupta A, Bhattacharya D, Pal G, Sinha NK. Comparative studies on Calotropin DI and DII from latex of *Calotropis gigantean*. *Arch Biochem Biophys* 1984; 232:17-25.

Subramanian SP, Saratha V. Evaluation of Antibacterial Activity of *Calotropis gigantea* Latex Extract on Selected Pathogenic Bacteria. *J Pharm Res* 2010; 3:14-17.

Calotropis procera R. Br.

English name: Roostertree.

Common name: Madar.

Ayurvedic names: *Arka, arkaparna, asphota, tulphala* and *vikirana*.

Family: Asclepidaceae.

Habitat: Throughout India, in dry waste places.

General Description: Erect, tall, large, much branched and perennial shrubs or small trees that grow to a height of 5.4 m., with milky latex throughout. Bark is soft and corky. Branches stout, terete with fine appressed cottony pubescence (especially on young). Leaves sub-sessile, opposite, decusate, broadly ovate-oblong, elliptic or obovate, acute, thick, glaucous, green, covered with fine cottony pubescent hair on young but glabrous later and base cordate. Flowers in umbellate-cymes and tomentose on young. Calyx glabrous, ovate and acute. Corolla glabrous, lobes errect, ovate, acute, coronal scales 5-6, latterly compressed and equally of exceeding the staminal column. Folicles are sub-globose or ellipsoid or ovoid. Seeds broadly ovate, acute, flattened, minutely tomentose, brown coloured and silky coma is 3.2 cm long.

Detailed Description of Stem Bark (Gupta, Sharma & Kapoor, 1971; Israili & Issar, 1977; Murti, Yogi & Pathak, 2010),

(a) *Macroscopic*

Drug occurs in channelled, quilled and fibrous pieces, upto 0.1 - 0.5 cm thick, external surface yellowish brown having longitudinal cracks, internal surface greenish, smooth, with an occasional wood tissue attached; fracture, fibrous; odour and taste not distinct.

(b) *Microscopic*

Stem bark shows exfoliated cork, consisting of 6-8 layers of tangentially elongated, thick-walled cells; where cork has not developed, epidermis present consisting of a single layered rectangular cells covered externally with striated cuticle; secondary cortex composed of tangentially elongated, oval, rounded or rectangular thinwalled, parenchymatous cells having intercellular spaces, some cells contain rosette crystals of calcium oxalate, a number of rounded, oval to elongated, single or groups of stone cells and latex cells also found scattered in this region; pericyclic fibres numerous, lignified; secondary phloem composed of sieve elements, phloem parenchyma, phloem fibres and phloem rays; phloem parenchyma rectangular to polygonal in shape having rosette crystals of calcium oxalate, latex cells and stone cells similar to those found in secondary cortex; phloem fibres aseptate with bordered pits; phloem rays mostly uniseriate and run straight.

Powder - Light yellowish-green; shows fibres, stone cells, rosette crystals of calcium oxalate and latex cells.

Chemical constituents: The leaves contain mudarine, the principal active constituent, a yellow bitter acid, resin and toxic glycosides calotropin, uscharin and calotoxin. In latex a powerful bacteriolytic enzyme, a very toxic glycoside calactin, calotropin D I, calotrapin D II, calotropin F I, calotropin F II and a non toxic protealytic enzyme calotopin (2-3%) had been identified. Calotopin is more proteolytic than papain, and bromelain and coagulates milk, digests meat, gelatin and casein. Whole plant contains α- and β-amyrin, teraxasterol, gigantin, giganteol, isogiganteol, β-sitosterol and a wax (Singh & Rastogi, 1972; Pant & Chaturvedi, 1989a; Pant & Chaturvedi, 1989b; Gupta, Sharma & Dixit, 1990; Ansari & Ali, 1999).

Actions: Drastic purgative.

Therapeutics: *C. procera* is used in the treatment of skin diseases and rheumatism. Powered drug is dusted cover unhealthy sores with benefit. The milk derived from the plant is used as an application in leucoderma. Root bark is emetic and purgative and given in helmenthiasis. It is also prescribed in malaria and snake poisoning.

The root bark is febrifuge, anthelmintic, depurative, expectorant, and laxative. The powdered root promotes gastric secretions and useful in asthma, bronchitis, and dyspepsia. Dried whole plant is good tonic, expectorant, depurative and anthelminthic.

Madar root-bark is very largely used in India as a treatment for elephantiasis, leprosy, and in chronic eczema. It also used as antidote substance and for abortive purposes. Leaves are useful in the treatment of paralysis, arthralgia, swellings and intermittent fevers. Flowers are useful in asthma, catarrh, anorexia, inflammations and tumours (Upadhyay, 1979; Moscolo *et al.*, 1988; Singh, 1992; Misra, Mohanty & Das, 1993).

Parts used: Root bark, flowers and leaves.

Bioactivity: Antimicrobial (Kumar & Chauhan, 1992), wound-healing (Rasik *et al.*, 1999), anti-inflammatory (Sangraula H, Kumar, 1999), antipyretic (Dewan, Kumar & Kumar, 2000), analgesic (Dewan, Sangraula & Kumar, 2000), anti-diarrhoeal (Kumar *et al.*, 2001), anthelmintic (Shivkar & Kumar, 2003) and hepatoprotective (Ramachandra *et al.*, 2007).

Further reading

Ansari SH, Ali M. Title: Phytochemical and pharmacological investigations on *Calotropis procera* (Ait.) R.Br. *Hamdard Medicus* 1999;42: 96-101.

Bharti S, Wahane VD, Kumar VL. Protective effect of *Calotropis procera* latex extracts on experimentally induced ulcers in rats. *J Ethnopharmacol* 2010; 127:440-444.

Dewan S, Kumar S, Kumar VL. Antipyretic effect of *Calotropis procera. Indian J Pharmacol* 2000; 32: 252.

Dewan S, Sangraula H, Kumar VL. Preliminary studies on the analgesic activity of latex of *Calotropis procera. J Ethnopharmacol* 2000; 73: 307-311.

Gupta RC, Sharma PC, Kapoor LD. Pharmacognostic study of the leaf of *Calotropis procera* (Ait.) R.Br. (arka). *J Res Indian Med* 1971; 6:167-172.

Gupta RS, Sharma N, Dixit VP. Calotropin - a novel compound for fertility control. *Ancient Sci Life* 1990; 9: 224-230.

Handa F, Sadna JK, Sharma PK. Allergic contact dermatitis due toplant *Calotropis procera* (vern. ark) - a case report. *Indian J Dermatol* 1984; 29: 27-29.

Israili AH, Issar RK. Pharmacognostical study of the Unani drug "poste-bekh-madar" (*Calotropis procera* (Ait.) R.Br. - root bark). *J Res Indian Med* 1977; 12: 41-48.

Kumar S, Chauhan UK. A study of antimicrobial activity of *Calotropis procera* leaf extract. *Geobios* 1992; 19:135-137.

Kumar S, Dewan S, Sangraula H, Kumar VL. Anti-diarrhoeal activity of the latex of *Calotropis procera. J Ethnopharmacol* 2001; 76:115-118.

Misra MK, Mohanty MK, Das PK. Studies on medico-ethnobotany of *Calotropis gigantea* and *C. procera. Ancient Sci Life* 1993; 13:40-56.

Moscolo N, Sharma R, Jain SC, Capasso P. Ethnopharmacology of *Calotropis procera* flowers. *J Ethnopharmacol* 1988; 22: 211-221.

Murti Y, Yogi B, Pathak D. Pharmacognostic standardization of leaves of *Calotropis procera* (Ait.) R.Br. (Asclepiadaceae). *Int J Ayur Res* 2010; 1:14-17.

Pant R, Chaturvedi K. Chemical analysis of *Calotropis procera* latex. *Curr Sci* 1989; 58:740-742.

Pant R, Chaturvedi K. Occurrence of lupeol in *Calotropis procera* latex. *Curr Sci* 1989; 58: 302-303.

Prasad G. Action of *Calotropis procera* on migraine. *J Nat Integr Med Assoc* 1985; 27: 7-10.

Rasik AM, Raghubir R, Gupta A, Shukla A, Dubey MP, Srivastava S, Jain HK, Kulshreshtha DK. Healing potential of *Calotropis procera* on dermal wounds in Guinea pigs. *J Ethnopharmacol* 1999; 68: 261-266.

Sangraula H, Kumar VL. Anti-inflammatory studies on latex of *Calotropis procera. Indian J Pharmacol* 1999; 31:79-80 (abstract No. 177).

Setty, R. Ramachandra, Afsar Ahmed Quereshi, A.H.M. Viswanath Swamy, Patil T, Prakash T, Prabhu K, Veeran Gouda A. Hepatoprotective activity of *Calotropis procera* flowers against paracetamol-induced hepatic injury in rats. *Fitoterapia* 2007; 78: 451-454.

Sharma P, Sharma JD. *In vitro* schizontocidal screening of *Calotropis procera. Fitoterapia* 2000; 71:77-79.

Shivkar YM, Kumar VL. Anthelmintic activity of latex of *Calotropis procera. Pharm Biol* 2003; 41:263-265.

Singh B, Rastogi RP. On *Calotropis procera* (Aiton) W.T. Aiton. *Phytochemistry* 1972; 11:757.

Singh S. On chewing the aak (*Calotropis procera*) - folk-beliefs and its scientific view. *Folklore* 1992; 32:36-39.

Upadhyay VP. Preliminary studies on the therapeutic effect of flowers of *Calotropis procera* (Ait.) R. Br., arka pushpa, on asthma - I. *J Sci Res Plant Med* 1979; 1:52-55.

11. ASHMANTAKA

Current representative: *Ficus rumphii*

Possible representative: *Bauhinia racemosa* and *Bauhinia tomentosa*

Ficus rumphii Blume

Syn: *Urostigma rumphii* (Blume) Miquel.

English name: Rumpfs Fig Tree.

Family: Moraceae.

Habitat: India.

General Description: Trees, ca. 15 m tall, usually epiphytic. Bark gray, wrinkled when dry. Stipules caducous, ovate-lanceolate, 1.5-2.5 cm, scar conspicuous. Petiole 6-8 cm, glabrous; leaf blade cordate to ovate-cordate, 6-13 × 6-11 cm, ± leathery, glabrous, base ± cordate to broadly cuneate, apex acuminate; basal lateral veins 4, outer 2 basal veins short and thin, secondary veins 5 or 6 on each side of midvein. Figs axillary on leafy branchlets, paired or in small clusters on leafless older branchlets, with dark spots when young, dark purple when mature, globose, 1-1.5 cm in diam., sessile; involucral bracts orbicular, small; apical bracts ± navel-like. Male, gall, and female flowers within same fig. Male flowers: few, scattered among other flowers; calyx lobes 3, spatulate; stamen 1. Gall and asexual flowers: calyx lobes 3, lanceolate. Female flowers: ovary white, ovoid, smooth; style persistent, long; stigma clavate. Achenes thin, tuberculate and with adherent liquid; style long; stigma clavate (Mangal & Das, 1979).

Chemical composition: β-amyrin, beta-sitosterol and flavonol glycoside (Baslas, 1979).

Action: Astringent.

Therapeutics: Diarrhoea, headache amd malaria.

Parts used: Bark, milk and fruit.

Further reading

Mangal AK, Das MN. Comparative powdered drug analysis of aerial and undergorund roots of *Ficus rumphii* Blume (*Ashmantaka*). *BMEBR* 2001; 22:99-106.

Baslas RK. Isolation and characterisation of beta-sitosterol and flavonol glycoside from the trunk bark of *Ficus rumphii*. *Curr Sci*1979; 48:113.

Bauhinia racemosa **Lam.**

Syn: *Bauhinia variegata* Linn.

English name: Mountain Ebony.

Ayurvedic **name:** *Kanchnara*.

Family: Fabaceae.

Habitat: India, Myanmar, Srilanka, Thailand, Vietnam.

General Description: This is a deciduous, 3 to 5 meters tall tree. The bark is dark brown, rough and longitudinally fissured. The leaves are sub orbicular, cordate at the base, bilobed, glabrous above and pubescent beneath, and petiolate. The flowers are creamy yellow in 4 to 7 cms. long terminals and racemes. The pods are linear oblong, woody slightly curved, dark brown or brownish black, constriction between all or few seeds. The seeds are oblong compressed, smooth, glabrous and brown. The flowering season is January to June.

Detailed Description of Stem Bark

(a) *Macroscopic*

Bark, dark brown, sometimes with silvery patches, rough, compact, exfoliating in oody strips and scales, outer surface with small transverse and longitudinal cracks, internal surface white, taste, astringent

(b) Microscopic

Transverse section of mature stem bark shows a wide stratified cork, outer cork composed of thin-wailed, slightly compressed, yellow brown cells followed by a number of layers of brown coloured cells, inner cork composed of transversely elongated orange brown cells, cork interrupted at certain places due to formation of rhytidoma, some secondary cortex composed or 15 or more rows or transversely elongated to circular, thin-walled, parenchymatous cells, some secondary cortex cells contain orange brown contents: groups of stone cells found scattered in this region occasionally arranged in 1-7 or more tangential rows, pericyclic fibres, thick-walled with narrow lumen, scattered in secondary cortex in singles or in groups, secondary phloem consists of sieve tubes, companion cells, phloem parenchyma and fibres traversed by funnel shaped medullary rays, phloem fibres arranged in radial rows throughout phloem region, prismatic and rhomboidal crystals or calcium, oxalate abundantly found in phloem and secondary cortex regions, very rarely found in cork cells, cluster crystals also present in secondary cortex and secondary phloem, crystal fibres also found in secondary phloem.

Powder - pinkish, under microscope showing abundant crystals of calcium oxalate, sclercids in singles or in groups with wide lumen, bits of fibres, cork and secondary cortex cells, containing coloured content, and numerous crystal fibres

Chemical constituents: Tannins, flavonoids: kaempferol and quercetin, coumarins: scopoletin and scopolin (Prakash *et al.,* 1976, Rajaram & Janardhanan, 1991), triterpenoids: β-amyrin and steroids: β-sitosterol (Dhar *et al.,* 1968).

Actions: Alterative.

Therapeutics: The stem bark of the plant is an astringent and is used in the treatment of goiter, headache, fever, skin diseases, tumors, diseases of the blood, dysentery and diarrhea

Parts used: Bark, root, leaves, flowers, seed and gum.

Bioactivity: Antitumor (Gupta *et al.,* 2004), antioxidant and hepatoprotective (Gupta *et al.,* 2004), antiulcer, analgesic and antipyretic (Borikar *et al.,* 2009) and anthelmintic (Kumar *et al.,* 2011).

Further reading

Borikar VI, Jangde CR, Philip P, Rekhe DS. Study of Analgesic activity of *Bauhinia racemosa lam* in Rats. *Veterinary World* 2009; 2:135-136.

Borikar VI, Jangde CR, Philip P, Rekhe DS. Study of Antipyretic Activity of *Bauhinia racemosa* lam in Rats. *Veterinary World* 2009; 2:215-216.

Dhar ML, Dhar MM, Dhawan BN, Mehrotra BN, Roy C. Screening of Indian plants for biological activity. *Indian J Exp Biol* 1968; 6: 232-247.

Gomathi P. Antioxidant and hepatoprotective effects of *Bauhinia racemosa* against paracetamol and carbon tetra chloride induced liver damage in rats. *Iranian J Pharm Ther* 2004; 3:12-20.

Gupta M, Mazumder UK, Sambath Kumar R, Siva Kumar T. Antitumor effect of *Bauhinia racemosa* against Ehrlich ascites carcinoma with reference to lipid peroxidation and antioxidant system in Swiss albino mice. *Acta Pharmacologica Sinica* 2004; 25:1070-1076.

Gupta M, Mazumder UK, Thangavel SK, Gomathi P, Sambath RK. Antioxidant and hepatoprotective effects of *Bauhinia racemosa* against paracetamol and carbon tetrachloride induced liver damage in rats. *Indian J Pharmacol Ther* 2004; 3:12-20.

Kumar T, Ajazuddin AA, Dewangan D, Khan J, Sharma M. Investigation of *in-vitro* anthelmintic activity of *Bauhinia racemosa Linn. J Pharm Sci* 2011; 1:73-75.

Prakash A, Khosa RL. Chemical studies on *Bauhini racemosa. Curr Sci* 1976; 45: 705-707.

Rajaram N, Janardhanan K. Chemical composition and nutritional potential of the tribal pulses *Bauhinia purpurea, B. racemosa* and *B. vahlii. J Sci Food Agric* 1991; 55:423-431.

Bauhinia tomentosa Linn.

English name: Camel foot tree, St. Thomas tree, yellow Bauhinia, yellow orchid tree.

Habitat : Tropical Africa and as far as India and Sri Lanka.

General Description: Medium to large shrub to a small tree, up to 4m in height. Leaves are divided into two lobes, light green in colour, with a leathery texture, carried on branches that are often drooping. It produces large bell-shaped, bright yellow flowers with a black to deep maroon coloured centre from December to March. The fruit are pea like, slender and velvety. They are light green, turning a pale brown with age and are produced from January to June or even later. Bark is gray or brown.

Chemical constituents: Tannins.

Therapeutics: In India, according to Kirtikar and Basu, the bruised bark is applied externally to tumors and wounds. On the coast of Malabar a decoction of the root-bark is administered for inflammation of the liver. The decoction of the root-bark is also used as a vermifuge. An infusion of the bark is also used as an astringent gargle.

Bioactivity: Anti-inflammatory (Kannan, Renitta & Guruvayoorappan, 2010).

Further reading

Dugasani S, Balijepalli MK, Tandra S, Pichika MR. Antimicrobial activity of *Bauhinia tomentosa* and *Bauhinia vahlii* roots.*Pharmacogn Mag* 2010;6:204-207.

Gupta SK. Phytopharmacognostic investigation of *Bauhinia tomentosa* Linn. *J Advanced Sci Res* 2011; 2:1-4.

Kannan N, Renitta RE, Guruvayoorappan C. *Bauhinia tomentosa* stimulates the immune system and scavenges free radical generation in vitro. *Basic Clin Physiol Pharmacol* 2010; 21:157-68.

Kumar VK, Kumar PS, Venkatachalam T, Selvi PK, Senthil NK. Pharmacognostic standardization and physico-chemical evaluations of *Bauhinia tomentosa* Linn. flower. *Int J Pharma Res Develop – Online* 2010; 2.

Mannangatti V, Ayyasamy B, Rangasamy M, Emin B, Natesan SK. Antihyperglycemic and anti-lipidemic activity of ethanolic extract of *Bauhinia tomentosa* Linn. Flower in normal and streptozotocin-induced diabetic rats. *J Global Pharma Tech* 2010; 2: 71-76.

12. ASHVA AND ASHVAKSURA

Current representative: *Clitoria ternatae*

Possible representative: *Medicago sativa*

Clitoria ternatea L.

Common name: Winged leaved Clitoria; butterfly pea.

Ayurvedic name: *Asphota, girikarni, vishnukranta, shankhapushpi, sephanda, sveta* and *maha sveta.*

Family: Fabaceae.

Habitat: India and Sri Lanka.

General Description: A good looking perennial herb twinner. Leaves compound, imparipinnate, leaflets 5 – 7, sub coriaceous, elliptic-oblong, obtuse, flowers white, solitary, axillary or in facicles, fruits nearly straight flattened pods. Seeds 6-10 smooth yellowish brown (Taur, Ravindra & Patil, 2010).

Chemical constituents: Resin, tannin, starch, β-sitosterol (Sinha, 1960, Joshi, Shrivastava RK, Shrivastava, 1988) and fatty acids in oil (Debnath *et al.,* 1975, Husain & Devi, 1988).

Actions: Demulcent and laxative.

Therapeutics: Constipation.

Parts used: Roots and seeds.

Bioactivity: Cholinergic (Kulkarni, Pattanashetty & Amrutharaj, 1988; Rai *et al.,* 2002), anti-inflammatory, analgesic and antipyretic (Mandal, Parimaladevi & Boominathan, 2003), antihyperglycemic and antihyperlipidemic (Daisy, Santosh & Rajathi, 2009), anthelmintic (Salhan *et al.,* 2011) and antihistaminic (Taur & Ravindra, 2011).

Further reading

Daisy P, Santosh K, Rajathi M. Antihyperglycemic and antihyperlipidemic effects of *Clitoria ternatea* Linn. in alloxan-induced diabetic rats. *African J Microbiol Res* 2009; 3:287-291.

Debnath NB, Chakravarti D, Ghosh A, *et al.,* Fatty acids of *Clitoria ternatea* seed oils. *J Institution Chem* (India) 1975; 47: 253-55.

Devi BP, Boominathan R, Mandal SC. Anti-inflammatory, analgesic and antipyretic properties of *Clitoria ternatea* root. *Fitoterapia* 2003; 74: 345-349.

Husain S, Devi KS. Fatty acid composition of three plant species: *Clitorea ternatea, Mandulea suberosa* and *Ruta chalapensis. J Oil Tech Assoc India* 1998; 30: 162-64.

Joshi SS, Shrivastava RK, Shrivastava DK. Chemical examination of *Clitoria ternatea* seeds. *J American Oil Chem Soc* 1981; 58: 714-15.

Kulkarni C, Pattanshetty JR, Amruthraj G. Effect of alcoholic extract of *Clitoria ternatea* Linn. on central nervous system in rodents. *Ind J Exp Biol* 1988; 26: 957-60.

Mukherjee PK, Kumar V, Kumar NS, Heinrich M. The Ayurvedic medicine *Clitoria ternatea*-from traditional use to scientific assessment. *J Ethnopharmacol* 2008; 120: 291-301.

Rai KS, Murthy KD, Karanth KS, et al., *Clitoria ternatea* root extract enhances acetylcholine content in rat hippocampus. *Fitoterapia* 2002; 73: 685-689.

Salhan M, Kumar B Tiwari I P, Sharma P, Sandhar HK. Comparative anthelmintic activity of aqueous and ethanolic leaf extracts of *Clitoria ternatea. Int J Drug Dev Res* 2011, 3:68-69.

Sinha A. β-Sitosterol from the seeds of *Clitoria ternatea. Curr Sci* 1960; 29: 180- 81.

Taur GJ, Ravindra Y, Patil RY. Pharmacognostic Evaluation of *Clitoria ternatea* Root. *J of Pharm Res* 2010; 3.

Taur GJ, Ravindra Y. Antihistaminic activity of *Clitoria ternatea* L. roots. *J Basic Clin Pharm* 2011; 2: 41-44.

Medicago sativa L.

Common name: Alfa alfa.

Ayurvedic names: *Ashwabala* and *ghodoaghasa*.

Family: Fabaceae.

Habitat: India.

General Description: An herbaceous perennial with deeply penetrating taproot. Stems procumbent, ascending to erect, arising from a woody base, 30–80 (–120) cm long. Leaf trifoliolate; stipules triangular, 5–15 mm long; pubescent on lower surface, glabrous on upper surface, coarsely toothed; petiole pubescent, 5–30 mm long, leaflets narrow, oblong to ovate or obovate, 8–28 mm x 3–15 mm, dentate near apex, glabrous on upper surface, slightly pubescent on lower surface. Inflorescences in dense racemes with 10–35 flowers, on peduncles 1–5 cm long; pedicel 1.5–2 mm long; calyx 5-lobed, 3–6 mm long, tube and pointed teeth about equal in length; corolla purple or blue, rarely white. Pod curled, 3–10 mm in diameter, indehiscent, not spined, containing 2–6 seeds. Seeds yellow to brown, kidney-shaped to ovoid, 1–2.5 mm x 1–1.5 mm.

Chemical constituents: Triterpene saponins, isoflavonoids, cyanogenic glycosides and fatty oil.

Actions: Diuretic.

Therapeutics: In Ayurveda, alfalfa is considered a mild tonic good for purifying the blood. Alfalfa is considered to have alterative, diuretic, antipyretic and hemostatic actions.

Parts used: Leaves.

Bioactivity: Hypolipidaemic and anti-fertility (Dixit & Jain, 1990).

Further reading

Dixit VP, Jain P. Hypolipidaemic effects of *Medicago sativa* seed extracts in rabbits under experimental conditions. *Ancient Sci Life* 1990; 10(1):52-55.

Dixit VP, Jain P. The role of castration in atherosclerosis and prevention with *Medicago sativa* seed extract. *Fitoterapia* 1990; 61:37-40.

13. ASHVGANDHA

Current representative: *Withania somnifera*

Withania somnifera (L.) Dunal.

English name: Winter Cherry.

Ayurvedic names: *Avarohaka*, *balya*, *gokarna*, *turangahava*, *vajikari*, *varada*, *varahakarni* and *vrsa*.

Family: Solanaceae.

Habitat: W. *somnifera* is found throughout the drier parts of India, into West Asia and northern Africa.

General Description: *W. somnifera* is an erect branching shrub that attains a height between 30-150 cm, covered in a wooly pubescence. The ovate leaves are up to 10 cm long and 2.5-5 cm wide, margins entire, arranged in an alternate fashion. The flowers are green or yellow, borne in axillary fascicles, giving rise to red globose fruits when mature. The roots are fleshy and cylindrical, the epidermis light brown and medulla white.

Detailed Description of Root

(a) *Macroscopic*

Roots straight, unbranched, thickness varying with age. roots bear fibre-like secondary roots, outer surface buff to grey-yellow with longitudinal wrinkles, crown consists of 2-6 remains of stem base, stem bases variously thickened, nodes prominent only on the side from where petiole arises, cylindrical, green with longitudinal wrinkles, fracture, short and uneven, odour, characteristic, taste, bitter and acrid.

(b) *Microscopic*

Transverse section of root shows cork exfoliated or crushed, when present isodiamatric and non-lignified, cork cambium of 2-4 diffused rows of cells, secondary cortex about twenty layers of compact parenchymatous cells, phloem consists of sieve tubes, companion cells, phloem parenchyma, cambium 4-5 rows of tangentially elongated cells, secondary xylem hard forming a closed vascular ring separated by multiseriate medullary rays, a few xylem parenchyma.

Chemical composition: Alkaloids: anagrine, anaferine, cuscohygrine, hygrine, isopelletriene, pseudopelletriene, 3- tropyltigloate, tropine and withasomnine, withanolides: withaferin-A, proteins and amino acids. Roots contain dulcitol and withaniol (a mixture of two withanolides), iron, potassium, magnesium and nickel.

Karma (Specific action): *Rasayana* (tonic).

Therapeutics: Depression, hypertension, general debility, oligozoospermia, stress, premature ejaculation and leucorrhoea. In asthma, ash of *Withania somnifera* is given with honey and vegetable oil.

Parts used: Root and leaves.

Bioactivity: Anti-inflammatory activity of methanolic fractions of the plant extracts (Begum and Sadique, 1987), antidepressant activity of a methanolic extract (Mehta *et al.,* 1991), *In vivo* growth inhibitory effect on a transplantable mouse tumor known as sarcoma 180 (Devi *et al.,* 1992), immunomodulator (Ziauddin, Phansalkar, Patki, *et al.,* 1996) antioxidant activity of glycowithanolides (Bhattacharya, Satyan and Ghosal, 1997), anti-stress (Archana and Namasivayam, 1999), hypoglycemic, diuretic and hypocholesterolemic effect of root (Andalluaa and Radhika, 2000) and adaptogenic activity of a novel, withanolide-free aqueous fraction from the roots (Singh, Saxena, Chandan, Gupta, Bhutani and Anand, 2001).

Further reading

Agrawal SS, Malik SS. Hepatoprotective effects of *Withnia somnifera. Indian J Pharmacol* 1995; 28: 45 (abstract No. 55).

Anand LV, Kuttan G. Use of *Withania somnifera* as an adjuvant during radiation therapy. *Amala Res Bull* 1995; 15: 83-87.

Aphale AA, Chhibba AD, Kumbhakarna NR, Mateenuddin M, Dahat SH. Subacute toxicity study of the combination of ginseng and ashwagandha in rats: A safety assessment. *Indian J Physiol Pharmacol* 1998; 42:99-302.

Begum VH, Sadique J. Effect of *Withania somnifera* on glycosaminoglycan synthesis in carrageenin-induced air pouch granuloma. *Biochem Med Metab Biol* 1987; 38:272-277.

Begum VH, Sadique J. Long term effect of herbal drug *Withania somnifera* on adjuvant induced arthritis in rats. *Indian J Exp Biol* 1988; 26: 877-882.

Bhattacharya A, Ghosal S, Bhattacharya SK. Anti-oxidant effect of *Withania somnifera* glycowithanolides in chronic footshock stress-induced perturbations of oxidative free radical scavenging enzymes and lipid peroxidation in rat frontal and striatum. *J Ethnopharmacol* 2001; 74: 1-6.

Bhattacharya A, Ghosal S, Bhattacharya SK. Effects of glycowithanolides from *Withania somnifera* on an animal model of Alzheimer's disease and perturbed central cholinergic markers of cognition in rats. *Phytother Res* 1995; 9:110-113.

Bhattacharya A, Ramanathan M, Ghosal S, Bhattacharya SK. Effect of *Withania somnifera* glycowithanolides on iron-induced hepatotoxicity in rats. *Phytother Res* 2000; 14: 568-570.

Bhattacharya SK, Goel RK, Ghoshal S. Antistress, antianxiety and CNS inhibitory properties of *Withania somnifera*. *Phytother Res* 1987; 1:32-37.

Bhattacharya SK, Kumar A, Ghosal S. Effects of glycowithanolides from Withania somnifera on an animal model of Alzheimer's disease and perturbed cholinergic markers of cognition in rats. *Phytother Res* 1995; 9:110-113.

Bhattacharya SK, Satyan KS, Ghosal S. Anxiolytic activity of glycowithanolides from *Withania somnifera*. *Indian J Exp Biol* 1997; 35:3, 236-239.

Bikshapathi T, Kumari K. Clinical evaluation of ashwagandha in the management of ama-vata. *J Res Ayur Siddha* 1999; 20, 46.

Chakraborti SK, De BK, Bandyopadhyay T. Variations in the antitumor constituents from *Withania somnifera*. *Experientia* 1974; 30:852-853.

Chatterjee S, Chakraborti SK. Antimicrobial activities of some antineoplastic and other withanolides. *Antonie Van Leeuwenhoek* 1980; 46:59-63.

Choudhary K, Neogy RK. Mode of action of Withaferin A and Withanolide D. *Biochem Pharmacol* 1975; 24:919-920; *Biol Abstr* 1975, 60, 62270.

Das H, Dutta SK, Bhattacharya B. Anti-neoplastic agents from plants. Antitumour activity of Withanolide D. *Indian J Cancer Chemother* 1985; 7:59-65.

Davis L, Kuttan G. Effect of *Withania somnifera* on cyclophosphamide-induced urotoxicity. *Cancer Lett* 2000; 148: 9-17.

Davis L, Kuttan G. Immunomodulatory activity of *Withania somnifera*. *J Ethnopharmacol* 2000; 7:193-200.

Devi PU *et al., In vivo* growth inhibitory effect of *Withania somnifera* (Ashwagandha) on a transplantable mouse tumor, Sarcoma 180. *Indian J Exp Biol* 1992; 30:169-72.

Devi PU. *Withania somnifera* Dunal (Ashwagandha): potential plant source of a promising drug for cancer chemotherapy and radiosensitization. *Indian J Exp Biol* 1996; 34:927-932.

Dhuley JN. Nootropic-like effect of ashwagandha (*Withania somnifera* L.) in mice. *Phytother Res* 2001; 15: 524-528.

Dhuley JN. Adaptogenic and cardioprotective action of ashwagandha in rats and frogs. *J Ethnopharmacol* 2000; 70: 57-63.

Gandhi SS. The potential of *Withania somnifera* as antistress agent. *Update Ayur* 1994; 94:16.

Ghosal S, Lal J, Srivastava RS. Immunomodulatory and CNS effect of sitoindosides IX and X, two regio glycowithanolides from *W somnifera. Indian J Nat Prod* 1988;4:12-13.

Grandhi A, Majumdar AM, Pathwardhan B. A comparative pharmacological investigation of Ashwagandha and Ginseng. *J Ethnopharmacol* 1944; 44: 131-135.

Gupta YK, Sharma SS, Rai K, Katiyar CK. Reversal of paclitaxel induced neutropenia by *Withania somnifera* in mice. *Indian J Physiol Pharmacol* 2001; 45: 253-257.

Jayaprakasam B, Zhang Y, Seeram NP, Nair MGGrowth inhibitor of human tumour cell lines by withanolides from *Withania somnifera* leaves. *Life Sci* 2003; 74:125-132.

Karnick CR. A double-blind, placebo-controlled clinical study on the effects of *Withania somnifera* and *Panax ginseng* extracts on psychomotor performance in healthy Indian volunteers. *Indian Med* (Vijayawada) 1991;3: 1-5.

Khanam S, Devi K. Antimutagenic activity of Ashwagandha. *J Nat Rem* 2005b 5:2.

Kulkarni SK, George B. Anticonvulsant action of *Withania somnifera* (aswagandha) root extract against pentylenetetrazol-induced kindling in mice. *Phytother Res* 1996; 10: 447-449.

Kulkarni SK, Ninan I. Inhibition of morphine tolerance and dependence by *Withania somnifera* in mice. *J Ethnopharmacol* 1997; 57: 213-217.

Kulkarni SK, Sharma A, Verma A, Ticku MK. GABA receptors mediated anticonvulsant action of *Withania somnifera* root extract. *Indian Drugs* 1993; 30:305-312.

Majumdar DN. Alkaloidal constituents of *Withania somnifera. Curr Sci* 1952; 21:46-47.

Majumdar DN. *Withania somnifera* Dunal, part II: Alkaloidal constituents and their chemical characterization. *Indian J Pharm* 1955; 17: 158-161.

Malhotra CL, Das PK, Dhalla NS. Studies of Withania ashawagandha. II Effect of total extracts on cardiovascular system, respiration, Skeletal muscles. *Indian J Physiol Pharmacol* 1960; 4: 46-49.

Malhotra CL, Mehta VL, Das PK, Dhalla NS. Studies on *Withania ashwagandha* Kaul V - The effect of total alkaloids (ashwagandholine) on the central nervous system. *Indian J Physiol Pharmacol* 1965; 9:127.

Malhotra CL, Mehta VL, Prasad K, Das PK. The effect of total alkaloids on the smooth muscles. *Indian J Physiol Pharmacol* 1965; 9: 9-15.

Mehta AK, Binkley P, Gandhi SS, Ticku MK. Pharmacological effects of *Withania somnifera* root extract on GABA receptor complex. *Indian J Med Res* 1991; 94:312-315.

Nagareddy PR, Lakshmana M. *Withania somnifera* improves bone calcification in calcium-deficient ovariectomized rats. *J Pharm Pharmacol* 2006; 58: 513-519.

Padmavathi B, Rath PC, Rao AR, Singh RP. Roots of *Withania somnifera* inhibit Fore stomach and skin carcinogenesis in mice. *Ecam* 2005; 2: 99-105.

Prasad S, Luthra SP. Pharmacognostic studies on *Withania somnifera* Dunal and *W. coagulans* Dunal. *Indian J Pharm* 1959; 21, 7:189-194.

Rasool M, Marylatha L, Varalakshmi P, Effect of *Withania somnifera* on lysosomal acid hydrolases in adjuvant-induced arthritis in rats. *Pharma Pharmacol Comm* 2000; 6: 187-190.

Sahni YP, Srivastava DN. Antiinflammatory and antiulcerogenic activity of *Withania somnifera. Update Ayur* 1994; 94:24-26.

Shah CS, Sukkawala BM, Vyas LS. Pharmacognosy of *Withania somnifera* Dunal. *Indian J Pharm* 1959; 21:7.

Sharada AC, Emerson SF, Uma Devi P. Toxicity of *Withania somnifera* root extracts in rats and mice. *Int J Pharmacog* 1993;31: 205-212.

Sharada AC, Solomon FE, Devi PU, Udupa N, Srinivasan KK. Antitumor and radiosensitizing effects of withaferin A on mouse Ehrlich ascites carcinoma *in vivo*. *Acta Oncologica Stockholm* 1996;35: 95-100.

Sharma K, Puri AS, Sannd BN. Role of aswagandha (*Withania somnifera* Dunal) in amvaat. *Nagarjun* 1968;11: 625-627.

Sharma lP, Singh VV, Kamath R, Uma Devi P, Udupa N. Niosomal Withaferin A with better antitumour efficacy. *Indian J Pharma Sci* 1998;60: 45-48.

Singh N, Nath R, Lata A, Singh SP, Kohli RP, Bhargava KP. *Withania somnifera* (ashvagandha), a rejuvenating herbal drug which enhances survival during stress (an adaptogen). *Pharm Biol* 1982;20: 29-35.

Srivastava GP, Krishnamurthy N. Standardization of ashwagandha (*Withania somnifera*). *Indian Pharm* 1952;8: 52.

Srivastava SK, Iyer SS, Ray GK. Estimation of the total alkaloids of *Withania somnifera* (Dunal). *Indian J Pharm* 1960; 22: 94-95.

Sudhir S, Budhiraja RD. Comparison of the protective effect of withaferin 'A' and hydrocortisone against carbon tetrachloride induced hepatotoxicity in rats. *Indian J Physiol Pharmacol* 1992;36: 127-129.

Thatte U, Pandit SA, Sha LP. Ashwagandha (*Withania somnifera*). Arch. Ind. Psychiat 1995;2:53-55.

Uma, D. P., K. Akagi, *et al.*, (1996). *International Journal of Radiation Biology*, 69(2): 193-197.

Venkataraghavan S, Seshadri C, Sundaresan TP, Revathi R, Rajagopalan V, Janaki K. The comparative effect of milk fortified with ashwagandha and punarnava in children - a double blind study. *J Res Ayur Siddha* 1980;1: 370-385.

Vidya Prabhu M, Rao A, Karanth KS. Neuropharmacological activity of Withania somnifera. *Fitoterapia* 1990;61: 237-240.

Ziauddin M, Phansalkar N, Patki P, Diwanay S, Patwardhan B. Studies on the immunomodulatory effects of Ashwagandha. *J Ethnopharmacol* 1996; 50:69-76.

14. ASOCA

Current representative: *Polyalthia longifolia*

Possible representative: *Saraca asoca*

Polyalthia longifolia (Sonn.) Thwaites

English name: Cemetery tree, Mast tree.

Ayurvedic name: *Kasthadaru*.

Family: Annonaceae.

Habitat: Throughout India cultivated as a garden tree.

General Description: A tall evergreen tree grows up to 45 meters in height. Leaves simple, alternate, acute, with undulate margins; flowers greenish yellow in axillary fascicles. Fruits are small globose berries (Rema Shree *et al.*, 2005; Dave, Nagani & Chanda, 2010).

Chemical composition: Tannins, quercetin), quercetin-3-O-β-glucopyranoside, kaempferol-3-O-α-rhamnopyranosyl-(1→6)-β-galactopyranoside, kaempferol-3-O-α-rhamnopyranosyl-(1→6)-β-glucopyranoside, rutin and allantoin (Sashidhara *et al.*, 2011).

Action: *Jvaraghana* (febrifuge).

Therapeutics: Plant pacifies vitiated *vata*, *pitta*, inflammation, fever, skin disease, diabetes, hypertension and worm infestation.

Parts used: Bark.

Bioactivity: Antibacterial (Goutam, Purohit, 1973b; Ali & Debnath, 1997), hypoglycemic and antihyperglycemic (Nair, Shukla & Chanda, 2007), hepatoprotective (Tanna, Nair &

Chanda, 2009), antitumor and antioxidant (Santhepete *et al.*, 2010; Sashidhara *et al.*, 2011); antiulcer and anti-inflammatory (Sharma *et al.*, 2011)

Note: Its bark is used as an adulterant for *Saraca asoka*.

Further reading

Ali MA, Debnath DC. Isolation and characterization of antibacterial constituent from devdaru (*Polyalthia longifolia* L.). *Bangladesh J Sci Indust Res* 1997; 32:20-24.

Dave R, Nagani K, Chanda S. Pharmacognostic Studies and Physicochemical Properties of the *Polyalthia longifolia* var. *pendula* Leaf. *Pharmacog J* 2010; 2:572-576.

Goutam MP, Purohit RM. Preliminary investigations on the antibacterial activity of leaves and oil of *Polyalthia longifolia*. *Indian J Pharm* 1973b; 35:118-119.

Santhepete M., Kenganora M, Parihar VK, Kumar S, Nayak PG, Kumar N, Pai KSR, Rao CM. Antitumor and antioxidant activity of *Polyalthia longifolia* stem bark ethanol extract. *Pharma Biol* 2010; 48:690-696.

Nair R, Shukla V, Chanda S. Assessment of *Polyalthia longifolia* var. pendula for hypoglycemic and antihyperglycemic activity. *J Clin Diag Res* 2007; 1:116-21.

Rema Shree AB, Sudhakar Raja S, Jayanthy A, Unnikrishnan KP, Balachandran I. Comparative anatomical and phytochemical markers to identify asoka from its common adulterant. *Aryavaidyan* 2005; 19: 13-24.

Sashidhara KV, Singh SP, Srivastava A, Puri A. Identification of the antioxidant principles of *Polyalthia longifolia* var. *pendula* using TEAC assay. *Nat Prod Res* 2011; 25:918-26.

Sharma RK, Mandal S, Rajani GP, Gupta N, Srivastava DP. Antiulcer and Anti-inflammatory Activity of Fresh Leave Extracts of *Polyalthia Longifolia* In Rats. *Int J Drug Develop Res* 2011;3:351-359.

Tanna A, Nair R, Chanda S. Assessment of anti-inflammatory and hepatoprotective potency of *Polyalthia longifolia* var. *pendula* leaf in Wistar albino rats. *J Nat Med* 2009; 63:80-85.

Saraca asoca (Roxb.) De Willd.

Syn: *S. indica* L. and *Jonesia asoca* Roxb.

English name: Asoka.

Family: Fabaceae.

Habitat: India and Sri Lanka.

General Description: *S. asoca* is a small evergreen tree 7-10 cm high. Leaves are parpinnate 15-20 cm long and the leaflets 6-12, oblong and rigidly sub-coriaceous. Leaves are narrowly lanceolate, cork like at the base and with a shot pestistipules are intra-petiolar and completely united. The bark is dark brown or grey or almost black with warty surface. Stem bark are rough and uneven due to the presence of rounded or projecting lenticles. Bark channeled, smooth with circular lenticles and traversely ridged, sometimes cracked. Fracture splinting exposing striated surface, a thin whitish and continuous layer is seen beneath the cork leaver. Flowers are fragrant. Flowers are polygamous apetalous, yellowish orange turning to scarlet, in short laterally placed corymbose, axillary panicles, bract small, deciduous, calyx petaloid. Seeds are 4-8, ellipsoid-oblong and compressed.

Microscopy: (Datta, 1948; Prasad, Natarajan & Wahi, 1961; Srivastava, Bagchi & Srivastava, 1988)

Bark: Transverse section of stem bark shows periderm consisting of a wide layer of cork, radially flattened narrow cork cambium, secondary cortex wide with one or two continuous layers of stone cells with many patches of sclereids, parenchymatous tissue contains yellow masses and prismatic crystals: secondary phloem consists of phloem parenchyma, sieve tubes with companion cells and phloem fibres occurring in groups, crystal fibres present

Stem: Transverse section of stem is circular. Small rounded to oval projecting lenticles are present on the surface. Epidermis is single layered with thin cuticle. Below the epidermis, 5-6 layers of cork are seen. Cortex is 12-16 layered. In the middle region of cortex, 3-5 layers of stone cells are clearly visible. Just above, the phloem region is very distinct and contains tannin cells. Cambium is very clear and is 2-3 layered. Xylem region is composed mostly of tracheids and a few vessels. Primary xylem is prominent. There is prominent pith, composed of thin walled parenchyma and many of the pith cells contain polygonal calcium oxalate crystals.

Root: In transverse section, the root appears somewhat circular in outline. The outermost zone is cork, composed of 8-10 layers of tangentially elongated thick walled cells. Phellogen is not distinct. Inner to the cork region, secondary cortex having two distinct zones are seen. The upper zone consists of 5-7 layers of thin walled parenchyma cells, some of them containing few small rounded starch grains. Below this parenchymatous one, 3-5 layers of mechanical cells are distinctly seen, of these the outer layer is schlerenchymatous and the inner layers are stone cells. Following this supporting region is a broad zone of primary and secondary phloem. The cells are parenchymatous, thin walled and polygonal. 4-6 cambial layers are very prominent below the bast zone. In secondary xylem region tracheids, vessels and parenchyma cells are arranged in a peculiar manner, i.e., xylem parenchyma and tracheids are in alternating patches.

Chemical composition: The bark contains (-) epicatechin, procyanidin p2,11'-deoxyprocyanidin B, (+) catechin, (24, £)- 24-methyl-cholesta-5-en-3p-ol (22 E, 21£)-24- ethycholesta-5,22 dien-33-ol,(24 £)-24- ethylcholesta-5-en-3-p-ol, leucopelargonidin-3-O-p-D glucoside, leucopelargonidin and leucocyanidin. The flower part of plant contain Oleic, linoleic, palmitic and stearic acids, ß-sitosterol, quercetin, kaempferol- 3-0-P-D- glucoside, quercetin- 3-0-P-D-glucoside, apigenin- 7-0-p-D-glucoside, pelargonidin- 3, 5- diglucoside, cyanidin-3, 5-diglucoside, palmitic, stearic, linolenic, linoleic, p and y sitosterols, leucocyanidin and gallic acid. Seed and Pod contains oleic, linoleic, palmitic and stearic acids, catechol, (-) epicatechol and leucocyanidin. Five lignan glyco-sides, lyoniside, nudiposide, 5-methoxy-9-ß-xylopyranosyl-(–)-isolariciresinol, icariside E3, and schizandriside, and three flavonoids, (–)-epicatechin, epiafzelechin-(4ß→8)-epicatechin and procyanidin B2, together with β-sitosterol glucoside, were isolated from dried bark (Sen, 1963).

Action as per Ayurevda: Astringent.

Therapeutics: The bark, which is the drug, is reported to have a stimulating effect on the endometrium and ovarian tissue and is useful in menorrhagia due to uterine fibroids, in leucorrhea and in internal bleeding. It is useful in all cases of uterine bleeding where ergot is indicated. The seeds are strengthening and the ash of plant is good for external application in rheumatoid arthritis (Pradhan *et al.*, 2009).

Parts used : Bark.

Bioactivity : Uterine-tonic (Satyavati *et al.*, 1969), antifungal (Singh & Bhatt, 1980), antibacterial (Pal *et al.*, 1985; Seethram *et al.*, 2003), uterine-tonic (Satyavati *et al.*, 1969), oxytocic (Satyavati *et al.*, 1970), chemoprotective (Varghese *et al.*, 1992), anticancer (Varghese *et al.*, 1992) and central nervous system depressant (Verma *et al.*, 2010)

Further reading

Datta DD. Comparative pharmacognostic studies of *Saraca indica* Linn. and *Trema orientalis* Bl. *Indian Pharm.*1948; 3:180.

Pal SC, Maiti A P, Chaterjee B P, Nandy A. Antibacterial activity of flower and buds of *Saraca indica* Linn. *Ind J Med Res* 1985; 82:188-189.

Pradhan P, Joseph L, Gupta V, Chulet R, Arya H, Verma R, Bajpai A. *Saraca asoca* (Ashoka): A review. *J Chem Pharm Res* 2009; 19:62-71.

Prasad S, Natarajan PN, Wahi SP. Pharmacognostic studies on ashoka bark: barks of *Saraca indica* Linn. and *Polyalthia longifolia* Benth. et Hook.f. *J Sci Indust Res* 1961; 20 B:125-131.

Satyavati GV, Prasad DN, Sen SP, Das PK. Investigations into the uterine activity of *Saraca indica* Linn. (ashoka). *J Res Indian Med* 1969; 4: 37-41.

Satyavati GV, Prasad DN, Sen SP, Das PK. Oxytocic activity of a pure phenolic glycoside (P2) from *Saraca indica* Linn. (ashoka): a short communication. *J Res Indian Med* 1970; 58: 660-663.

Seethram N, Sujeetha H, Jyothiswaran G, Barad A, Sharanbasappa G, Praveen S. Antibacterial activity of *Saraca asoka* Bark. *Indian J Pharm Sci* 2003; 65: 658-659.

Sen SP. Chemical study of the indigenous plant *Saraca indica* (asoka). *Curr Sci* 1963; 32:502-503.

Singh KV, Bhatt SK. Efficacy of seed extracts of *Saraca indica* against some dermatophytic fungi. *Indian Drugs Pharma Ind* 1980; 15: 39-40.

Srivastava GN, Bagchi GD, Srivastava AK. Pharmacognosy of Ashoka stem bark and its adulterants. *Int J Crude Drug Res* 1988; 26:65-72.

Varghese CD. *et al.,* Chemoprotection of *Saraca indica* extracts on cisplatin induced toxicity in mice. *Amla Rese Bull* 1992; 12:86-91.

Varghese CD. *et al.,* Potential anticancer activity of *Saraca indica* extracts towards transplantable tumours in mice. *Indian J Pharma Sci* 1992; 54: 37-40.

Verma A, Jana GK, Sen S, Chakraborty R, Sachan S, Mishra A. Pharmacological Evaluation of *Saraca indica* Leaves for Central Nervous System depressant activity of mice. *J Pharm Sci Res* 2010; 2:338-343.

15. BALA

Current representative: *Sida cordifolia*

Other representative (s): *Abutilon indicum (Atibala), Grewia hirsuta (Nagabala), Sida veronicaefolia, Sida rhombifolia and Grewia tenax.*

Sida cordifolia Linn.

Ayurvedic names: *Vatyalaka, sitapaki, vatyodarahva, bhadraudani, samanga, samamsa* and *svarayastika.*

Common name: Country mallow.

Family: Malvaceae.

Habitat: *S. cordifolia* is a perennial subshrub native to India (Kolammal, 1972).

General Description: *S. cordifolia* is an erect perennial that reaches 50 to 200 cm (20 to 79 in) tall, with the entire plant covered with soft white felt-like hair that is responsible for one of its common names, "flannel weed". The stems are yellow-green, hairy, long and slender. The yellow-green leaves are oblong-ovate, covered with hairs and 3.5 to 7.5 cm (1.4 to 3.0 in) long by 2.5 to 6 cm (0.98 to 2.4 in) wide. The flowers are dark yellow, sometimes with a darker orange center, with a hairy 5-lobed calyx and 5-lobed corolla.

Detailed Description of Root

(a) *Macroscopic*

Drug occurs as entire root or cut pieces of varying lengths, 7-8 mm in thickness, with wavy lateral roots comparatively thinner than main roots having numerous rootlets, brownish-yellow, surface, rough due to scars of small rootlets and lenticels; fracture, hard and splintery.

(b) *Microscopic*

Mature root shows cork consisting of 3-10 rows of narrow, rectangular, tangentially elongated, thin-walled, parenchymatous cells, a few containing rosette crystals of calcium oxalate; secondary phloem composed of phloem fibres in wedgeshaped patches with thin-walled parenchyma in between; phloem rays thin-walled, tangentially elongated towards secondary cortex; a few rosette crystals of calcium oxalate found scattered in phloem parenchyma; secondary xylem composed of vessels, fibre, parenchyma and rays; vessels arranged in radial rows, fibres moderately long, thick-walled, lignified with wide lumen and pointed apex; xylem rays 2-3 cells wide, a few containing rosette crystals of calcium oxalate

Powder - Creamish-grey; shows moderately large, thick-walled, lignified fibres, with wide lumen and pointed tips, fragments of cork cells simple, pitted vessels and a few rosette crystals of calcium oxalate.

Chemical constituents: Asparagin, alkaloids (hypaphorine, ephedrine and vasicine), phytosterols, mucin, gelatin, potassium nitrate and rutin.

Action: *Balya* (tonic).

Therapeutics: *S. cordifolia* is used in the treatment of leucorrhoea, gonorrhea, general debility and rheumatism. Expressed juice of the whole plant is useful in premature ejaculation. The juice obtained from the roots is applied to unhealthy sores. Decoction of the root bark is given in sciatica and rheumatism.

Rajanighantu describes this herb as extremely bitter (*atitikta*), yet madhura, and it is beneficial in deranged pitta. It cures diarrhea and is invigorating and nutritive. It is also efficacious in diseases caused by deranged *kapha*. The rejuvenating action of this herb extends to the nervous, circulatory, and urinary systems. It has a diuretic effect and is useful in urinary problems, including cystitis. Being cooling and astringent, it is used in inflammations and bleeding disorders also.

The herb has rasayana (rejuvenating) properties and assists in convalescence and debility. It is very effective as medicated oil. The popular *Mahanarayana taila, Balati taila, Prabhanjana Vimardhana, Ksheera-bala taila* contain this herb. The oils are used topically to massage the sore muscles, sore joints in arthritis and rheumatism, in sciatica and neuritis of legs. Another oil called *Dhanvantari tailam* (21 and 101 times boiled) which contains *Bala* along with 47 other substances, and is prepared in milk and given for all disorders produced by the derangement of '*Vata*', such as emaciation, weakness, diseases of reproductive organs.

Parts used: Roots and seeds.

Bioactivity: Imnnunomodulator (Dixit, Tewar & Gupta, 1987), antihepatotoxic (Rao & Mishra, 1998), analgesic, anti-inflammatory and hypoglycemic (Muzaffer, 1998; Kanth & Diwan 1999; Sutradhar *et al.,* 2006; Swathy *et al.,* 2010), antipyretic and anti-ulcerogenic (Philip *et al.,* 2008) and antiperoxidative (Swathy *et al.,* 2010).

Further reading

Dixit SP, Tewar PV, Gupta RM. Experimental studies on the immunological aspects of atibala (*Abutilon indicum* (Linn.) Sw.), mahabala (*Sida rhombifolia* Linn.), bala (*Sida cordifolia* Linn.) and bhumibala (*Sida veronicaefolia* Lam.) *J Res Ind Med* 1987; 13:50-66.

Kanth VR, Diwan PV. Analgesic, anti-inflammatory and hypoglycaemic activities of *Sida cordifolia*. *Phytother Res* 1999; 13:75-77.

Kolammal M. The botanical sources of bala. *J Res Ind Med* 1972; 7: 37-45.

Muzaffer Alam KK, Dasan S, Susan T, Joy S. Anti-inflammatory potential of balarishta and dhanvantara gutika in albino rats *Ancient Sci Life* 1998; 17: 305-312.

Philip BK, Muralidharan A, Natarajan B, Varadamurthy S, Venkataraman S. Preliminary evaluation of anti-pyretic and anti-ulcerogenic activities of *Sida cordifolia* methanolic extract. *Fitoterapia* 2008; 79:229-231.

Rao KS, Mishra SH. Antihepatotoxic activity of *Sida cordifolia* whole plant. *Indian Perfumer* 1998; 3: 99-100.

Singh KK, Singh OP, Dubey SD. Clinical evaluation of bala (*Sida cordifolia* Linn) as a brimhaniya (anabolic drug). *Sachitra Ayur* 1998; 50:754-756.

Sumanth M, Mustafa SS. Antistress, adoptogenic activity of *Sida cordifolia* roots in mice. *Indian J Pharm Sci* 2009; 71: 323-324.

Sutradhar RK, Matior Rahman AKM, Ahmad M, Bachar SC, Saha A, Guha SK. Bioactive alkaloid from *Sida cordifolia* Linn. with analgesic and anti-inflammatory activities. *Iranian J Pharmacol Ther* 2006; 5:175-178.

Swathy SS, Panicker S, Nithya RS, Anuja MM, Rejitha S, Indira M. Antiperoxidative and antiinflammatory effect of *Sida cordifolia* Linn. on quinolinic acid induced neurotoxicity. *Neurochem Res* 2010; 35: 1361-1367.

Abutilon indicum G.Don.

Ayurvedic name: *Atibala, Balika, Vatyapushpy* and *Bhuribala.*

Common name: Country mallow.

Family: Malvaceae.

Habitat: The plant grows as a weed, and is found abundantly in wastelands from the seashore to 1,200 meters high in India and in the sub-Himalayan tracts.

General Descrption: It is an herbaceous or shrubby, softly tomentose plant. Stem is round, often tinged with purple color. The leaves are petiolate, ovate to orbicular-cordate, acuminate and toothed. Flowers are borne solitary in long, jointed and axillary pedicels. Calyx lobes divided in the middle, ovate and apiculate. Corolla is yellow or orange-yellow and opens in the evening. Carpels are 15-20 in number. Fruits are hispid, scarcely longer than the calyx and the awns are erect. Seeds are three to five, kidney-shaped, dark brown or black, tubercled or with minutely stellate hairs (Lakshmayya *et al.,* 2003).

Microscopy: Transverse Section of root shows outermost tissue the cork, which is thin and composed of 4 to 7 or more rows of rectangular or slightly tangentially elongated cells. Occasionally, lenticels may be seen. Inner to the cork is the phellogen single row of narrow, thin walled, elongated cells. This is followed by the cortex made up of 3 or 4 layers; some cortical cells contain starch.

Bast is present in the form of linear conical bands. In between these conical bands are present, the phloem elements formed mostly of phloem parenchyma. Some of these contain small calcium oxalate crystals while others have starch grains. Inner to the phloem, a cambium consisting of 5 to 7 rows of narrow thin-walled rectangular cells is present. Wood is composed of many vessel elements, wood fibres, wood parenchyma and medullary rays. Wood parenchyma cells are thickwalled. Medullary rays are long, 1 or 2 seriate radiating from centre of the wood extending up to the cortex. All the rays widen towards periphery. Medullary rays also have calcium oxalate crystals and starch grains (Chakraborthy, 2009).

Chemical composition: Asparagin, mucilage and minerals (Govind & Chopra, 1974). The flower contains flavonoid glycosides and flavonoids: gossypin, gossypitrin, cyanidin-3-rutinoside (Subramaniam & Nair, 1972), luteolin, chrysoeriol, cynaroside, termopsoside, apigetrin, hirsutrin, rutin. The whole herb contains monoterpenoids: eugenol (Ahmed *et al.,* 2004), sesquiterpenoids: alantolactone, isoalantolactone (Ahmed *et al.,* 2004), triterpenoids: ß-amyrin [Dennis & Kumar, 1984); organic acids: gallic acid (Shivhare *et al.,* 2010). It also contains essential oils: β-pinene, caryophyllene, caryophyllene oxide, cineole, geraniol, geranyl acetate, elemene, farnesol, borneol and eudesmol (Jain, Sharma & Bokadia, 1982).

Action: Tonic.

Therapeutics: In traditional medicine, *A. indicum* is used as a demulcent, aphrodisiac, laxative, diuretic, pulmonary and sedative (leaves). The bark is astringent and diuretic; laxative, expectorant and demulcent (seeds); laxative and tonic, anti-inflammatory and anthelmintic (plant); analgesic (fixed oil); diuretic and for leprosy (roots).

Parts used: Leaves and seeds.

Biaoctivity: Analgesic (Bagi *et al.,* 1984, Sharma, Ahmed & Sharma, 1989), antimicrobial (Mehta, Neogi & Kotra, 1997), hypoglycemic (Seetharam *et al.,* 2002), antidiarrhoeal (Chandrashekhar *et al.,* 2004) and bronchodilator (Sharma Ahmed & Sharma, 2008).

Further reading

Bagi MK, Kalyani GA, Dennis TJ, Akshaya Kumar K, Kakrani HK. A preliminary pharmacological screening of *Abutilon indicum*: analgesic activity. *Indian Drugs* 1984; 22: 69-72.

Chakraborthy GS. Pharmacognostical and phytochemical evaluation of leaf of *Abutilon indicum* (Linn.). *Int J Pharm Sci Drug Res* 2009; 1: 28-31.

Chandrashekhar, V.M., A.M. Nagappa, T.S. Channesh, P.V. Habbu and K.P. Rao. Antidiarrhoeal activity of *Abutilon indicum* Linn leaf extract. *J Nat Rem* 2004; 4:1.

Dennis TJ, Kumar KA. Chemical examination of the roots of *Abutilon indicum* Linn. *J Oil Technol Assoc India* 1984, 15: 82-83.

Dixit SP, Tewar PV, Gupta RM. Experimental studies on the immunological aspects of atibala (*Abutilon indicum* (Linn.) Sw.), mahabala (*Sida rhombifolia* Linn.), bala (*Sida cordifolia* Linn.) and bhumibala (*Sida veronicaefolia* Lam.) *J Res Ind Med* 1987; 13:50-66.

Govind KN, Chopra KS. on *Abutilon indicum* (Linn.) Sw. *Planta Med* 1976; 30:174.

Lakshmayya K, Nelluri NR, Kumar P, Agarwal NK, Gouda TS, Setty RS. Phytochemical and pharmacological evaluation of leaves of *Abutilon indicum*. *Indian J Trad Knowl* 2003; 2: 79-83.

Mehta BK, Neogi R, Kotra S. Antimicrobial activity of *Abutilon indicum*. *Fitoterapia* 1997; 68:273-274.

Paranjape AN, Mehta AA. A study on the clinical efficacy of *Abutilon indicum* in the treatment of bronchial asthma. *Oriental Pharm Exper Med* 2006; 6: 330-335.

Paranjape AN, Mehta AA. Investigation into the Mechanism of Action of *Abutilon indicum* in the Treatment of Bronchial Asthma. *Global J Pharmacol* 2008; 2:23-30.

PK Jain, TC Sharma, MM Bokadia. Chemical investigation of the essential oil of *Abutilon indicum*. *Acta Ciencia Indica Chem* 1982, 8:136-139.

Poonkothai, M. Antibacterial activity of leaf extract of *Abutilon indicum*. *Ancient Sci Life* 2008; 26: 39-41.

Seetharam YN, Chalageri G, Setty SR, Bheemachar H. Hypoglycemic activity of *Abutilon indicum* leaf extracts in rats. *Fitoterapia* 2002; 73:156-159.

Sharma PV, Ahmed ZA, Sharma VV. Analgesic constituents of *Abutilon indicum*. *Indian Drugs* 1989; 26: 333.

Shivhare Y, Singh P, Singh S, Tiwari R, Bharti PK, Upmanyu N . Ethanobotanical and Ethanopharmacological profile of *Abutilon indicum* Linn: A Review. *Der Pharmacia Sinica* 2010, 1: 40-45.

Subramanian SS, Nair AGR. Flavonoids of four malvaceous plants. *Phytochemistry* 1972, 11:1518-1519.

Grewia hirsuta Vahl.

Ayurvedic name: Nagbala.

Family: Tiliaceae.

Habitat: Rajasthan and Bihar.

General Description: Large shrubs. Leaves hairy, ovate, cordate at base. Flowers white, buds globose, in axillary cymes. Drupe entire.

Action: Tonic.

Therapeutics: Nagbala has properties to pacify provocation of *Vata* and *Pitta* humor. It is useful as nervine tonic, brain tonic, demulcent, anti-acidic, expectorant, antipyretic, diuretic, aphrodisiac, carminative and cardiac tonic.

Parts used: Roots.

Sida veronicaefolia Lam.

Ayurvedic name: *Bhumibala* and *rajbala.*

Syn: *S. javensis* Cav., *S. humilis* Cav.

Common name: Penghu Sida.

Family: Malvaceae.

Habitat: Throughout hotter parts of India and Nepal.

General Description: *S. veronicaefolia* is a hairy herb with cordate, ovate leaves, pale yellow axillary flowers and fruits containing brown seeds (Kolammal, 1972).

Action: Tonic.

Therapeutics: Decoction of the root of *S. veronicaefolia* is a nervine and brain tonic so it's useful in loss of memory due to *vata* disorders.

Bioactivity: Immunomodulator (Dixit, Tewar & Gupta, 1987) and antioxidant (Pandey *et al.,* 2009).

Further reading

Dixit SP, Tewar PV, Gupta RM. Experimental studies on the immunological aspects of atibala (*Abutilon indicum* (Linn.) Sw.), mahabala (*Sida rhombifolia* Linn.), bala (*Sida cordifolia* Linn.) and bhumibala (*Sida veronicaefolia* Lam.) *J Res Ind Med* 1987; 13:50-66.

Kolammal M. The botanical sources of bala. *J Res Ind Med* 1972; 7: 37-45.

Pandey M, Sonker K, Kanoujia J, Koshy MK, Saraf SA. *Sida veronicaefolia* as a source of natural antioxidant. *Int J Pharm Sci Drug Res* 2009; 1:180-182.

Sida rhombifolia L.

Ayurvedic names: *Mahabala.*

Syn: *Sida retusa.*

Common name: Arrow leaf Sida, Paddy's lucern, Jelly leaf.

Family: Malvaceae.

Habitat: Found as a weed, all around India.

General Description: A semi woody, erect annual or perennial undershrub grows up to 1 meter in height. Leaves are variable, Rhomboid-lanceolate, serrated at apical half; flowers yellowish white, solitary, found in leaf axils. Fruits schizocarp, enclosed within the calyx. Seeds brown colored and smooth.

Chemical composition: Mucilage and alkaloids (Anand, Varma & Ghosal, 1981).

Action: Tonic.

Therapeutics: Plant pacifies vitiated *vata, pitta*, pain, arthritis, asthma, bronchitis, burning sensation, and urinary retention.

Bioactivity: Anti-inflammatory (Kumar & Mishra, 1997; Venkatesh *et al.,* 1999), antinociceptive (Venkatesh *et al.,* 1999), antioxidant (Dhalwal *et al.,* 2007), and hypolipidemic (Thounaojam *et al.,* 2009, Dhalwal *et al.,* 2010).

Further reading

Anand P, Varma RK, Ghosal S. Alkaloidal constituents of *Sida acuta, S. humilis, S. rhombifolia* and *S. spinosa. Planta Med* 1981; 43:384-388.

Dhalwal K, Shinde VM, Singh B, Mahadik KR. Hypoglycemic and hypolipiodemic effect of *Sida rhombifolia* ssp. retusa in diabetic induced animals. *Int J Phytomed* 2010; 2:160-165.

Dhalwal, Kamlesh, Yogesh S. Deshpande and Ajit P. Purohit. Evaluation of the *in vitro* antioxidant activity of *Sida rhombifolia* (L.) ssp. *retusa* (L.). *J Med Food* 2007; 10: 683-688.

Kumar RS, Mishra SH. Anti-inflammatory and hepatoprotective activities of *Sida rhombifolia* Linn. *Indian J Pharmacol* 1997; 29:110-116.

Thounaojam MC, Jadeja RN, Patel VB, Devkar RV, Ramachandran AV. Potential of *Sida rhomboidea* Roxb. leaf extract in controlling hypertriglyceridemia in experimental models. *Pharmacog Res* 2009; 1:208-212.

Venkatesh S, Siva Rami Reddy Y, Suresh B, Madhava Reddy B and Ramesh M. Antinociceptive and anti-inflammatory activity of *Sida rhomboidea* leaves. *J Ethnopharmacol* 1999; 67: 229-232.

16. BANSHALOCHANA

Current representative: *Bambusa arundiacea*

Other representative: *Dendrocalamus strictus*

Bambusa arundinacea (Retz.) Willd.

Syn: *Bambusa bambos* Druce.

Common name: Spiny bamboo, Thorny bamboo.

Ayurvedic names: *Vamsa, Venu, Rujasaha, Vamsalekhana.*

Family: Poaceae.

Habitat: Throughout India growing wild, also cultivated.

General Description: A large grass grows up to 40 meters in height. Leaves alternate, long, base chordate, lanceolate, acute, and very rough; plant flowers once in its lifetime. Flowers are seen in large branched panicles, bisexual, small yellowish; fruits are oblong grains, grooved at one side. Stems are numerous, nodes prominent, internodes up to 50 cm long, greenish or yellowish. Very rarely some edible substance called Bamboo manna can be seen inside the hollow stems.

Chemical composition: The stem consists almost entirely of cellulose and hemicellulose (xylans, arabans, polyuronides, etc.) and lignins, with a small amount of resins. Oven-dried stems contain 3.3% ash, 1.8% silica, 6.0% hot water solubles (see above), 19.6% pentosans, 30.1% lignin, and 57.6% cellulose. A cyanogenic glucoside like anti-thyroidal substance has been studied (Chandra *et al.*, 2004).

Action: Expectorant.

Therapeutics: *B. arundinacea* pacifies vitiates kapha, pitta, indigestion, diarrhea, amenorrhea, Ulcer, muscle spasm and general weakness. It is used in chronic bronchitis with success. Tabasheer, a siliceous secretion (up to 97% SiO_2), considered aphrodisiac, cooling, and tonic, is used in asthma, cough and debilitating diseases (C.S.I.R., 1948–1976). An Ayurvedic remedy, *Sitopaladi Churna*, was used traditionally for tuberculosis and other wasting diseases and has been adopted as a popular remedy for common cold, sore throat, sinus congestion, and cough. It is a powder made with tabasheer as the main ingredient, plus small amounts of long pepper, cardamom, and cinnamon in a base of sugar (Pattanayak *et al.*, 2010).

Parts used: Bamboo manna (tabashir), roots, leaves and fruit.

Bioactivity: Antibacterial (Basu, 1975), antifertility (Vanithakumari *et al.*, 1989) antiinflammatory and antiulcer (Muniappan & Sundararaj, 2003) and thyrotropic (Chandra *et al.*, 2004).

Further reading

Basu AP. Studies on antibacterial activity of *Bambusa arundinacea* Willd. *Indian J Pharm* 1975; 37: 176.

C.S.I.R. (Council of Scientific and Industrial Research). 1948–1976. The wealth of India. 11 vols. New Delhi.

Chandra AK, Ghosh D, Mukhopadhyay S, Tripathy S. Effect of bamboo shoot, *Bambusa arundinacea* (Retz.) Willd. on thyroid status under conditions of varying iodine intake in rats. *Indian J Exp Biol* 2004; 42:781-786.

Dharmananda S. Bamboo as medicine. http://www.itmonline.org/arts/bamboo.htm.

Muniappan M, Sundararaj T. Antiinflammatory and antiulcer activities of *Bambusa arundinacea. J Ethnopharmacol* 2003; 88: 161-167.

Pattanayak P, Panda SK, Dash S, Behera M, Mishra SK. Study of antitussive activity of sitopaladi churna: a poly-herbal formulation. *Int J Pharma Sci Rev Res* 2010; 4:65-67.

Vanithakumari G, Manonayagi S, Padma S, Malini T. Antifertility effect of *Bambusa arundinacea* shoot extracts in male. *J Ethnopharmacol* 1989; 25:173-80.

Dendrocalamus strictus (Roxb.) Nees

Common name: Iron bamboo, Calcutta bamboo, male bamboo.

Family: Poaceae.

Habitat: It is native to South Asia.

General Description: A deciduous densely tufted bamboo. Culms 8-16 m high, 2.5-8 cm diameter, pale blue green when young, dull green or yellow on maturity, much curved above half of its height; nodes somewhat swollen, basal nodes often rooting, lower nodes often with branches; internodes 30-45 cm long, thick-walled. Culms-sheaths variable, lower ones shorter, 8-30 cm long with golden brown stiff hairs on the back, sometimes glabrous in dry localities, striate, rounded at the top, margin hairy; ligule 2-3 mm high, toothed; auricles small, blade triangular, awl-shaped, hairy on both sides. Leaves linear-lanceolate, small in dry localities, up to 25 cm long and 3 cm broad in moist areas, rounded at the base into a short petiole, tip sharply acuminate with twisted point, rough and often hairy above, softly hairy beneath; ligule very short. Inflorescence a large panicle of large dense globular heads 4-5 cm apart; rachis rounded, smooth. Spikelets spine scent, usually hairy, the fertile intermixed with many sterile smaller ones, 7.5-12 mm long and 2.5-5.0 mm broad, with 2 or 3 fertile flowers; empty glumes 2 or more, ovate spine scent, many-nerved; flowering glumes ovate, ending in a sharp spine surrounded by ciliate tufts of hair; palea ovate or obovate, emarginate, lower ones 2-keeled, ciliate on the keels and 2-nerved between them, uppermost not keeled, often nearly glabrous, 6 to 8-nerved. Stamens long-exerted; turbinate, stalked, hairy above and surmounted by a long style ending in a purple feathery stigma.

Note: Decoction of leaves and nodes and silicious matter is used in the traditional medicine.

17. BHARANGI

Current representative: *Clerodendron serratum*

Possible representative (s): *Clerodendron siphonanthus, Gardenia turgida, Gardenia latifolia, Premna corymbosa Premna herbacea* and *Picrasma quassioides.*

Clerodendron serratum Linn.

Ayurvedic names: *Bharangi, brgubhava, padma, kasaghani, gandhaparvani, sukramata, bhanji* and *brahmana yastika.*

Family: Verbenaceae.

Habitat: India and Nepal.

General Description: *C. serratum* is a perennial shrub, growing 0.3-3.0 metres in height. The leaves are 7-15 cm long and 5 cm. broad, rough, serrate and tapering at the apes. The flowers are blue, fragrant and the plant looks beautiful during flowering. The fruits are purple drupes and succulent. The flowering and fruiting occurs during April and November (Gantait, 2010).

Detailed Description of Root (Prasad *et al.*, 1967)

(a) *Macroscopic*:

Mature root hard, woody, cylindrical, upto 5 cm thick, external surface light brown having elongated lenticels; bark, thin and easily separated from a broad wood which shows marked medullary rays and concentric growth rings in a transversely cut surface; fracture, short; taste, acrid.

(b) *Microscopic*:

Mature root shows stratified cork composed of 14-20 layers of thin-walled, tangentially elongated cells; each stratification consists of 3-5 layers of cells; secondary cortex wide, outer 2 or 3 layers radially arranged and tangentially elongated, inner cells polyhedral or circular to ellipsoidal with intercellular spaces; a few cells modified into stone cells with greatly thickened wall having concentric striations and radiating canals with narrow lumen; some cells contain acicular crystals of calcium oxalate and a few contain brown colouring matter; secondary phloem consists of sieve elements and parenchyma mostly collapsed in outer region, forming ceratenchyma; some phloem parenchymatous cells modified into stone cells similar to those in secondary cortex but somewhat smaller and with greater thickening' of walls; secondary xylem diffused porous consisting of vessels, tracheids, fibres and xylem parenchyma traversed by xylem rays; macerated preparation show wider vessels cylindrical, drum-shaped, some being elongated at one end having bordered pits, rarely reticulate or pitted, while narrower ones elongated with spiral to reticulate thicken- tracheids long, cylindrical with tapering ends and bordered pits; xylem fibres moderately thick-walled with mostly tapering, pointed ends and oblique bordered pits; xylem parenchyma square to rectangular with simple pits on their walls; medullary rays 1-4 cells wide and 2-50 cells high, 2 or 3 cell wide rays more common, having simple pits on their walls; acicular crystals and abundant simple and compound starch grains measuring up to 20 μ in dia. present in a number of cells throughout the region.

Powder - Light-brown; shows vessels reticulate, spiral and with bordered pits, starch grains simple and compound, round to oval, measuring upto 20 μ in dia. and acicular crystals; stone cells as describes under microscopy present.

Chemical composition: Triterpenoid constituents: oleanolic acid, queretaroic acid, and a new acid, named serratagenic acid (Rangaswami & Sarangan, 1969), D-mannitol (Garg & Verma, 1967), gamma-sitosterol (Vasavada, Banerjee SK, Chakravarti, 1967), serratin and lupeol (Ravikumar, Lakshmanan & Ravi, 2008).

Action: Antiasthamtic.

Therapeutics: Bharngi is bitter, pungent and astringent in taste, pungent in the post digestive effect and has hot potency (virya). According to *Raja Nighantu*, it is useful in asthma, cough, fever, worms, burning sensation of the body and wounds. Susruta has mentioned it as a panacea for epilepsy and also as *stanyasodhaka* (lactodepurant). *Vagbhata* has cited its usefulness in cough due to *kapha* and *vata*.

Parts used: Roots.

Bioactivity: Antihistaminic (Sachdev, Vasavada & Joseph, 1964, Gupta, 1968) and hepatoprotective (Vidya *et al.,* 2007).

Further reading

Gantait A, Roy P, Nema NK, Dutta PK, Mukherjee PK. A validated method for standardization of the bark of *Clerodendron serratum. Nat Prod Commun* 2010; 5:863-866.

Garg VP, Verma SC. Chemical examination of *Clerodendron serratum.* Isolation and characterization of D-mannitol. *J Pharm Sci* 1967; 56:639-40.

Gupta SS. Development of antihistamine and antiallergic activity after prolonged administration of a plant saponin from *Clerodendron serratum. J Pharm Pharmacol* 1968; 20: 801-802.

Narayanan N, Thirugnanasambantham P, Viswanathan S, Vijayasekaran V, Sukumar E. Antinociceptive, anti-inflammatory and antipyretic effects of ethanol extract of *Clerodendron serratum* roots in experimental animals. *J Ethnopharmacol* 1998; 65: 237-241.

Rangaswami S, Sarangan S. Sapogenins of *Clerodendron serratum*: Constitution of a new pentacyclic triterpene acid, serratagenic acid. *Tetrahedron* 1969; 25:3701-3705.

Ravikumar R, Lakshmanan AJ, Ravi S. Chemical constituents from *Clerodendron serratum. J Asian Nat Prod Res* 2008; 10:659-62.

Sachdev KC, Vasavada SA, Joseph AD. Antihistaminic activity of *Clerodendron serratum* (Linn) Moon. *Indian J Pharm* 1964; 26:105-106.

Sachdev KS, Banerjee SK, Charkarvarti RN. Preliminary chemical examination of the root-bark of *Clerodendron serratum* (Linn) Moon. *Bull Calcutta Sch Trop Med* 1965; 13:17-18.

Vasavada SA, Banerjee SK, Chakravarti RN. Gamma-Sitosterol from *Clerodendron serratum. Bull Calcutta Sch Trop Med* 1967; 15:61.

Vidya SM, Krishna V, Manjunatha BK, Mankani KL, Ahmed M, Singh SD. Evaluation of hepatoprotective activity of *Clerodendrum serratum. Indian J Exp Biol* 2007; 45:538-542.

Clerodendron siphonanthus (R.Br.) CB Clarke N.O.: Verbenaceae

Syn: *C. indicum* (Linn.) Kuntze.

English name: Peacock's tail, Tube flower.

Family: Verbenaceae.

Habitat: Cultivated as an ornamental throughout India.

General Description: A small shrub grows up to 2 meters in height with soft succulent branches. Leaves simple, 3-4 in a whorl, lanceolate, sessile, acute upper side glabrous, lower coarsely pubescent, with prominent veins. Flowers arises from terminal spikes, whitish, with long and curved corolla tube. Flowers are usually sterile (Prasad *et al.*, 1967).

Therapeutics: According to M.C. Dutt, the plant is used as *bharangi* in Bengal. Plant pacifies vitiated *vata, kapha*, asthma, cough, intestinal worms, fever, arthritis, atrophy, emaciation, consumption, and general weakness.

Bioactivity: Antifertility (Shukla & Pandey, 2008).

Further reading

Prasad S, Khosa RL, Wahi AK, Bhardwaj RC. Pharmacognostic studies on bharangi root (Part I): roots of *Clerodendron serratum* (Linn) and *Clerodendron indicum* Kuntze. *J Res Indian Med* 1967; 1:223-229.

Shukla M, Pandey SK. Antifertility with histopathological effect of *Clerodendron siphonanthus* (R. Br.) on the ovary of female albino rats. *J Ecophysiol Occupat Heal* 2008;8.

Gardenia turgida Roxb. N.O.: Rubiaceae

Syn: *Ceriscoides turgida* (Roxb.) Tirveng.

Habitat: India.

Family: Rubiaceae.

General Description: *G. turgida* is an evergreen shrub growing to 1–15 metres tall. The leaves are opposite or in whorls of three or four, 5–50 centimetres (2.0–20 in) long and 3–25 centimetres (1.2–9.8 in) broad, dark green and glossy with a leathery texture. The flowers are solitary or in small clusters, white, or pale yellow, with a tubular-based corolla with 5-12 lobes (petals) from 5–12 centimetres (2.0–4.7 in) diameter (Nair *et al.*, 1971; Bhatnagar, Singh VK & Pandey, 1973b).

Chemical composition: *Triterpenoids* (Reddy, Ayengar & Rangaswami, 1973, Singh & Paradasani, 1979).

Therapeutics: Bark of the plant is used in the treatment of conjunctivitis.

Bioctivity: Antiasthmatic (Chaturvedi, Tiwari & Sen, 1976).

Further reading

Bhatnagar LS, Singh VK, Pandey G. Investigations on the new botanical source of bharangi in Madhya Pradesh – I. *J Res Indian Med* 1973b;8: 86-94.

Chaturvedi GN, Tiwari SK, Sen SP. Experimental studies on bharangi (*Gardenia turgida*), an indigenous drug used for the treatment of bronchial asthma. *J Res Indian Med* 1976; 11: 100-103.

Nair NK, Yoganarasimhan SN, Keshavamurthy KR, Mary Z. Studies on some South Indian market samples of Ayurvedic drugs - I. *Ancient Sci Life* 1971; 2: 71-78.

Reddy GCS, Ayengar KNN, Rangaswami S. *Triterpenoids of Gardenia turgida Phytochemistry* 1973; 12: 1831.

Singh P, Paradasani RT. Chemical examination of the roots of *Gardenia turgida* Roxb. *J Indian Chem Soc* 1979; 56:327-328.

Gardenia latifolia Aiton

Ayurvedic name: *Parpataki.*

English name: Ceylon Boxwood, Indian Boxwood.

Family: Rubiaceae.

Habitat: *G. latifolia* is found in the forests of Madhya Pradesh in India, and has been widely cultivated elsewhere, to the point of naturalization, especially in Nigeria, West Africa where the tree is highly valued for both its fruit and shade.

General Description: *G. latifolia* is medium-sized to large, long-lived tree, 5-10 m high. The berry is globose to oblong, smooth to lightly net-ridged, 2-4.5 cm long x 2-3 cm wide, often with 5-9 persistent calyx lobes, initially leathery, drying to a hard, woody consistency.

Chemical composition: Triterpenoids: oleanolic acid, hederagenin, siaresinolic acid, spinosic acid, D-mannitol and sitosterol (Reddy, Ayengar & Rangaswami, 1975).

Further reading

Reddy GCS, Ayengar KNN, Rangaswami S. Triterpenoids of *Gardenia latifolia*. *Phytochemistry* 1975; 14: 307

Premna herbacea Roxb.

Syn: *P. timoriana* H. Lam., *Pygmaeopremna humilis* Merr.

Family: Verbenaceae.

General Description: *P. herbacea* is a small, inconspicuous undershrub up to 15 centimeters in height; it is unusal in that it is produced from stout, elongated, woody roots, with hardly any stems. The roots are about as thick as a crowquill with numerous, almost-globular, woody knots. The leaves are simple, obovate-oblong or obovate, up to 13 centimeters long, 6.5 centimeters wide, and pointed at both ends, with entire and irregularly toothed margins. The stalks are very short, being 2 to 4 millimeters long. The flowers are greenish-white, 4-parted, and borne on short, terminal inflorescences about 1 centimeter long. The fruit is black, broadly obovoid, and 4 to 5 millimeters diameter.

Chemical composition: Diterpenoid: sirutekkone (Sandhya, Krishnamurthy & Chandrakumar, 1988).

Bioactivity: Antipyretic, antinociceptive and anti-inflammatory (Narayanan *et al.,* 2000), antimicrobial (Marthanda Murthy *et al.,* 2006) and antitumor (Kumar Ravishankar *et al.,* 2008.)

Further reading

Kumar Ravishankar, Pai KSR, Dhamija Isha, Setty MM, Manjula SN, Ramalingayya GV. An appraisal of the antitumor activity of alcoholic extract of *Premna herbacea* roxb in Ehrlich.s ascetic carcinoma model. *Indian J Pharmacol* 2008; 4: 567 (abstract 107).

Marthanda Murthy M, Subramanyam M, Giridhar KV, Jetty A. Antimicrobial activities of bharangin from *Premna herbacea* Roxb. and bharangin monoacetate. *J Ethnopharmacol* 2006; 104: 290-292.

N Narayanan, P Thirugnanasambantham, S Viswanathan, S Rajarajan, E Sukumar. Comparative Antibacterial Activities of *Clerodendrum serratum* and *Premna herbacea* Roots. *Indian J Pharm Sci* 2004; 66:453-454.

Narayanan N, Thirugnanasambantham P, Viswanathan S, Kannappa Reddy M, Vijayasekaran V, Sukumar E. Antipyretic, antinociceptive and anti-inflammatory activity of *Premna herbacea* roots. *Fitoterapia* 2000; 71:147-153.

Sandhya G, Krishnamurthy R, Chandrakumar N. Structure of sirutekkone, a diterpenoid from *Premna herbacea. Phytochemistry* 1988; 27:2249-2250.

Vasudevan Nair K, Yoganarasimhan SN, Keshavamurthy KR, Mary Z. Studies on some South Indian market samples of Ayurvedic drugs – I. *Ancient Sci Life* 1971;2: 71-78.

Picrasma quassioides (D.Don) Benn.

Syn: *Simaruba quassioides.*

Family: Simaroubaceae.

General Description: *P. quassioides* is a deciduous shrub or small tree growing to 10–15 m (rarely 20 m) tall with a trunk up to 50 cm diameter. The bark is smooth and dark grey-brown. The leaves are 15–40 cm long, pinnate, with 7–15 leaflets 2.5–10 cm long and 1.5–4.5 cm broad, with a coarsely and irregularly toothed margin. The flowers are green to yellow-green with four or five sepals and petals, produced in cymes 8–15 cm long in mid to late spring. The fruit is an ovoid to globose red to black drupe 6–7 mm diameter.

Further reading

Vasudevan Nair K, Yoganarasimhan SN, Keshavamurthy KR, Mary Z. Studies on some South Indian market samples of Ayurvedic drugs – I. *Ancient Sci Life* 1971;2: 71-78.

Bhatnagar LS, Singh VK, Pandey G. Investigations on the new botanical source of bharangi in Madhya Pradesh – I. *J Res Indian Med* 1973b;8: 86-94.

18. BHRINGRAJA

Current representative: *Eclipta alba*

Possible representative (s): *Wedelia calendulaceae*

There is some confusion between the Sanskrit and vernacular names of *Wedelia calendulacea*, and *Eclipta alba*. Both these plants are called kesaraj or kesuria in Bengali, and the two Sanskrit names of bhringaraja and kesaraja are used as synonyms. The Hindustani term bhangra, derived from the Sanskrit bhringardja is however generally applied to *Wedelia calendulacea*, and this is the plant used in medicine at present.

Eclipta alba (L.) Hassk.

Family: Asteraceae.

Common name: Trailing eclipta.

Ayurvedic names: *Angaraka, bhekaraja, kesharanjaka, marvaka* and *suryavallabha*.

Habitat: India.

Description

(a) *Macroscopic*

Root - Well developed, a number of secondary branches arise from main root, upto about 7 mm in dia., cylindrical, greyish.

Stem - Herbaceous, branched, occasionally rooting at nodes, cylindrical or flat, rough due to oppressed white hairs, node distinct, greenish, occasionally brownish.

Leaf - Opposite, sessile to subsessile, 2.2 - 8.5 cm long, 1.2 - 2.3 cm wide, usually oblong, lanceolate, sub-entire, sub-acute or acute, strigose with appressed hairs on both surfaces.

Flower - Solitary or 2, together on unequal axillary peduncles; involucral bracts about 8, ovate, obtuse or acute, herbaceous, strigose with oppressed hairs; ray flowers ligulate, ligule small, spreading, scarcely as long as bracts, not toothed, white; disc flowers tubular, corolla often 4 toothed; pappus absent, except occasionally very minute teeth on the top of achene; stamen 5, filaments epipetalous, free, anthers united into a tube with base obtuse; pistil bicarpellary; ovary inferior, unilocular with one basal ovule.

Fruit - Achenial cypsella, one seeded, cuneate, with a narrow wing, covered with warty excrescences, brown.

Seed - 0.2 - 0.25 cm long, 0.1 cm wide, dark brown, hairy and non endospermic.

(b) *Microscopic*

Root - Mature root shows poorly developed cork, consisting of 3-5 rows of thin-walled, tangentially elongated cells; secondary cortex consists of outer one or two rows of tangentially elongated or rounded cells with air cavities, inner secondary cortex of tangentially elongated to irregular shaped, parenchymatous cells with conspicuous air cavities; stone cells found scattered in secondary cortex and cork, in singles or in groups of various shape and size; pericyclic fibres in tangentially arranged bands of many cells or in singles; secondary phloem consists of sieve elements including phloem fibres traversed by multiseriate phloem rays; phloem rays broader towards periphery, consisting of rounded cells; xylem composed of vessels, fibre tracheids, fibres and xylem parenchyma, traversed by xylem rays; vessels numerous, found scattered throughout wood, in macerated preparation vessels small, drum-shaped, cylindrical elongated with pitted walls and perforations, simple, rarely slightly oblique; fibre tracheids, pitted, with very pointed tips, xylem fibres long with pointed tapering ends and short lumen, a few fibres show peg-like outgrowths towards the tapering ends; xylem parenchyma sparse usually squarish to rectangular having simple pits on their walls, xylem ray distinct, run straight in tangential section, generally 5-32 cells in height and 3-5 cells in width although very rarely uniseriate and biseriate rays also found, ray cells pitted.

Leaf-

Petiole - shows single layered upper and lower epidermis consisting of tubular cells, covered with striated cuticle; trichomes of two types, non-glandular, uniseriate, 1-5 celled, warty, and with pointed apical cell; epidermis followed by wide cortex, consisting of 2-5 layered collenchyma on both, upper arid lower side with distinct angular thickening; parenchyma 4-6 layered on upper side and 5-8 layered on lower side consisting of isodiametric, thin-walled cells with intercellular spaces; five vascular bundles central one largest while four others small flanking to either side of central bundle, consists of xylem on dorsal side and phloem on ventral side; xylem vessels arranged in radial rows traversed by xylem rays.

Midrib - cut at basal region shows both upper and lower single layered epidermis, externally covered with cuticle, a few epidermal cells elongate outwards to form uniseriate hairs; epidermis followed by cortex, consisting of 3-5 layered collenehymatous cells on both sides; section cut at middle region shows 3-4 layered collenchymatous cells on dorsal and 1-3 layered on ventral side, while the section cut at apical region, shows 2 layered collenchymatous cells on both sides, similarly transverse section cut at a basal, middle and apical regions shows 4-6 layered parenchymatous cells on dorsal side and 6-9 layered parenchyma on ventral side, in section cut at basal region 4-6 layered parenchyma on both the sides in the middle region with thin-walled cells and intercellular spaces, 2-3 layered parenchymatous cells on both side in the apical region; in the basal region section shows vascular bundle similar to that of petiole while in the section cut at middle and apical region section shows 4 smaller bundles shifting towards lamina.

Lamina - shows a dorsi ventral structure, epidermis single layered, externally covered with cuticle, followed by single layered palisade parenchyma containing chlorophyll contents; spongy parenchyma irregularly arranged with distinct intercellular spaces and filled with chlorophyll contents; mesophyll traversed by number of veins; anisocytic and anomocytic stomata present on both surface, more abundant on lower surfaces; stomatal index 20.0-22.5 on upper and 23.5 -26.0 on lower surface; palisade ratio 3.8 -4.5; hairs stiff, pointed, wide at the base, about 3 celled, uniseriate, middle cells longest, uppermost generally not exceeding the basal cell in length, septa thick-walled.

Stem - Mature stem shows single layered epidermis, externally covered with cuticle, a few epidermal cells elongate to form characteristic non-glandular trichomes, the cork where formed, poorly developed consistsing of rectangular cells; secondary cortex composed of large, rounded or irregular shaped parenchymatous cells having wide air spaces; endodermis single layered consists of tangentially elongated cells; pericyclic fibres distinct, arranged in tangential strands; vascular bundles in a ring, collateral, endarch, of varying sizes traversed by medullary rays; phloem a narrow strip composed of sieve elements and phloem parenchyma; xylem consists of large number of vessels, xylem fibres and xylem parenchyma; xylem vessels appear evenly distributed throughout the xylem; in macerated preparation vessels barrel-shaped, some elongated with simple perforations, pitted with spiral thickening; xylem fibres with wide lumen, pointed tips and pitted walls, a few often bifurcate and a few other large, peg-like outgrowth; xylem parenchyma rectangular with pitted thickening; xylem rays triseriate to pentaseriate, normally biseriate and uniseriate, 8-15 cells in height and 3-5 cells in width; centre occupied by a wide pith consisting of isodiametric cells of parenchyma.

Powder - Dark green; shows vessels in large groups or single broken pieces with pitted walls, numerous fibres entire or in pieces, trichomes entire or in pieces, warty, a few attached with epidermal and subsidiary cells, anomocytic and anisocytic stomata.

In Ayurveda, three varieties of *E. alba* have been mentioned. The description is based on the color of the flowers. White variety: It is the common variety and distributed widely. Yellow variety, according to experts, is represented by *Wedelia chinesis* Merrill. Blue variety is of controversial origin (Santapau, 1957; Gopalakrishnan & Johnson Solomon, 1992).

Chemical constituents: Resin, alkaloid: ecliptine (as per old literature) and 25-β-hydroxyverazine, coumarins: wedelolactone and demethylwedelolactone and triterpenoid glucosides: eclabatin, α-amyrin, ursolic acid and oleanolic acid (Brindha, Sasikala & Purushothaman, 1981).

Specific action as per Ayurveda: Beneficial for hairs.

Therapeutics: *Eclipta alba* is used in the treatment of hepatomegaly, jaundice, and hypertension. Expressed juice of *Eclipta alba* mixed with goat's milk is used in frontal sinusitis. Expressed juice is used in baldness and earache. Seeds are used for increasing vitality. Expressed juice of the leaves mixed with honey is given in nasal catarrh in children. *Eclipta alba* is given with buttermilk for correcting digestive power.

Parts used: Whole plant and roots.

Bioactivity: Analgesic (Pandey *et al.,* 1997; Swant, Issac and Narayanan, 2004), anti-inflammatory (Sureshkumar *et al.,* 2005), hepatoprotective (Dixit and Achara, 1979; Saxena, Singh and Anand, 1993; Chandan, Agarwal and Anand 2001), wedelolactone and demethylwedelolactone: trypsin inhibitory effect (Syed *et al.,* 2004), immunomodulator (Jayatirtha & Mishra, 2004) and wound-healing (Sharma & Sikarwar, 2008)

Clinical studies: Hyperchlorhydria (Chaturvedi & Singh, 1968), infective hepatitis (Dixit & Achar, 1979) and jaundice (Dixit & Achar, 1981) mild hypertension (Rangineni, Sharada & Saxena, 2007).

Further reading

Brindha P, Sasikala B, Purushothaman K. Phytochemical analysis of *E. alba. BMEBR* 1981; 3: 84-96.

Chaturvedi GN, Singh RN. Treatment of hyperchlorhydria with an indigenous drug *Eclipa alba. Curr Med Prac* 1968; 8: 283.

Dixit SP, Achar MP. Study of bhringaraja (*Eclipta alba*) therapy in jaundice in children. *J Sci Res Plants Med* 1981; 2: 96-100.

Dixit SP, Achar MP. Bhringaraja in the treatment of infective hepatitis. *Curr Med Pract* 1979; 23:237-242.

Gopalakrishnan S, Johnson Solomon M. Comparative pharmacognostic studies on the leaves of *Eclipta alba* and *Wedelia calendulacea. Pharma Biol* 1992; 30: 33-38.

Jayatirtha MG, Mishra SH. Preliminary immunomodulatory activities of methanol extracts of *Eclipta alba* and *Centella asiatica. Phyto-medicine* 2004; 11: 361-365.

Pandey PS, Pandey KK, Upadhyay OP, Pandey DN. Experimental evaluation of the analgesic property of *Eclipta alba* (L) Hassk. *Ancient Sci Life* 1997; 17:36-40.

Rangineni, V, Sharada D, Saxena S. Diuretic, hypotensive and hypocholesterolemic effects *Eclipta alba* in mild hypertensive subjects: a pilot study. *J Med Food* 2007; 10: 143-148.

Santapau H. *Eclipta prostrata, E. erecta* or *E. alba*: which is the correct name? JBNHS 1957; 54, 476-477.

Saxena, A.K.; Singh, B. & Anand, K.K. (1993). Hepatoprotective effects of *Eclipta alba* on subcellular levels in rats. *J Ethnopharmacol* 1993; 40: 155-161.

Sharma S, Sikarwar MS. Wound healing activity of ethanolic extract of leaves of *Eclipta alba. Pharmacog Mag* 2008; 4: 108-111.

Singh B, Saxena AK, Chandan BK, Agarwal SG, Anand KK. *In vivo* hepatoprotective activity of active fraction from ethanolic extract of *Eclipta alba* leaves. *Indian J Physiol Pharmacol* 2001; 45: 435-441.

Sureshkumar S, Sivakumar T, Chandrasekar MJN, Suresh B. Evaluation of anti-inflammatory activity of *Eclipta alba* in rats. *Ancient Sci Life* 2005;24:112-118.

Swant M, Issac JC, Narayanan S. Analgesic studies on total alkaloids and alcohol extracts of *Eclipta alba* (Linn.) Hassk. *Phytother Res* 2004; 18:111-113.

Syed SD, Deepak M, Yogisha S, Chandrashekar AP, Muddarachappa KA, D'Souza P, Agarwal A, Venkataraman BV. Trypsin inhibitory effect of wedelolactone and demethylwedelolactone. *Phytother Res* 2003; 17: 420-421.

Syed, S.D., Deepak, M., Yogisha, S., Chandrashekar, A.P., Muddarachappa, K.A., D'Souza P, Agarwal A, Venkataraman BV. Trypsin inhibitory effect of wedelolactone and demethylwedelolactone. *Phytother Res* 2003;17: 420-421.

Thakur VD, Mengi SA. Neuropharmacological profile of *Eclipta alba* (Linn.) Hassk. *J Ethnopharmacol* 2005;102: 23-31.

Wedelia calendulacea Less.

Syn: *Wedelia chinensis* (Osbeck) Merr., *Verbesina calendulacea* Roxb.

Family: Asteraceae.

Common name: Chinese wedelia.

Habitat: India.

General Description: A perennial herb, 0.3-0.9 m long; stem procumbent at the base and rooting at the lower nodes, terate, more or less appressed hairy. Leaves opposite, subsessile, oblong, entire or irregularly crenate-serrate, scabrous with short white hairs. Heads 2-3.2 cm diam., solitary; peduncles 2.5-15 cm long, erect, slender. Ray-florets ligulate, yellow, 2-3 toothed (Gopalakrishnan & Johnson, 1992).

Chemical constituents: Coumestans derivatives: wedelolactone and demethylwedelolactone (Govindachari, Nagarajan & Pai, 1956), benzofuran-3-carboxylic acid: norwedelic acid (Govindachari and Permila, 1985) and isoflavanoids (Annie, Prabhu & Malini, 2006).

Therapeutics: The leaves, which have a slight camphoraceous taste, are considered tonic, alterative, and useful in cough, cephalalgia, skin diseases and alopecia. The juice of the leaves is much used as a snuff in cephalalgia, and in soaking various sorts of powders for the preparation of pills.

Bioactivity: Wound-healing (Hegde, Khosa & Chansouria, 1994), antiulcer and cytoprotective (Hegde, Khosa & Goel, 1994; Manikandan & Senthil, 2007), post- menopausal osteoporosis (Annie, Prabhu & Malini, 2006), hepatoprotective (Sharma *et al.*, 1999) and antidepressant (Prakash, Rao & Viswanatha Swamy, 2008; Verma & Khosa, 2009).

Further reading

Annie S, Prabhu RG, Malini S. Activity of *Wedelia calendulaceae* in post-menopausal osteoporosis. *Phytomedicine* 2006; 13:43-48.

Gopalakrishnan S, Johnson MS. Comparative pharmacognostic studies on the leaves of *Eclipta alba* and *Wedelia calendulacea. Int J Pharmacog* 1992; 30: 33-38.

Govindachari TR, Nagarajan K, Pai BR. Chemical examination of *Wedelia calendulacea*. Part I. Structure of wedelolactone. *J Chem Soc* 1956; 629-632.

Govindacharia TR, Premila MS. The benzofuran norwedelic acid from *Wedelia calendulaceae. Phytochemistry* 1985; 24: 3068-3069.

Hegde DA, Khosa RL, Chansouria JPN. A Study of the effect of *Wedelia calendulacea* less. On wound healing in rats. *Phytother Res* 1994; 8: 439-440.

Hegde DA, Khosa RL, Goel RK. Antiulcer and cytoprotective action of *Wedelia calendulacea* Less. *Ancient Sci Life* 1994; 14:77-81.

Manikandan L, Senthil PG. Anticancer activity of *Indigofera aspalathoides* and *Wedelia calendulacea* in Swiss albino mice. *Iranian J Pharm Res* 2007; 6: 141-145.

Prakash T, Rao NR, Viswanatha Swamy AH. Neuropharmacological studies on *Wedelia calendulacea* Less stem extract. *Phytomedicine* 2008. [Epub ahead of print].

Sharma AK, Anand KK, Pushpangadan P, Chandan BK, Chopra CL, Prabhakar YS, Damodaran NP. Hepatoprotective effects of *Wedelia calendulacea. J Ethnopharmacol* 1989; 25: 93-102.

Verma N, Khosa RL. Effect of *Costus speciosus and Wedelia chinensis* on Brain Neurotransmitters and Enzyme Monoamine Oxidase Following Cold Immobilization Stress. *J Pharm Sci Res* 2009; 1: 22-25.

20. BRAHAMDANDI

Current representative: *Tricholepis glaberrima*

Possible representative (s): *Argemone mexicana, Echinops echinatus, Lamprachenium microcephalum, Voluntarella divaricata* and *Xanthuim sturmarium.*

Tricholepis glaberrima DC.

Ayurvedic name: *Ajadandi.*

Family: Asteraceae.

Habitat: India.

General Description: Herb.

Chemical constituents: Betulin, spinasterol, stigmasterol (Manjrekar & Kulkarni, 1978), triterpenoid: cynaropicrin (Bhattacharya, Barua & Ghosh, 1995), guaianolides (Singhal, Chowdhury & Sharma, 1982), C_{23}-C_{27} n-alkanes and C_{28}-C_{32} n-alkanols (Kapoor *et al.,* 1976; Kapoor & Sangal, 1978).

Actions: Aphrodisiac

Therapeutics: Aerial parts of this plant are used as a nervine tonic, and its infusion was prescribed in seminal debility and impotence (Issar, 1976b).

Parts used: Whole plant.

Bioactivity: Aphrodisiac (Padashetty & Mishra, 2007) and antioxidant (Naphade *et al.,* 2009).

Further reading

Bhattacharya PR, Barua NC, Ghosh AC. Cynaropicrin from *Tricholepis glaberrima*: a potential insect feeding deterrent compound. *Ind Crops Prod* 1995; 4: 291-294.

Chawla AS, Kapoor VK, Sangal PK, Gupta AK, Evans FJ. Chemical constituents of *Tricholepis glaberrima. Planta Med* 1976; 30:151-153.

Issar RK. The botanical identification of market sample of brahamdandi. *J Res Indian Med* 1974b 9: 94-95.

Kapoor S *et al.,* Transactions of the Illinois State Academy of Science. 1976; 69: Abstract.

Kapoor VK, Sangal PK. Cycloart 23-3n3-3beta, 25 diol from *Tricholepis glaberrima. Plant Med* 1978; 34:109-110.

Manjrekar SV, Kulkarni AB. Chemical investigation of *Tricholepis glaberrima. Ind J Chem SB,* 1978; 16:439-440.

Naphade SS, Khadabadi SS, Deore SL, Jagtap NS, Hadke SP. Antioxidant activity of different extrcats of plant *Tricholepis glaberrima* DC (Asteraceae). *Int J Pharm Tech Res* 2009; 1:502-505.

Padashetty SA, Mishra SH. Aphrodisiac studies of *Tricholepis glaberrima* with Supportive Action from Antioxidant Enzymes. *Pharm Biol* 2007; 45: 580-586.

Satpute, Ashok D. Rasaratnasamuchchaya (Chapters 1-11) (Sanskrit text with English translation) 2003: The Chaukhamba Ayurvijnan Studies 41.

Singhal AK, Chowdhury PK, Sharma R.P. Guainolides from *Tricholepis glaberrima*. *Phytochemistry* 1982; 21:462-463.

Echinops echinatus Roxb.

Ayurvedic name: *Utkantaka*.

Englih name: Globe thistle.

Family: Asteraceae.

Habitat: India.

General Description: *E. echinatus* is an annual prostrate herb upto 70cm high. Leaves sessile, oblong, 7-12 cm long, pinnatifid; lobes triangular, spinescent; spines 2-3 cm long, pale, scabrous above, white arachnoid beneath. heads compound, forming a spherical ball, 3-4 cm in diameter, having single bisexual white floret. Involucre surrounded by strong white bristles, resembling pappus; bracts tri cuspidate, sharp spinose; inner bracts 6-7 mm long, connate; white, tubular, 5-lobed. Achenes obconic, densely villous, pappus bristly deciduous (Somashekar & Mishra, 2007).

Chemical constituents: Alkaoid echinozolinone (Chaudhuri, 1987), flavones, flavone glycosides (Singh *et al.*, 1988) and flavonoids: apigenin, apigenin-7-O-glucoside, echinacin (Singh *et al.*, 1988), echinosidesoflavone glycoside, echinoside, together with 7-hydroxyisoflavone, kaempferol-4'-methylether, kaempferol-7-methylether, myrecetin-3-O-alpha-L-rhamnoside, kaempferol and kaempferol-3-O-alpha-L-rhamnoside (Singh *et al.*, 2006) and triterpenoid: taraxasterol (Singh *et al.*, 1999).

Parts used: Whole plant.

Bioactivity: Antifertility (Sharma, Mishara & Mehta, 1988), antifungal (Singh *et al.*, 1988) and anti-inflammatory (Singh *et al.*, 1989) and antimicrobial (Sharma & Mehta, 1989).

Further reading

Chaudhuri PK. Echinozolinone, an alkaloid from *Echinops echinatus*. *Phytochemistry* 1987; 26: 587-589.

Padashetty SA, Mishra SH. Effect of terpenoidal fraction of *Echinops echinatus* roots on reproductive parameters of male rats. *J Nat Med* 2007; 61:453-453.

Ram SN, Roy R, Singh B, Singh RP, Pandey VB. An acylflavone glucoside of *Echinops echinatus* flowers. *Planta Med* 1996; 62:187.

Sharma KS, Mishra S, Mehta BK. Antifertility activity of *Echinops echinatus* in albino rats. *Indian J Med Sci* 1988; 42:23-26.

Singh B, Gambhir SS, Pandey VB, Joshi VK. Anti-inflammatory activity of *Echinops echinatus*. *J Ethnopharmacol* 1989; 25:189-199.

Singh B, Ram SN, Pandey VB, Joshi VK, Gambhir SS. Studies on antiinflammatory activity of taraxasterol acetate from *Echinops echinatus* in rats and mice. *Phytother Res* 1991; 5: 103-106.

Singh S, Upadhyay RK, Pandey MB, Singh JP, Pandey VB. Flavonoids from *Echinops echinatus*. *J Asian Nat Prod Res* 2006; 8:197-200.

Singh UP, Pandey VB, Singh KN, Singh RDN. Antifungal activity of some new fiavones and fiavone glycosides of *Echinops echinatus*. *Canadian J Bot* 1988; 66:1901-1903.

Somashekar AP, Mishra SH. Pharmacognostic Parameters for Evaluation of the Roots of *Echinops echinatus* marketed as Brahmadandi. *Pharmacog Mag* 2007; 3:196-202.

Sharma S, Mehta BK. Antimicrobial activity of *Echinops echinatus* root extract. *Fitoterapia* 1989; 60: 82-83.

Argemone mexicana Linn.

English name: Mexican poppy, Mexican prickly poppy, cardo or cardosanto.

Ayurvedic names: *Kanchana kshiri*, *katiparni*, *pitadugdha* and *svarna kshiri*.

Family: Papaveraceae.

Habitat: *A. mexicana* is a species of poppy found in Mexico and now widely naturalized in the United States, India and Ethiopia.

General Description: A strong branched prickly annual, 60-90 cm in height with yellow latex; leaves simple, sessile and spiny. Flowers large, bright yellow, terminal on the short leafy branches; fruits prickly capsules, oblong-ovoid, opening by 4-6 valves; seeds numerous (Issar, 1975, Sukumar, Nambi & Sulochana, 1984, Dwivedi, Pandey & Jha, 1998).

Chemical constituents: Alkaloids: protopine, berberine and argemonine, Four quaternary isoquinoline alkaloids, dehydrocorydalmine, jatrorrhizine, columbamine, and oxyberberine (Santra & Saoji, 1971). Sanguinarine, a hepatotoxic alkaloid is present in the seeds (Dalvi, 1985).

Actions: *Sukhavirechana* (mild laxative).

Therapeutics: *A. mexicana* pacifies vitiated pitta, guinea-worm infestation, skin diseases, itching, inflammations, all types of poisoning, constipation, flatulence, colic, malarial fever, and constipation. Root powder is used in tapeworm infestation. Latex obtained from the plant is used as application to corns.

Parts used: Milk, seeds, roots and oil.

Bioactivity: Wound-healing (Patil, Jalalpuri & Ali, 2001).

Further reading

Bose BC, Vijayvargiya R, Saifi AQ, Sharma SK. Chemical and pharmacological studies on *Argemone mexicana* Linn. *J Pharm Sci* 1972; 521: 1172-1175.

Dalvi RR. Sanguinarine: it's potential as a liver toxic alkaloid present in the seeds of *Argemone mexicana. Experientia* 1985; 41: 77-78.

Dwivedi PK, Pandey S, Jha CB. Some facts about *Argemone mexicana. Sach Ayur* 1998; 1: 370-376.

Issar RK. The botanical identification of market sample of brahamdandi. *J Res Indian Med* 1974b; 9: 94-95.

Kumar L, Chugh KS, Singhal PC, Sharma BK, Walia BN. *Argemone mexicana* poisoning in North India. *J Assoc Physi India* 1975; 23: 205-208.

Patil MB, Jalalpuri SS, Ali A. On *Argemone mexicana* Linn. and wound healing. *Indian Drugs* 2001; 38: 288.

Santra DK, Saoji AN. Phytochemical study of *Argemone mexicana* Linn. Latex. *Curr Sci* 1971; 40:548-549.

Singh S, Singh TD, Singh VP, Pandey VB. Quaternary alkaloids of *Argemone mexicana. Pharm Biol* 2010; 48:158-60.

Sukumar D, Nambi RA, Sulochana N. Studies on the leaves of *Argemone mexicana. Fitoterapia* 1984; 55: 352-353.

Vaidya AD, Pandita N, Godse C, Meshram M. *Argemone mexicana* and epidemic dropsy: phytopharmacology and putative sequelae and therapy. *J Assoc Physi India* 2001; 49:1041-1042.

Xanthium strurmarium Linn.

Syn: *Xanthium canadense* (Mill.)

Ayurvedic names: *Aristha, itara, kambu-vanamalini, sarpakshi.*

Common name: Rough cocklebur.

Family: Asteraceae.

Habitat: The plant probably originated in North America and has been extensively naturalized elsewhere.

General Description: *X. strumarium* is an annual herb with a short, stout, hairy stem. Leaves broadly triangular-ovate or suborbicular; flower heads in terminal and axillary racemes; white or green; numerous; male upper most; female ovoid, covered with hooked bristles; Fruit obovoid, enclosed in the hardened involucre, with 2 hooked beaks and hooked bristles. Flowering time in India is August-September. It can be propagated through seeds. This weed is easily dispersed through animals as the fruits have hooked bristles and 2 strong hooked beaks (DasGupta & Kundu, 1963, Reeta *et al.,* 2010).

Chemical constituents: Toxic glucoside: xanthostrumarin. C_{23}-C_{35} n-alkanes, C_{22}-C_{30} n-alkanols, and sterols: sitosterol, stigmasterol and campesterol (Kapoor, *et al.,* 1976; Kamboj & Saluja, 2010). Dried leaves contain tannins (Singh & Kachroo, 1976).

Actions: Anthelmintic, antipyretic, cooling, digestive, laxative and tonic.

Therapeutics: *X. strumarium* cures leucoderma, biliousness, and poisonous bites of insects, epilepsy, salivation and fever (Chopra, Nayar & Chopra, 1986). Oil obtained from seeds is used in herpes zoster.

Parts used: Whole plant, roots and fruit.

Bioactivity: Antitrypanosomal (Talakal, Dwivedi & Sharma, 1995; Chandel *et al.,* 2011).

Further reading

Chandel S, Bagai U, Vashishat N. Antiplasmodial activity of *Xanthium strumarium* against Plasmodium berghei-infected BALB/c mice. *Parasitol Res* 2011. [Epub ahead of print].

Chopra. R.N., Nayar. S.L. and Chopra. I.C. (1986). *Glossary of Indian Medicinal Plants* (Including the Supplement). Council of Scientific and Industrial Research, New Delhi.

DasGupta J, Kundu BC. Pharmacognostic studies on *Xanthium strumarium* Linn. *Planta Med* 1963; 11: 471-478.

Kamboj Ao, Saluja AK. Phytopharmacological review of *Xanthium strumarium* L. *Int J Green Pharm* 2010; 4: 129-139.

Kapoor VK, Chawla AS, Gupta AK, Bedi KL. Studies on the oil of *Xanthium strumarium* Linn. *Indian J Pharm* 1975; 37: 170.

Mandal SC, Dhara AK, Kumar A, Maiti BC. Neuropharmacological activity of *Xanthium strumarium* Linn. extract. *J Herbs Spices Med Plants* 2001; 8: 69-77.

Reeta S, Kavita G, Arjun P, Rimpal J. Pharmacognostical Standardizati on of Leaves of *Xanthium strumarium* Linn. *Phcog J* 2010; 2: 492-497.

Singh G, Kachroo P. *Forest Flora of Srinagar.* 1976. Bishen Singh Mahendra Pal Singh.

Talakal TS, Dwivedi SK, Sharma SR. *In vitro* and *in vivo* antitrypanosomal activity of *Xanthium strumarium* leaves. *J Ethnopharmacol* 1995; 49:141-145.

***Voluntarella divaricata* C.B. Clarke N.O.: Asteraceae**

The plant is used in the treatment of insanity.

21. BRAHAMI

Current representative: *Centella asiatica*

Possible representative: *Bacopa monniera*

Centella (Hydrocotyle) asiatica Linn.

English name: Indian pennywort.

Ayurvedic names: *Divya, lavanya, mahaushadi, mandukparni, manduki, munika, somavalari* and *tvasti*.

Family: Apiaceae.

Habitat: India.

Description:

(a) *Macroscopic*

Small creeping herb with slender stem, rooting at nodes giving rise to thin, brownish-grey, roots of about 2.5 to 6.0 cm in length; leaves 1 to 3 from each node, orbicular-reniform, crenate, base cordate, petioles channelled with adnate stipules; flowers fascicled umbels each carrying 3 or 4 flowers, short stalked; fruits cremocarp, ovoid, with laterally compressed seeds.

(b) *Microscopic*

Root - Shows wavy outline, consisting of 3 to 5 layered, rectangular, cork cells having exfoliated cells, followed by 3 or 4 layers of parenchyma cells containing oval to round, simple, starch grains measuring 8 to 16 μ in dia., having centric hilum and microsphenoidal crystals of calcium oxalate; secondary cortex composed of thin-walled, oval to polygonal, parenchymatous cells; secretory cells present, scattered towards periphery region; secondary phloem and secondary xylem consisting of usual elements; vessels lignified with reticulate and spiral thickening; pith nearly obliterated.

Stem - More or less concave-convex outline, shows single layered epidermis composed of round to cubical cells covered by striated cuticle; below this 2 or 3 layers of collenchymatous cells, followed by 6 to 8 layers of thin-walled, isodiametric, parenchymatous cells with intercellular spaces present; vascular bundles collateral, open, arranged in a ring, capped by patches of sclerenchyma and traversed by wide medullary rays; vessels with spiral thickening present, resin duct present in parenchymatous cells of cortex and generally one in between vascular bundles; pith of isodiametric, parenchyma with intercellular spaces.

Leaf-

Petiole - shows a characteristic outline due to two projections adjacent to ventral groove; epidermis single layered, cells cubical covered by a thick cuticle; inner walls of epidermal cells adjoining the cortex much thickened; hairs absent; collenchyma 2 or 3 layered, absent on the projections, a broad zone of more or less rounded parenchyma cells present with intercellular spaces, and a few containing rosette crystals of calcium oxalate; resin canal present on dorsal side of each vascular bundle except in the vascular bundles occurring projecting arms; vascular bundles seven in number, two of which less developed and present in projections.

Midrib - show a single layered epidermis, 2 or 3 layered collenchyma on both surfaces, 4 or 5 layered parenchyma, mostly devoid of chloroplasts; central zone occupied by vascular bundles differentiated into xylem towards ventral side and phloem towards dorsal side; phloem consisting of sieve tubes, companion cells and phloem parenchyma; xylem consisting of radial rows of vessels with xylem parenchyma in between.

Lamina -shows an epidermis of tangentially elongated cells on both surfaces, larger on the upper surface, covered by striated cuticle; mesophyll differentiated into 2 or 3 layers of palisade cells, 5 to 7 layers of loosely arranged, somewhat isodiametric spongy parenchyma; rosette crystals of calcium oxalate present in a few cells; stomata more on the lower surface, anisocytic in general, but anomocytic type also occurs on both surfaces, palisade ratio 3 to 5, stomatal index on upper surface, 9 to 12, and lower surface 11 to 17.

Fruit - Shows several ridges in outline; epicarp consists of single layered epidermis covered externally with thick cuticle; mesocarp consists of polygonal, thin walled parenchymatous cells having patches of sclerenchymatous cells on both lateral side; each ridge having a vittae and patch of sclerenchyma; endocarp consists of columnar shaped sclereids arranged in wavy layers; endosperm and embryo composed of oval topolygonal, thin-walled parenchymatous cells.

Powder - Green to greenish-brown, shows fragments of epidermal cells polygonal in surface view with stomata, palisade cells, vessels with spiral, reticulate and annular thickening; microsphenoidal and rosette crystals of calcium oxalate; simple, oval to round starch grains measuring 8 to 16 μ in dia (Jelani *et al.*, 1993).

Chemical constituents: Triterpenoid acids (asiatic acid and madecassic acid), Triterpene saponins (asiaticoside, medecassoside, brahmoside and brahminoside) and alkaloid (hydrocotyline), flavonoids and 0.1% volatile oil (Rastogi & Dhar, 1963; Datta & Basu, 1967, Srivastava, Shukla & Kumar, 1997; Shukla *et al.*, 2000).

Specific action as per Ayurveda: *Medya rasayana* (nootropic or brain tonic).

Therapeutics: *C. asiatica* is used in the treatment of loss of memory, psoriasis and eczema, fevers and bloody diarrhea. The decoction of the leaves is given in leprosy.

Parts used: Whole plant.

Bioactivity: Anticancer (Babu & Padikkala, 1993), cytotoxic and anti-tumour (Babu TD, Kuttan G, Padikkala, 1995), psychotropic (Deshpande *et al.*, 1980), antianxiety (Diwan, Karawande & Singh, 1991), antifertility (Dutta & Basu, 1968), anti-oxidant (Jayashree *et al.*, 2003), antiamnesic (Rao, Chetana & Uma Devi, 2005), wound healing (Suguna L, Sivakumar P, Chandrakasan; Rasik *et al.*, 1997; Shetty *et al.*, 2006), gastroprotective (Sairam Rao & Goel, 2001), radioprotective (Sharma & Sharma, 2002), antibacterial (Srivastava, Shukla & Darokar, 1997)

Clinical studies: A study repoted efficiacy of *C. asiatica* on the general mental ability of mentally retarded children (Apparao *et al.*, 1973). *C. asiatica* is repoted to be useful in varicose veins (Khandelwal, Wortikar & Mahajani, 1999).

Further reading

Apparao MVR, Srinivasan K, Rao K, *et al.*, The effect of mandookparni (*Centella asiatica*) on the general mental ability (Medhya) of mentally retarded children. *J Res Indian Med* 1973, 8: 9-16.

Babu TD and Padikkala J. Anticancer activity of an active principle from *Centella asiatica*. *Amla Res Bull* 1993; 13: 46-49.

Babu TD, Kuttan G, Padikkala J. Cytotoxic and anti-tumour properties of certain taxa of Umbelliferae with special reference to *Centella asiatica* (L.) Urban. *J Ethnopharmacol* 1995; 48: 53-57.

Datta T, Basu UP. Terpenoids IV: Isolation & identification of Asiatic acid from *Centella asiatica* Linn. *Indian J Chem* 1967; 5:586.

Deshpande S, Gupta SS, Shinde S, Iyengar VL, Shastry S. Psychotropic effects of *Centella asiatica*. *Indian J Pharmacol* 1980; 12: 64.

Diwan PV, Karawande I, Singh AK. Antianxiety profile of mandukparni (*Centella asiatica*) in animals. *Fitoterapia* 1991; 62:253-257.

Dutta T, Basu UP. Crude extract of *Centella asiatica* and Products derived from its Glycosides as Oral Antifertility Agents. *Indian J Exp Biol* 1968; 6: 181-182.

Jayashree G, Kurup Muraleedhara G, Sudarslal S, Jacob VB. Antioxidant activity of *Centella asiatica* on lymphoma-bearing mice. *Fitoterapia* 2003;74 431-434.

Jelani S, Jabeen F, Prabhakar M, Leelavathi P. Pharmacognostic studies on *Centella asiatica* (L.) Urban. *Ancient Sci Life* 1993; 12:439-450.

Joseph GVR, Chaturvedi S, Deokule SS. Standardisation and quality evaluation of *Centella asiatica* Linn. *Ancient Sci Life* 2001;20: 99-110.

Khandelwal S, Wortikar P, Mahajani SS. Effects of *Centella asiatica* on venous parameters. *Indian J Pharmacol* 1999; 31: 52 (abstract No. 53).

Rao SB, Chetana M, Uma Devi P. *Centella asiatica* treatment during postnatal period enhances learning and memory in mice. *Physiol Behav* 2005, 86: 449-457.

Rasik AM, Shukla A, Patnaik GK, Kulshrestha DK, Dhawan BN. Wound healing: effect of *Centella asiatica*. *Indian J Pharmacol* 1997; 29: 28 (abstract No. 19).

Rastogi RP, Dhar ML. Chemical examination of *Centella asiatica* Linn. Part II. Brahmoside and brahminoside. *Indian J Chem* 1963; 1: 267.

Sairam K, Rao CV, Goel RK. Effect of *Centella asiatica* Linn on physical and chemical factors induced gastric ulceration and secretion in rats. *Indian J Exp Biol* 2001; 39:137-142.

Sharma J, Sharma R. Radioprotection of Swiss albino mouse by *Centella asiatica* extract. *Phytother Res* 2002; 16:785-786.

Shetty BS, Udupa SL, Udupa AL, *et al.*, Effect of *Centella asiatica* L (Umbelliferae) on normal and dexamethasone-suppressed wound healing in Wistar albino rats. *Int J Low Extrem Wounds* 2006; 5: 137-143.

Shukla YN, Srivastava R, Tripathi AK, *et al.*, Characterization of an ursane triterpenoid from *Centella asiatica* with growth inhibitory activity against *Spilarctia bliqua*. *Pharm Biol* 2000; 38: 262-267.

Srivastava R , Shukla YN, Darokar MP. Antibacterial activity of *Centella asiatica*. *Fitoterapia* 1997; 68: 466-467.

Srivastava R, Shukla YN, Kumar S. Chemistry and pharmacology of *Centella asiatica*: a review. *J Med Arom Plants Sci* 1997; 19:1049-1056.

Suguna L, Sivakumar P, Chandrakasan G. Effects of *Centella asiatica* extract on dermal wound healing in rats. *Indian J Exp Biol* 1996, 34:1208-1211.

Bacopa monniera Linn.

Syn: *Herpestis monniera* Linn.

English name: Thyme leaved Gratiola.

Ayurvedic names: *Brahmacaroni, sarsavati, satyahva* and *soma.*

Family: Scophulariaceae.

Habitat: Himalayas.

Description:

(a) *Macroscopic*

Root - Thin, wiry, small, branched creamish-yellow. Stem - Thin, green or purplish green, about 1-2 mm thick, soft, nodes and internodes prominent, glabrous; taste, slightly bitter. Leaf - Simple, opposite, decussate, green, sessile, 1-2 cm long, obovate-oblong; taste, slightly bitter. Flower - Small, axillary and solitary, pedicels 6-30 mm long, bracteoles shorter than pedicels. Fruit - Capsules upto 5 mm long, ovoid and glabrous.

(b) *Microscopic*

Root - Shows a single layer of epidermis, cortex having large air cavities; endodermis single layered; pericycle not distinct; stele consists of a thin layer of phloem with a few sieve elements and isolated material from xylem shows vessels with reticulate thickenings.

Stem - Shows single layer of epidermis followed by a wide cortex of thin-walled cells with very large intercellular spaces; endodermis single layered; pericycle 3 consisting of 1-2 layers; vascular ring continuous, composed of a narrow zone of phloem towards periphery and a wide ring of xylem towards centre; centre occupied by a small pith with distinct intercellular spaces; starch grains simple, round to oval, present in a few cells of cortex and endodermis, measuring 4-14 μ in dia., and 8.0-14.0 x 2.5-9.0 μ in dia. respectively.

Leaf -Shows a single layer of upper and lower epidermis covered with thin cuticle; glandular hairs sessile, subsidiary cells present on both surfaces; a few prismatic crystals of calcium oxalate occasionally found distributed in mesophyll cells; mesophyll traversed by small veins surrounded by bundle sheath; no distinct midrib present.

Powder - Yellowish-brown; shows xylem vessels with reticulate thickening, glandular hairs, simple, round and oval starch grains, measuring 4-14 μ in diameter.

Chemical constituents: Triterpenoid saponins (bacoside A-B, monnierin), alkaloids (herpestine and brahmine), dammarane-type saponins (bacopasaponins), flavonoid glycosides, betulinic acid, stigmastanol and phytosterols (Chatterji, Rastogi & Dhar, 1963; Dutta & Basu, 1963; Chatterji, Rastogi & Dhar, 1965; Kulshreshtha & Rastogi, 1973; Chandel, Kulshreshtha & Rastogi, 1977; Jain & Kulshreshtha, 1993; Garai *et al.*, 1996a; Garai *et al.*, 1996b).

Specific action as per Ayurveda: *Medya rasāyana* (Nootropic or brain tonic).

Therapeutics: Loss of memory and anxiety neurosis.

Parts used: Whole plant.

Bioactivity: Nootropic (Singh and Dhawan, 1982), muscle-relaxant (Dar & Channa, 1997), cytotoxic (Elangovan *et al.*, 1995), antioxidant (Tripathi *et al.*, 1996; Bhattacharya *et al.*, 2000), anxiolytic (Singh & Singh, 1979; Bhattacharya and Ghosal, 1998), cognition-enhancing (Vohora, Pal & Pillai, 2000), hepatoprotective (Sumathy *et al.*, 2001; Sumathi & Nongbri, 2008), gastroptotective (Sairam *et al.*, 2001; Goel *et al.*, 2003), neuroprotective (Amar, Sethi & Sharma, 2007) and antidepressant (Sairam *et al.*, 2002).

Clinical studies: In an uncontrolled study, the effect of *Bacopa monniera* was studied in patients suffering from epilepsy, mental retardation. It was observed that a defatted extract of the dried plant was more potent than a decoction of the fresh plant. However nothing special was concluded after the study (Mukherjee and Dey 1966).

Kidd (1999) in his article wrote about 4-week study of *Bacopa monniera* in 35 patients with anxiety neurosis. The daily dose of dried *Bacopa monniera* herb was 12 g, as syrup. The patients showed significant improvements in anxiety concentration and memory. Work-related mental fatigue and other symptoms like insomnia, cephalgia and palpitation were reduced. A randomized controlled trial of standardized *B. monniera* extract reported efficiacy in age-associated memory impairment (Raghav *et al.*, 2006).

Further reading

Anonymous (1998), *Indian Herbal Pharmacopoeia Volume 1.* Worli, Mumbai: Indian Drug Manufacturers Association.

Bhattacharya SK, Bhattacharya A, Kumar A, Ghosal S. Antioxidant activity of *Bacopa monniera* in rat frontal cortex, striatum and hippocampus. *Phytother Res* 2000;14: 174-179.

Chandel RS, Kulshreshtha DK, Rastogi RP. Bacogenin-{A}{3}: a new sapogenin from *Bacopa monniera*. *Phytochemistry* 1977; 16: 141-143.

Chatterji N, Rastogi RP, Dhar ML. Chemical examination of *Bacopa monniera* Wettst. - Part I: Isolation of chemical constituents *Indian J Chem* 1963;1:212-215.

Chatterji N, Rastogi RP, Dhar ML. Chemical examination of *Bacopa monniera* Wettst. - Part II: The constitution of Bacoside A. *Indian J Chem* 1965; 3: 24-29.

Dutta T, Basu UP. Terpenoids: Part II: Isolation of a new triterpene saponin, monnierin, from *Bacopa monniera* Wettst. *Indian J Chem* 1963;1: 408-409.

Garai S, Mahato SB, Ohtani K, Yamasaki K Bacopasaponin D-A pseudojujubogenin glycoside from *Bacopa monniera*. *Phytochemistry* 1996b; 42: 815-820.

Garai S, Mahato SB, Ohtani K, Yamasaki K. Bacosaponin D - A pseudoju-jubogenin glycoside from *Bacopa monniera*. *Phytochemistry* 1996a 43: 447-449.

Goel RK. Sairam E, Dora Babu M, Tavares IA, Raman A. *In vitro* evaluation of *Bacopa monniera* on anti-Helicobacter pylori activity and accumulation of prostaglandins. *Phytomedicine* 2003;10:523-527.

Jain P, Kulshreshtha DK. Bacoside A1, a minor saponin from *Bacopa monniera*. *Phytochemistry* 1993; 33: 449-451.

Jyoti A, Sethi P, Sharma D. *Bacopa monniera* prevents from aluminium neurotoxicity in the cerebral cortex of rat brain. *J Ethnopharmacol* 2007; 111: 56-62.

Kidd PM. A Review of Nutrients and Botanicals in the Integrative Management of Cognitive Dysfunction. *Altern Med Rev* 1999;4:144-161.

Kulshreshtha DK, Rastogi RP. Bacogenin-{A}: a novel dammarane triterpene sapogenin from *Bacopa monniera*. *Phytochemistry* 1973; 12:887-892.

Mukherjee GD, Dey CD. Clinical trial on Brahmi. I. *J Exper Med Sci* 1966; 10:5-11.

Raghav S, Singh H, Dalal PK, Srivastava JS, Asthana OP. Randomized controlled trial of standardized *Bacopa monniera* extract in age-associated memory impairment. *Indian J Psych* 2006;48, 4, 238-242.

Sairam K, Dorababu M, Goel RK, Bhattacharya SK. Antidepressant activity of standardized extract of *Bacopa monniera* in experimental models of depression in rats. *Phytomedicine* 2002;9: 207-211.

Sairam K, Rao CV, Babu MD, Goel RK. Prophylactic and curative effects of *Bacopa monniera* in gastric ulcer models. *Phytomedicine* 2001;8: 423-430.

Singh HK. Dhawan BN. Effect of *Bacopa monniera* Linn. (brahmi) extract on avoidance responses in rats. *J Ethnopharmacol* 1982; 5: 205-214.

Singh RH, Singh L. Studies on the anti-anxiety effect of the medhya rasayana drug brahmi (*Bacopa monniera* Linn.) *J Res Indian Med* 1979; 14:3, 1-6.

Sumathi T, Nongbri A. Hepatoprotective effect of bacoside-A, a major constituent of *Bacopa monniera* Linn. *Phytomedicine* 2008; 15: 901-905.

Sumathy T. Subramanian S, Govindasamy S, Balakrishna K, Veluchamy G. Protective role of *Bacopa monniera* on morphine induced hepatotoxicity in rats. *Phytother Res* 2001; 15: 643-645.

Vohora D, Pal SN, Pillai KK. Protection from phenytoin-induced cognitive deficit by *Bacopa monniera*- a reputed Indian nootropic plant. *J Ethnopharmacol* 2000;71:383-390.

22. CHAMARA MUSALI

Current representative: *Vitis araneosa* Laws

23. CHAVYA

Current representative: *Piper chaba*

Possible representative: *Plantago amplexicaulis, Scindapsus scandens* and *Tetranthere apatela.*

Piper chaba W. Hunter.

English name: Chavica.

Ayurvedic names: *Chavana, chavika, kolavallika* and *ucchista.*

Family: Piperaceae.

Habitat: It is found throughout India especially in the warmer places. It is also found in Malaysia, Indonesia, Singapore, Sri Lanka and south Asian regions.

General Description: Climber.

Chemical constituents: Resin (Mishra & Tewari, 1964), lignan (Bhandari *et al.,* 1998) and chabamide (Jyothi *et al.,* 2009).

Specific action as per Ayurveda: *Anulomaka* (carminative).

Therapeutics: Indigestion and chronic cough.

Parts used: Roots.

Further reading

Bhandari SPS, Babu UV, Garg HS. A lignan from *Piper chaba* stems. *Phytochemistry* 1998; 47:1435-1436.

Jyothi D, Vanathi P, Mangala Gowri P, Rama Subba Rao V, Madhusudana Rao J, Sreedhar AS. Diferuloylmethane augments the cytotoxic effects of piplartine isolated from *Piper chaba*. *Toxicol In Vitro* 2009; 23:1085-1091.

Mishra SS, Tewari JP. Phytochemical investigation of *Piper chaba*. *J Pharm Sci* 1964; 53:1423-1424.

Rukachaisirikul T, Prabpai S, Champung P, Suksamrarn A. Chabamide, a novel piperine dimer from stems of *Piper chaba*. *Planta Med* 2002; 68: 853-855.

Plantago amplexicaulis Cav.

Family: Plantiginaceae.

Habitat: India.

General Description: Annual herb, sparsely pilose to nearly glabrous, stemless or shortly caulescent, often branched at base, up to 30 cm tall. Leaves close or distant, arranged spirally, narrow lanceolate, 4-15 cm long, 0.3-1 (-1.5) cm broad, entire or sparingly toothed, acuminate, at bases attenuated into narrow, long, amplexicaul, vaginate, 3-5 nerved petioles; vagina 4-5 mm long. Scapes many, axillary, erect to slightly arcuate, 10-20 (-25) cm long. Spikes ellipsoid, 1.5-4 cm long, dense. Bracts equalling the calyx, broadly ovate-oblong, 5-6 cm long, obtuse, glabrous, membranous except the green midrib. Sepals broad ovate, 3.5 mm long, subacute membranous, with green midrib or the inner ones entirely membranous, posterior sepals at midrib glabrous or hairy. Corolla glabrous, ovate, 2-3.75 mm long, acute. Capsule ellipsoid, ± 6 mm long, subobtuse. Seeds 2, 4 mm long, nearly black (Skyrme, 1935).

Chemical constituents: Mucilage.

Actions: Demulcent, febrifuge and pectoral.

Therapeutics: The dried seeds are demulcent, febrifuge, ophthalmic and pectoral. They are used in the treatment of intestinal complaints and as a demulcent in cases of dysentery. Plantain seeds contain up to 30% mucilage which swells up in the gut, acting as a bulk laxative and soothing irritated membranes. Sometimes the seed husks are used without the seeds. The leaves are astringent. They are said to be useful in the treatment of intermittent fevers and pulmonary affections. An infusion is applied to the eyes in the treatment of ophthalmia.

Parts used: Leaves.

Further reading

Chopra. R N, Nayar SL, Chopra IC. *Glossary of Indian Medicinal Plants* (Including the Supplement). Council of Scientific and Industrial Research, New Delhi, 1986.

Skyrme EW. Drugs derived from the genus Plantago: their botanical sources (*Plantago ovata, P. psyllium, P. amplexicaulis, P. arenaria*). *Quarterly J Pharm* 1935; 7: 1-12.

24. CHITRAKA

Current representative: *Plumbago zeylanica*

Plumbago zeylanica L.

English name: White flowered lead wort.

Ayurvedic names: *Hutabuk, vyala, daruna, dahana, aruna, agnimali, havihpachi* and *vahni.*

Family: Plumbaginaceae.

Habitat: India.

General Description: *P. zeylanica* is a perennial, sub-scandent shrub. The leaves are ovate and glabrous. The flowers are white, in elongated spikes. The capsules are oblong, pointed, contained in a viscid glandular persistent calyx.

Detailed Description of the Root (Iyengar & Pendse, 1962).

(a) *Macroscopic*

Roots 30 cm or more in length, 6 mm or more in diameter as also as short stout pieces, including root stocks reddish to deep brown, scars of rootlets present, bark thin and brown, internal structure striated, odour, disagreeable, taste, acrid.

(b) *Microscopic*

Transverse section of root shows outer most tissue of cork consisting of 5 -7 row, of cubical to rectangular dark brown cells, secondary cortex consists of 2-3 rows of thinwalled rectangular, light brown cells, most of the cortex cells contain starch grains, secondary cortex followed by a wide zone of cortex, composed of large polygonal to tangentially elongated parenchymatous cells varying in size and shape, containing starch grains and some cells with yellow contents, fibres scattered singly or in groups of 2-6, phloem a narrow zone of polygonal, thin-walled cells, consisting of usual elements and phloem fibres, similar to cortical zone, phloem fibres usually in groups of 2-5 or more but occasionally occurring singly, lignified with pointed ends and narrow lumen, similar in shape and size to those of secondary cortex, cambium indistinct, xylem light yellow to whitish, vessels radially arranged with pitted thickenings, medullary rays straight, 1-6 seriate, cells radially elongated starch filled with starch grains, stone cells absent.

Chemical constituents: Napthoquinone: plumbagin (Vijayalaksmi, Rajan, & Srinivasan, 1987; Satheesh, & Seeni, S. (2002), glucoside: hydroplumbagin, sterol: β-sitosterol, plumbagic acid.

Specific action as per Ayurveda: Appetizer and digestive.

Therapeutics: Chronic diarrhoea and dyspepsia.

Parts used: Roots.

Bioactivity: Antimutagenic (Durga, Sridhar & Polasa, 1992), hypolipidemic (Dwivedi, 1997), antitumour (Singh *et al.*, 1997; Devi, Solmon, & Sharda, 1999), radimodifying (Devi, Solmon, & Sharda, 1999) and central nervous system stimulant (Bopaiah, & Prahdan, 2001).

Further reading

Bopaiah CP. Prahdan N. Central nervous system stimulatory action from the root extract of *Plumbago zeylanica* in rats. *Phytother Res* 2001; 15:153-156.

Devi U, Solmon FE, Sharda AC. Plumbagin, A Plant Napthoquinone with Antitumour and Radimodifying Properties. *Pharm Biol* 1999; 37: 231-236.

Durga R, Sridhar P, Polasa H. Effect of plumbagin on antibiotic resistance in bacteria. *Indian J Med Res* 1992;96:143-145

Dwivedi S. Effect of *Plumbago zeylanica* in hyperlipidemic rabbits and its modification by vitamin E *Indian J Pharmacol* 1997; 29:138.

Gujar GT. Plumbagin, a naturally occurring naphthoquinone- its pharmacological and pesticidal activity. *Fitoterapia* 1990;61:387-394.

Iyengar MA, Pendse GS. Pharmacognosy of the root of *Plumbago zeylanica. Indian J Pharm* 1962; 24:290-291.

Karnick CR, Tiwari KC, Majumdar R. Cultivation trials, pharmacognosy and ethno-botanical investigation of *Plumbago zeylanica* L. (Chitraka) of the Indian system of medicine. *Int J Crude Drug Res* 1982; 20: 193-199.

Premkumari P, Rathinam K, Santhakumari G. Antifertility activity of plumbagin. *Indian J Med Res* 1977; 65, 829.

Singh U.V. *et al.,* Reduced toxicity and enhanced antitumor efficacy of plumbagin using poly (Lactic-co-glycolic) biodegradable injectable implant. *Indian J Pharmacol* 1997; 29: 168-172.

Tilak JC, Devasagayam TP. Antioxidant properties of *Plumbago zeylanica*, an Indian medicinal plant and its active ingredient, plumbagin. *Redox Rep* 2004; 9:223-227.

Vijver LM. The constituents in the roots of *Plumbago auriculata* Lam. and *Plumbago zeylanica* L. responsible for antibacterial activity. *Planta Med* 1971; 20: 8-13.

25. DANTI

Current representative: *Baliospermum montanum*

Baliospermum montanum (Willd.) Müll. Arg.

Ayurvedic names: *Nikumbha, mukulaka, pratyaka shreni* and *udumburparni.*

Family: Euphorbiaceae.

Habitat: India.

Description (Shah & Shah, 1971)

(a) *Macroscopic*: A perennial and woody undershrub grows up to 1.5 meters in height. Leaves simple, sinuate-toothed, upper ones small, lower ones are large, flowers are numerous, in axillary recemes with male flowers above and female below. Fruits capsule, 12 mm long, obovoid, seeds ellipsoid and smooth (Shah & Shah, 1971; Venkataraman & Gopalakrishnan, 2003).

(b) *Microscopic*: Shows 5-18 layered cork, consisting of brown coloured, suberised or lignified brick-shaped cells, a few cells containing tannin and red colouring matter; secondary cortex consists of 2-7 layers of oval to elliptical, tangentially elongated cells, a few cortical fibres are also present in this region; secondary phloem consists of usual elements, traversed by uni to biseriate phloem rays; secondary xylem consists of usual elements; vessels and tracheids, bordered pits, a few having reticulate thickening; fibres slightly thick-walled, narrow lumen and blunt tips; xylem rays 1 or 2 cells wide; rosette crystals of calcium oxalate and starch grains, present only in secondary cortex and phloem; starch grains solitary and in groups, simple, round to oval measuring 6-17 μ in dia.

Powder - Brown; shows fragments of cork more or less rectangular, thick-walled in surface view; rosette crystals of calcium oxalate; numerous phloem fibres with narrow lumen and blunt tips, border pitted- and reticulate vessels, tracheid and tannin cells, round to oval simple starch grains measuring 6-17 μ in diameter, and in groups occasionally.

Chemical constituents: Starch, β-sitosterol and phorbol ester derivatives (baliospermin and montanin) and resin (Nayak & Singhai, 2003, Mali & Wadekar, 2008).

Specific action as per Ayurveda: Laxative.

Therapeutics: Plant pacifies vitiated vata, dropsy, constipation, flatulence, jaundice, hemorrhoids, skin diseases, calculi, wounds, splenomegali, leukoderma, and fever. Seeds are drastic purgative.

Parts used: Leaves, Root, Seeds.

Bioactivity: Analgesic (Nayak, Sahai & Singhai, 2003), anthelmintic (Mali & Wadekar, 2008), and hepatoprotective (Wadekar *et al.,* 2008; Mishra & Kumar, 2009).

Further reading

Mali RG, Wadekar RR. *In vitro* Anthelmintic Activity of *Baliospermum montanum* Muell. Arg roots. *Indian J Pharm Sci* 2008; 70: 131–133.

Mali, RG, Wadekar RR. *Baliospermum montanum* (Danti): ethnobotany, phytochemistry and pharmacology - a review. *Int J Green Pharm* 2008; 2:194-199.

Mishra SH, Kumar VS. Hepatoprotective activity of *Baliospermum montanum* (Willd) Muell.-Arg. in rats treated with carbon tetrachloride: *In vivo* and *in vitro* studies. *Pharmacog Mag* 2009; 5:196-202.

Nambiar VP, Krishnan TK, Jayanthi SA, Rajendra KK. Pharmacognostical studies on *Baliospermum montanum* (Willd.) Muell.-Arg. *Aryavaidyan* 1997/1998; 11: 95-108.

Nayak S, Sahai A, Singhai AK. Analgesic activity of the roots of *Baliospermum montanum* Linn. *Ancient Sci Life* 2003; 23:108-113.

Nayak S, Singhai AK. Preliminary phytochemical studies on the roots of *Baliospermum montanum* Linn. *Sach Ayur* 2003; 56: 455-458.

Shah CS, Shah NS. Pharmacognostic study of *Baliospermum montanum* Muell-Arg Publication date: *J Res Indian Med* 1971; 6: 199-208.

Venkataraman R, Gopalakrishnan S. Comparative pharmacognostic studies on the roots of genuine and commercial samples of *Baliospermum montanum.* Muell. J. Joseph Thas (Ed.), 2001; 68-69 (abstract).

Wadekar RR, Agrawal SV, Tewari KM, Shinde RD, Mate S, Patil K. Effect of *Baliospermum montanum* root extract on phagocytosis by human neutrophils. *Int J Green Pharm* 2008; 2:111-113

Wadekar RR, Supale RC, Tewari KM, Patil KS, Jalalpure SS. Screening of roots of *Baliospermum montanum* for hepatoprotective activity against paracetamol induced liver damage in albino rats. *Int J Green Pharm* 2008; 2:220-223.

26. DARUHARIDRA

Current representative: *Berberis aristata*

Possible representative: *Berberis aquifolium*, *Berberis lyceum* and *Coscinum fenestratum.*

Berberis aristata DC.

Syn: *Berberis coriaria* (Lindl.) and *Berberis chitria* (Hort.)

English name: Indian Barberry.

Other Ayurvedic names: *Darvi, katankati, suvranavarna, panchadha* and *pittadaru.*

Family: Berberidaceae.

Habitat: Himalaya and Sri Lanka.

General Description: A small evergreen shrub grows up to the height of 3 meters. The leaves are strong with fine whorled venation, straight, but with dentate margins. Inflorescence 5 to 8 cm long, with large yellow colored flowers. The fruits are bluish purple and small (Tiwari, 1991).

Chemical constituents: Alkaloids (berberine, berbamine and oxycanthine) and tannins.

Specific action as per Ayurveda: *Ykkrita uttejaka* (Hepatic stimulant).

Therapeutics: Used in fever, loss of appetite liver diseases, internal hemorrhoids, and conjunctivitis. It is especially indicated in malaria. *Rasaunt* (aqueous extract of *B. aristata*) mixed with honey is applied over aphthous ulcers. Decoction of *B. aristata* along with *amalaki* (*Emblica officinalis*) is given in urinary system ailments. Decoction of the root bark is used for washing unhealthy sores.

Parts used: Root, stem and flower and extract (*Rasaunt* or *rasanjana*).

Bioactivity: Antiperiodic, anticancer and antibacterial (Porwal *et al.*, 2010), platelet aggregation prevention (Tripathi & Shukla, 1996), antihyperglycemic (Semwal *et al.*, 2009; Singh & Kakkar, 2009).

Clinical Studies: Cholrea (Dutta & Panse, 1962), psoriasis (Singh, Rai & Pathak, 2002) and infective hepatitis (Rai, Mishra & Seth, 2005).

Further reading

Dutta NK, Panse MV. Usefulness of berberine (an alkaloid from *Berberis aristata*) in the treatment of cholera (experimental). *Indian J Med Res* 1962; 732-736.

Porwal KP, Kumar K, Jain N, Pathak KA, Jain P. Evaluation of antitumour activity of *Berberis aristata* DC. *Indian Drugs* 2010; 47, (article 3).

Rai NP, Mishra BK, Seth V. Role of *Berberis aristata* in the treatment of infective hepatitis with special reference to its antioxidant properties. *Aryavaidyan* 2005; 18: 157-161.

Semwal, BC Gupta J, Singh S, Kumar Y, Giri M. Antihyperglycemic activity of root of *Berberis aristata* DC. in alloxan-induced diabetic rats. *Int J Green Pharm* 2009; 3: 259-262.

Singh OP, Rai NO, Pathak NN. A comparative clinical evaluation of daruharidra in the management of kitibh (psoriasis) *J Res Ayur Siddha* 2002; 23: 31-37.

Singh, J, Kakkar P. Antihyperglycemic and antioxidant effect of *Berberis aristata* root extract and its role in regulating carbohydrate metabolism in diabetic rats. *J Ethnopharmacol* 2009; 123:1.

Tiwari, NN. Some crude drugs: source, substitute and adulterant with special reference to KTM crude drug market. *Sac Ayur* 1991; 44:284-290.

Tripathi YB, Shukla SD. *Berberis aristata* inhibits PAF induced aggregation of rabbit platelets. *Phytother Res* 1996; 10: 628-630.

Berberis lycium Royle

Syn: *Berberis pseudoumbellata.*

English name: Barberry.

Family: Berberidaceae.

Habitat: E. Asia – Himalayas

General Description: A semi deciduous shrub, 2 to 4 meter high, leaves lanceolate or narrowly obovate-oblong, entire or with a few large spinous teeth, arranged alternately on stem. Inflorescence a racemes, flowers yellow born in axillary clusters longer then the leaves. Fruit, berries black (Kapahi, Srivastava & Sood, 1997; Modgil & Sood, 2010).

Chemical constituents: Alkaloids (berberine, berbamine and oxycanthine) and tannins.

Therapeutics: Eye complaints, menorrhagia, chronic diarrhea and piles. Locally, the plant is used for the treatment of internal injuries. An ointment made from root powder is mixed with oil and applied on broken bones.

Parts used: Fruit and root bark.

Bioactivity: Hypolipidemic (Chand *et al.,* 2007) and antimicrobial (Singh, Srivastava & Rawat, 2007).

Further reading:

Chand N, Durrani F, Qureshi F, Durrani Z (2007). Role of Berberis *lycium* in reducing serum cholesterol in broilers. *Asian-Australien J Pl Sci* 2007; 21:563-568.

Chauhan KC, PujarGS, Ramesh C. Variation in *Berberis lycium* royle populations for berberene contents. *Indian J Plant Gen Res* 2000; 13.

Kapahi BK, Srivastava TN. Resource survey and pharmacognostic studies of commercially important Berberis species of north-western Himalaya. *J Plant Anat Morphol* 1997; 7:133-139.

Singh M, Srivastava S, Rawat AKS. Antimicrobial activities of Indian Berberis species. *Fitoterapia* 2007; 78: 574-576.

Sood P, Modgil R, Sood M. Physico-chemical and nutritional evaluation of indigenous wild fruit Kasmal, *Berberis lyceum. Indian J Nat Prod Resources* 2010; 1:362-366.

Coscinium fenestratum (Gaertn) Coleb.

English name: Tree turmeric.

Ayurvedic *name:* *Valli haridra.*

Family: Menispermaceae.

Habitat: India, Singapore, Sri Lanka and Viet Nam.

General Description: A large woody climber grows with cylindrical and yellowish stem. Leaves simple, oblong, deltoid, alternate, minutely tomentose beneath, smooth above, with yellowish tint. Flowers yellow, unisexual in supra axillary inflorescence. Fruits globose drupes, smooth contain one globose seed (Singh & Singh, 1967; Mehrotra & Mitra, 1997; Tushar *et al.,* 2008).

Chemical constituents: Alkaloid: berberine and saponin (Varier & Pillai, 1943, Child & Natheniel, 1943).

Therapeutics: *C. fenestratum* pacifies vitiates kapha, vata, skin diseases, and diseases of the eye, inflammations, wounds, jaundice, diabetes, fever and general debility.

Parts used: Stem.

Dose: Decoction (5-10 ml); extract (1-3G).

Bioactivity: Hypotensive (Singh, 1990), hepatoprotective (Venukumar & Latha, 2004), Antidiabetic (Shirwaikar, Rajendran &

Punitha, 2005; Nair *et al.*, 2005), and antidiabetic (Shirwaikar, Rajendran & Punitha, 2005).

Further reading

Child R, Natheniel WRN. The berberine content of *Coscinium fenestratum* (Colebr.) *Curr Sci* 1943; 12:255-256.

Mehrotra S, Mitra R. Pharmacognostical study of *Coscinium fenestratum* (Gaertn.) Colebr. Stem. *Indian J Pharm* 1997; 39: 6 (Abstract).

Nair GM, Narasimhan S, Shiburaj S, Abraham TK. Antibacterial effects of *Coscinium fenestratum. Fitoterapia* 2005; 76: 585-587.

Punitha ISR, Rajendran K, Shirwaikar A, Shirwaikar A. Alcoholic stem extract of *Coscinium fenestratum* regulates carbohydrate metabolism and improves antioxidant status in streptozotocin-nicotinamide induced diabetic rats. *Evidence-based Comp Alter Med* 2005; 2: 375-381.

Shirwaikar A, Rajendran K, Punitha ISR. Antidiabetic activity of alcoholic stem extract of *Coscinium fenestratum* in streptozotocin-nicotinamide induced type 2 diabetic rats. *J Ethnopharmacol* 2005; 97: 369-374.

Singh GB, Singh S, Bani S, Malhotra S. Hypotensive action of a *Coscinium fenestratum* stem extract. *J Ethnopharmacol* 1990; 30:151-155.

Singh RS, Singh BN. Identification of kaleeyak and some comments about the misapplication of *Coscinium fenestratum* (Gaertn.) Colebr. (Menispermaceae). *Indian Drugs* 1967; 5:15-16.

Tushar KV, George S, Remashree AB, Balachandran I. *Coscinium fenestratum* (Gaertn.) Colebr.-A Review on this Rare, Critically Endangered and Highly-Traded Medicinal Species. *J Plant Sci* 2008; 3: 133-145.

Varier NS, Pillai PP. A note on the alkaloids of *Coscinium fenestratum* (Colebr.). *Curr Sci* 1943; 12:228-229.

Venukumar MR, Latha MS. Effect of *Coscinium fenestratum* on hepatotoxicity in rats. *Indian J Exp Biol* 2004; 42: 798-802.

27: DUGDHIKA

Current representative: *Euphorbia hirta*

Possible representatives: *Euphorbia hypericifolia, Euphorbia thymifolia, Euphorbia onicrophylla, Euphorbia parviflora, Euphorbia prostrata* and *Daemia extensa.*

***Euphorbia hirta* L.**

Syn: *Euphorbia pilulifera* L., *Chamaesyce hirta* ((L.) Millsp.)

English name: Asthma weed and cat's hair.

Ayurvedic name: *Nagarjuni*

Family: Euphorbiaceae.

Habitat: East Asia.

General Description: The hairy plant grows up to 2 m in height; it has numerous small flowers clustered together with opposite oblong leaves, which have a toothed margin. The young yellow fruit is a small hairy capsule with 3 reddish - brown wrinkled seeds. There is milky latex in all parts of the plant. The plant flowers and fruits all year long (Sharma, 1976; Bose, 1994).

Chemical constituents: Latex (euphorbon), flavonoids and sterols (Baslas & Agarwal, 1980).

Parts used: Whole plant.

Bioactivity: Bronchodilator (Karpagam *et al.,* 2004), ulcer-protective (Rao *et al.,* 2004), wound-healing (Chandramohan & Reddy, 2006), antimicrobial (Sudhakar *et al.,* 2006), anti-allergic (Singh *et al.,* 2006), antidiabetic (Rashmi & Kumar, 2010).

Further reading

Baslas RK, Agarwal R. Isolation and characterisation of different constituents of *Euphorbia hirta* Linn. *Curr Sci* 1980; 49: 311-312.

Bose AB. Pharmacognostic study of *Euphorbia pilulifera. Sci Cul* 1994; 9: 301.

Chandramohan JB, Reddy DN. Burn wound healing activity of *Euphorbia hirta Ancient Sci Life* 2006;25: 16-18.

Karpagam KS, Kumarappan CT, Jaswanth A, Valarmathy R. Bronchodilator effect of alcoholic extract of *Euphorbia hirta* Linn. *Ancient Sci Life* 2004; 23: 1-5.

Rao ChV, Rao GMM, Kartik R, Sudhakar M, Mehrotra S, Goel RK. Ulcer protective effect of *Euphorbia hirta* extract in rats. *J Trop Med Plants* 2004; 4:2.

Rashmi SK Kumar D. Evaluation of antidiabetic activity of *Euphorbia hirta* Linn. in streptozotocin induced diabetic mice. *Indian J Nat Prod Res* 2010; 1:200-203.

Sharma GP. Synonyms, its significance and botanical description of dugdhika. *Sach Ayur* 1976; 28: 455-461.

Singh GD, Kaiser P, Youssouf MS, Singh S, Khajuria A, Koul A, Bani S, Kapahi BK, Satti NK, Suri KA, Johri RK. Inhibition of early and late phase allergic reactions by *Euphorbia hirta* L. *Phytother Res* 2006; 20:316-321.

Sudhakar M, Rao Ch.V, Rao PM, Raju DB, Venkateswarlu Y. Antimicrobial activity of *Caesalpinia pulcherrima*, *Euphorbia hirta* and *Asystasia gangeticum*. *Fitoterapia* 2006; 77: 378-380.

Euphorbia hypericifolia Linn.

Syn: *Chamaescye hypericifolia* (L.) Millsp., *Euphorbia bifida* auct. non Hook. & Ar., *Euphorbia glomerifera* (Millsp.) L.C. Wheeler, *Chamaescye glomerifera* Millsp.

English name: Black purslane, milk purslane and spurge.

Family: Euphorbiaceae.

Habitat: India.

General Description: The stem is from 1 to 2 feet in length, ascending, and much branched; it is smooth, and when the plant grows in sunny situations, is of a purple color. The branches are alternate, and proceed from opposite sides of the stem, giving the plant a flat appearance. The leaves are about an inch long, opposite, unequal at the base, and supported on very short leaf-stalks; they are oblong, obtuse, triple-veined from the base and serrulate, with numerous small, appressed teeth. The larger leaves have large purple spots near the center, which is very characteristic of the plant. The flowers are small, inconspicuous, and appear late in summer. They have the peculiar structure of the genus *Euphorbia,* and the involucrate clusters are borne from the forks of the branches on slender stalks about 1/4 inch long. The fruit is a 3-lobed carpel, containing 3 wrinkled, blackish seeds.

Chemical constituents: Caoutchouc, resin, tannin and gallic acid.

Actions: Astringent, sedative and central nervous system stimulant.

Therapeutics: Diarrhea and dysentery.

Parts used: Whole plant.

Sharma (1994), while translating and explaining *Bhavprakash Nighantu* included *Euphorbia prostrata* in the *Dugdhika*.

Euphorbia prostrata Aiton

Syn: *Anisophyllum prostrata* (Aiton.) Haw.

Common name: Blue weed, caustic weed, ground spurge and prostrate spurge.

Family: Euphorbiaceae.

Habitat: India.

General Description: *E. prostrata* is an annual herb producing slender prostrate stems up to about 20 centimeters long, sometimes purple-tinted in color. The oval-shaped leaves are up to a centimeter long with finely toothed edges. The inflorescence is a cyathium less than 2 millimeters wide, with white petal-like appendages surrounding the actual flowers. There are four male flowers and a single female flower, the latter developing into a lobed, hairy fruit one to two millimeters wide (Sinha, Prasad & Wahi, 2004).

Chemical constituents: Flavonoids: apigenin, apigenin-7-glucoside, luteolin and luteolin-7-glucoside. Phenolics: Gallic acid, ellagic acid and tannins.

Actions: Alterative.

Therapeutics: Gastritis, constipation, worm infestation, heart-weakness, chronic cough, bronchial asthma, dysuria, pyuria, spermaturia and dysmenorrhoea. *Charaka* mentions medicated ghee prepared from the plant in the treatment of bleeding hemorrhoids.

Parts used: Whole plant.

Bioactivity: Antifungal (Pal & Gupta, 1971), antispasmodic and bronchodilator (Sharma & Sharma, 1972), anti-inflammatory (Vasudev, Tripathi & Sinha, 1999).

Clinical studies: Bronchial asthma (Sharma, Upadhyay & Tripathi, 1982; Sharma & Tripathi, 1984).

Further reading

Pal S, Gupta I. *In vitro* studies on the antifungal activity of chotidudhi plant (*Euphorbia prostrata* Ait. and *Euphorbia thymifolia* Linn.). *Indian J Pharmacol* 1971; 3:29.

Sharma GD, Tripathi SN. Experimental evaluation of dugdhika (*Euphorbia prostrata* W.Ait.) for the treatment of `tamaka svasa' (bronchial asthma). *Ancient Sci Life* 1984;3: 143-150.

Sharma GD, Upadhyay BN, Tripathi SN. A clinical trial of *Euphorbia prostrata* W.Ait. and *E. thymifolia* L. in the treatment of bronchial asthma (tamakaswasa). *J Res Ayur Siddha* 1982; 3: 109-118.

Sharma GP, Sharma PV. Experimental studies on antispasmodic and bronchodilator actions of dugdhika (*Euphorbia thymifolia* Linn. and *Euphorbia prostrata* W.Ait.). *J Res Indian Med* 1972; 7: 24-28.

Sinha P, Prasad RR, Wahi SP. Pharmacognostical studies on *Euphorbia prostrata* W. Ait. *Indian Bot Reptr* 1904; 3: 48-55.

Vasudev A, Tripathi CD, Sinha VK. Evaluation of anti-inflammatory activity of bioflavonoids extracted from *Euphorbia prostrate. Indian J Pharmacol* 1999; 31:56-57 (abstract No. 72).

***Daemia extensa* (Jacq.) R.Br.**

Syn: *Pergularia daemia* (Forssk.) Choiv., *Pergularia extensa* (R.Br.) N.E. Br.

Ayurvedic names: *Kurutakah, Uttamarani, Visanika* and *Yugaphala*.

Common name: Blue weed, caustic weed, ground spurge and prostrate spurge.

Family: Asclepiadaceae.

Habitat: Throughout India.

General Description: A slender, hispid, fetid-smelling perennial climber. Leaves opposite, membranous, 3-9 cm long and about as wide, broadly ovate, orbicular or deeply Cordate, Acute or short acuminate at apex, pubescent beneath, petioles 2-9 cm. Long. Flowers greenish-yellow or dull white tinged with purple, borne in axillary, long-peduncled, drooping clusters. Fruits (follicles) lanceolate, long pointed, about 5cm. long, covered with soft spines and seeds are pubescent, broadly ovate (Shah, Shah & Bhatt, 1973).

Chemical constituents: Triterpenes and sterol (Gupta *et al.*, 1950; Raman & Barua, 1958; Seshadri & Vydeeswaran, 1971).

Actions: Stomachic, laxative and diuretic.

Therapeutics: The plant is useful in the diseases of *vata*, convulsion, asthma, poisoning; the root is useful in mental disorder, anaemia, leprosy, piles and gynacological diseases. Leaf paste mixed with castor oil is applied to joints in inflammation, liver complaints and spleen enlargement. The juice of the leaves is given in asthma and applied to rheumatic swellings in combination with lime or ginger; it is also used in the preparation of medicinal oil given in rheumatism, amenorrhoea, and dysmenorrhoea.

Parts used: Whole plant.

Bioactivity: Antifertility (Raman & Barua, 1958), antibacterial (Patel & Trivedi, 1958), hepatoprotective (Kumar & Mishra, 2008a; Kumar & Mishra, 2008b), anti-inflammatory (Hukkeri *et al.*, 20001), and anti-diabetic (Wahi *et al.*, 2002).

Further reading

Gupta JC, Roy PK, Ray GK, Dutta A. Pharmacological action of an active-constituent isolated from *Daemia extensa* Linn. (syn. *Pergularia extensa*). *Indian J Med Res* 1950; 38:75-82.

Jabalpure SS, Habbu PV, Patil MB, Kulkarni RV, Simpi CC, Patil CC. Analgesic and antipyretic activity of *Pergularia extensa* in rats. *Indian J Pharm Sci* 2002; 64: 493-495.

Hukkeri VI, Patil MB, Jalalpure SS, Ali A. Anti-inflammatory activity of various leaves extracts of *Pergularia extensa* N.E.B. *Indian J Pharm* Sci 2001; 63: 429.

Patel RP, Trivedi BM. Antibacterial activity of *Daemia extensa. Indian J Pharm* 1958; 20: 328-330.

Raman SP, Barua AK. Chemical in vestigation of *Daemia extensa* R. Br. J Am *Pharm Assoc Am Pharm Assoc* (Baltim). 1958; 47:559-560.

Seshadri TR, Vydeeswaran S. Chemical constituents of *Daemia extensa* (roots). *Curr Sci* 1971; 40: 594-595.

Shah CS, Shah N, Bhatt JG. Pharmacognostic study of *Daemia extensa* {R}.Br. *J Res Indian Med* 1973; 8: 75-85.

Suresh Kumar SV, Mishra SH. Hepatoprotective Effect of *Pergularia daemia* (Forsk.) Ethanol extract and it's fraction. *Indian J Exper Biol* 2008; 46:447-452.

Suresh Kumar SV, Mishra SH. *In-vitro* Evaluation of Hepatoprotective activity of *Pergularia daemia* (Forsk.). *Pharma Mag* 2008(b); 48:298-302.

Wahi AK, Ravi J, Hemalatha S, Singh PN. Anti-diabetic activity of *Daemia extensa* R.Br. *J Nat Rem* 2002; 2:1.

28. GAMBHARI

Current representative: *Gmelina arborea*

Possible representatives: *Trewia nudiflora*

Gmelina arborea Roxb.

Ayurvedic names: *Bhadra parnika, hira, kashmari, kashmarya, krsna varnika, madhuparnika, pitarohini, sarvatobhadra and shriparni.*

Family: Verbenaceae.

Habitat: Native to India, Bangladesh, Sri Lanka, Myanmar, Thailand and Southern China.

General Description: *G. arborea* is a medium-sized deciduous tree up to 40 m tall and 140 cm in diameter, but usually smaller than this (Jensen, 1995). The tree form is fair to good, with 6–9 m of branchless, often crooked trunk and a large, low-branched crown. The bark is thin and grey. Leaves are simple, opposite, more or less heart-shaped, 10–25 cm long, and 5– 18 cm wide. The yellow or brown flowers are arranged in panicled cymes 15–30 cm long, which appear after leaf-fall. The trumpet-shaped flowers are 4 cm long and are hairy and short-stalked. The fruit is a drupe 2–2.5 cm long and contains 1–4 seeds (Natarajan, 2003).

Detailed Description of Dried Stem

(a) *Macroscopic*:

Stem occurs as longitudinally and transversely cut pieces having varying length and thickness; hard, woody, smooth except for a few scars of branches; yellowish-grey externally and cream coloured internally.

(b) *Microscopic*:

Thin stem shows 10-15 or more layers of lignified cork, consisting of tangentially elongated, rectangular cells; secondary cortex 5-10 layers, oval to elliptical, thin-walled cells with tangential groups of fibres; pericycle present in the form of continuous ring consisting of patches of fibres alternating with stone cells: secondary phloem composed of usual elements, phloem fibres absent; in thick stem secondary cortex almost absent; secondary phloem well developed, consisting of usual elements; groups of stone cells and fibres scattered throughout this region; secondary xylem consists of usual elements; vessels solitary or 2-4 in groups having spiral thickening and bordered pits; fibres mostly aseptate but some septate with wide lumen; parenchyma paratracheal, a few in number; medullary rays 3-22 cells high and 1-4 cells wide; starch grains, simple as well ascompound having 2-4 components measuring 3-11μ in dia., present in secondary cortex, phloem and xylem parenchyma and ray cells.

Powder - Crearnish-grey; shows fragments of lignified cork cells, thin-walled, parenchymatous cells, aseptate and a few septate fibre with wide lumen; vessels with spiral thickening and bordered pits, stone cells, simple, round to oval starch grains, measuring 3-1 μ in dia.

Chemical constituents: Oil, alkaloid, resin and gmelinol (Unnikrishnan *et al.*, 2007).

Specific action as per Ayurveda: *Shothahara* (anti-inflammatory).

Therapeutics: Arthritis and rheumatism.

Parts used: Root bark.

Dose: Decoction (50-100 ml).

Bioactivity: Antioxidant (Shukla, Saluja & Pandya, 2000), immunomodulator (Sinha *et al.*, 2000), anthelmintic and wound-healing (Shirwaikar, Ghosh & Rao, 2005; Ambujakshi & Thakkar, 2009).

Further reading

Ambujakshi HR, Thakkar H. Anthelmintic activity of *Gmelina arborea* Roxb. Leaves extracts. *Int J Pharm Res Devlop* 2009; 9:1-5.

Natarajan M. Micromorphological characterization of two stimul-ating root-drugs: *Gmelina arborea* Roxb. and *G. asiatica* L. (Verbenaceae). J. Joseph Thas (Ed.), 2003; 75-76.

Shirwaikar A, Ghosh S, G.M. Rao P. Effect of *Gmelina arborea* Roxb. leaves on wound healing in rats. *J Nat Rem* 2005; 3:1.

Shukla SH, Saluja AK, Pandya SS. Modulating effect of *Gmelina arborea* Linn. on immunosuppresed albino rats. *Pharmacog Res* 2000; 2: 359-363.

Sinha, S, Dixit P, Bhargava S, Devasagayam DPA, Ghaskadhi S. Bark and fruit extracts of *Gmelina arborea* protect liver cells from oxidative stress. *Pharm Biol* 2006; 44: 237-243.

Unnikrishnan KP, Raja S, Remashree AB, Balachandran I. Occurrence of gmelinol in the stem bark of *Gmelina arborea* Roxb. *Aryavaidyan* 2007; 21: 34-37.

Trewia nudiflora Linn.

Family: Euphorbiaceae.

General Description: Deciduous tree, branchlets woody. Leaves opposite, ovate. 11-20 cm by 7-12 cm. long pointed, hairy beneath when young, glabrous later; stalks 2-7.5 cm long. Male and female flowers on separate trees, males yellow in long lax drooping inflorescences, females green, solitary or 2-3 together in the leaf axils. Fruit fleshy, depressed globose, greyish-green, 3.5 cm by 3 cm.

Chemical constituents: Phenol (Prakash *et al.*, 2007).

Bioactivity: Antioxidant (Prakash *et al.*, 2007).

Further reading

Dhan P, Suri S, Upadhyay G, Singh BN. Total phenol, antioxidant and free radical scavenging activities of some medicinal plants. *Int J Food Sci Nut* 2007;58:18-28.

Maheshwari JK, Sikarwar RLS, Painuli RM. Plants used in tribal medicine in the Dangs District of Gujarat. *Aryavaidyan* 1994, 7: 201-205.

29. GANDA DURVA

Family: Poaceae.

Botanical name: It is not established.

Ayurvedic names: *Matsyagandha, matsyakshi and sakuladani.*

Specific action as per Ayurveda: Anti diarrhoeal.

Therapeutics: Polydipsia, burning syndrome, skin diseases and fevers.

30. GANDIRA

Current representative: *Coleus barbatus*

Possible representatives: *Amaranthus viridis*

Coleus barbatus (Andrews) Benth.

Syn: *Coleus forskohlii* Briq., *Plectranthus forskohlii* Willd., *P. barbatus* Andr. and *P. comosus* Willemse

Common name: Coleus.

Family: Lamiaceae.

Habitat: Western Ghats in India.

Description : Herb (Shah, 1989).

Chemical constituents: A new labdane diterpenoid: (Bhat *et al.,* 1977)

Clinical research: *C. forskohlii* is found to be effective in skin conditions as eczema and psoriasis. Coleus is primarily indicated for cardiovascular diseases including hypertension, congestive heart failure and angina. Studies have indicated the use of coleus for asthma, intestinal colic, uterine cramps, as well as painful urination. Coleus can aid in weight loss due to its ability to breakdown stored fat as well as inhibit the synthesis of adipose tissue, additionally, it increases thyroid hormone production and release thereby increasing metabolism. Ophthalmic preparation of forskolin to the eyes lowers eye pressure thus reducing the risk of Glaucoma. Coleus can aid in digestive and malabsorption disorders through its ability to stimulate secretion of saliva, hydrochloric acid, pepsin, amalyase and pancreatic enzymes and increase nutrient absorption in the small intestine (Rupp, de Souza & Dohadwalla, 1985).

Parts used: Roots.

Further reading

Bhat SV, Bajwa BS, Dornauer H, Souza NJ, & Fehlhaber HW. Structure and stereochemistry of a new labdane diterpenoid from *Coleus forskohlii* Briq. *Tetrahed Lett* 1977;19:1669-1672.

Rupp RH, de Souza NJ, Dohadwalla AN., eds. 1985. Proceedings of the International Symposium on Forskolin: its chemical, biological and medical potential. Hoechst India Limited, Bombay 400 080. p 19-30.

Shah VC. Biosystematic studies on *Coleus barbatus* (Andr.) Benth. Ph.D. thesis. University of Bombay, Bombay, India,1989.

Amaranthus viridis L.

Syn: *Amaranthus gracilis* Desf.

Common name: Slender amaranth.

Family: Amaranthaceae.

Habitat: India.

Description: *A. viridis* is an annual herb, erect, 10 to 75 (-100) cm stem; slender, branched, angular, glabrous leaves; glabrous, long petiolate, 10 cm, lamina deltoidovate to rhomboid-oblong, 2 to 7 × 1.5 to 5.5 cm flowers; green, axillary or terminal, often paniculate spikes, 2.5 to 12 cm long and 25 mm wide. Bracts and bracteoles ovate to lanceolate-ovate, whitish, pale or reddish awn, bracteoles shorter than the perianth (1 mm); Perianth, male flowers, oblong-oval, acute, concave, 1.5 mm, female flowers narrowly oblong to narrowly spathulate, finally 1.25 to 1.75 mm, midrib green and thickened above. Stigmas 2 to 3, short, erect. Capsule subglobose, 1.25 to 1.5 mm. Seed, 1 to 1.25 mm, round, compressed, dark brown to black, reticulate. Flowering summer-fall.

Chemical constituents: 24-Alkyl-DELTA-7-sterols (Behari, Shri *et al.,* 1986) and flavonoids: Rutin and Quercetin (Ashok *et al.,* 2009). Seeds contain fixed oil (Singhal & Kulkarni, 1983).

Actions: Emollient and vermifuge.

Therapeutics: A decoction of the entire plant is used to stop dysentery and inflammation. Root juice is used to treat inflammation during urination. In India, stem used as antidote for snake bites. Leaves used for scorpion stings. Traditionally used for constipation, inflammation, eczema, bronchitis, anemia and leprosy.

Parts used: Roots.

Bioactivity: Antiproliferative and antifungal (Kaur *et al.*, 2006), anti-inflammatory (Macharla *et al.*, 2010), anthelmintic (Ashok *et al.*, 2010), antidiabetic, antihyperlipidemic and antioxidant (Ashok *et al.*, 2011).

Further reading

Ashok KBS, Lakshman K, Jayaveea KN, Sheshadri SD, Saleemulla K, Thippeswamy BS, Veerapur VP (2010). Antidiabetic, antihyperlipidemic and antioxidant activities of methanolic extract of *Amaranthus viridis* Linn in alloxan induced diabetic rats. *Exp Toxicol Pathol* 2011; 3:11-16.

Ashok Kumar BS, Lakshman K, Jayaveera KN, Nandeesh R, Mani Tripathi SN, Vamshi Krishna N, Manjunath M, Suresh MV. Estimation of Rutin and Quercetin in *Amaranthus viridis* L by High Performance Layer Chromatography (HPLC). *Ethnobot Leaflets* 2009; 13: 437-442.

Ashok Kumar BS, Lakshman K, Jayaveera KN, Ranganayakulu D, Manoj B. *In vitro* anthelmintic properties of methanol extract of *Amaranthus viridis* Linn. *Electronic J Envir Agri Food Chem* 2010; 9:1093-1097.

Behari M, Shri V *et al.*, 24-Alkyl-DELTA-7-sterols of the herb *Amaranthus viridis*. *Fitoterapia* 1986; 57: 276-277.

Kaur N, Dhuna V, Kamboj SS, Agrewala JN, Singh J. A novel antiproliferative and antifungal lectin from *Amaranthus viridis* Linn seeds. *Protein Pept Lett* 2006; 13: 897-905.

Macharla SP, Goli V, Bhasker KV, Devi SP, Dhanalakshmi Ch, Sanjusha Ch. Effects of anti-inflammatory activity of *Amaranthus viridis* Linn. *Annals Biol Res* 2011, 2: 435-438.

Singhal RS, Kulkarni PR. Composition of the seeds of some Amaranthus species. *J Sci Food Agricul* 1983; 42: 325-331.

31. GOJIHVA

Current representative: *Elephantus scaber*

Possible representatives: *Onosma bracteatum*

Elephantopus scaber L.

Common name: Elephant's foot

Family: Asteraceae.

Habitat: Asia and Australia.

Description: A perennial rigid herb found in dry localities. Leaves obovate, mostly spreading rosette on the ground and hairy on both surfaces, usually 10-20 cm long. Flowers purple on head inflorescence, fruits achenes of 4-6 bristles dilated at the base.

Chemical constituents: Sesquiterpene lactones (main is deoxyelephantopin) and steroid (Daisy *et al.*, 2009).

Specific action as per Ayurveda: Astringent.

Therapeutics: *E. scaber* pacifies vitiated *vata, pitta*, toxins, cough, cardiac ailments, diarrhoea and dysentery.

Parts used: Leaves.

Bioactivity: Hepatoprotective (Rajesh & Latha, 2001), anti-inflammatory (Sankar *et al.,* 2001), antitumor (Rajkapoor, Jayakar and Anandan, 2002), antimicrobial (Suresh, Perumal & Suresh 2004; Avani & Neeta, 2005; Daisy *et al.,* 2007) wound healing (Singh *et al.,* 2005), and hypoglycemic (Daisy *et al.,* 2008).

Further reading

Avani K, Neeta S. A study of the antimicrobial activity of *Elephantopus scaber*. *Indian J Pharmacol* 2005; 37:126-127.

Daisy P, Jasmine R, Ignacimuthu S, Murugan E. A novel steroid from *Elephantopus scaber* L., an ethnomedicinal plant with antidiabetic activity. *Phytomedicine* 2009;16: 252-257.

Daisy P, Jasmine R. Role of *Elephantopus scaber* on the glucose oxidation in liver and skeletal muscles of streptozotocin (STZ) induced diabetic adult male rats. *Res J Med Plant* 2008;2: 22-27.

Jasmine R, Daisy P, Selvakumar BN. Evaluating the antibacterial activity of *Elephantopus scaber* extracts on clinical isolates of beta-lactamase producing methicillin resistant *Staphylococcus aureus* from UTI patients. *Int J Pharmacol* 2007;3:165-169.

Rajesh MG, Latha MS. Hepatoprotection by *Elephantopus scaber* Linn. in CCl4-induced liver injury. *Indian J Physiol Pharmacol* 2001; 45: 481-486.

Rajkapoor B, Jayakar B, Anandan R. Antitumor activity of *Elephantopus scaber* Linn. against Dalton's ascitis lymphoma. *Indian J Pharm Sci* 2002;64: 71-73.

Sankar V, Kalirajan R, Sales SV, Raghuraman S. Anti-inflammatory activity of *Elephantopus scaber* in albino rats. *Indian J Pharm Sci* 2001; 63, 523-525.

Singh SDJ, Krishna V, Mankani KL, Manjunatha BK, Vidya SM, Manohara YN. Wound healing activity of the leaf extracts and deoxyelephantopin isolated from *Elephantopus scaber* Linn. *Indian J Pharmacol* 2005; 37: 238-242.

Suresh K, Perumal SP, Suresh B. Antibacterial studies on leaf extract of *Elephantopus scaber* Linn. *Ancient Sci Life* 2004;23:6-8.

Onosma bracteatum Wall.

Syn: *Coccinia glauca* Savi.

Ayurvedic names: *Davipatra* and *kharapatra*.

Family: Boraginaceae.

Habitat: It is distributed in North Western Himalayas at altitudes of 3,500-4,500 m.

Description

(a) *Macroscopic*: Stem - Cut pieces available in 5-9 cm long and 3.2 to 4.7 cm in dia., flattened, erect, stout; rough due to white, hard, hispid hairs and cicatrices, and longitudinal wrinkles; colour greenish-yellow; fracture, short; odour and taste not characteristic. Leaf - Lanceolate to ovate-lanceolate, 12-30 cm long, 1.5-3.5 cm broad, acuminate tubercle-based hispid hairs present on both surfaces; greenish to light yellow on top and white beneath.

(b) *Microscopic*: Stem - shows single-layered epidermis, covered with thick cuticle, some epidermal cells elongate to form long, warty, tubercle-based unicellular hairs, cortex differentiated in two zones, 5-7 layered outer collenchyrna, 3-4 layered inner parenchymatous cells, consisting of thin-walled, round to oval cells; phloem composed of usual elements; phloem fibres absent; xylem consisting of usual elements, vessels mostly solitary or rarely 2-3 in groups having spiral thickening, and fibres and tracheids having blunt tips and simple pits; xylem ray not distinct: pith consisting of round, thin-walled, parenchymatous cells.

Leaf -Midrib -single layered epidermis with thick cuticle and long warty, tubercle-based unicellular hairs present on both surfaces followed by 5-7 layers of collenchymatous and 3-4 layers parenchymatous cortical cells; vascular bundle situated centrally.

Lamina - isobilateral, single layered epidermis on either surface covered with thick cuticle, long warty, tubercle-based, simple, unicellular hairs present on both surfaces; palisade 2 layered, spongy parenchyma 8-10 layered, stomata paracytic. Powder - Greenish-brown; shows groups of oval to polygonal, thin-walled straight epidermal cells; spiral vessels; a few fibres entire or in pieces, elongated with blunt tips; long warty, tubercle-based unicellular hairs and a few paracytic stomata.

Chemical constituents: Alkaloids, resins, tannins, glycosides and saponins (Dandiya & Arora, 1957; Tandale, Parasar & Shilaskar, 1986).

Specific action as per Ayurveda: *Chedena* (expectorant).

Therapeutics: Common cold, bronchitis, urinary tract infection and general debility.

Parts used: Aerial parts.

Bioacticvity: Tranquilizing (Tandale, Parasar & Shilaskar, 1986), mast cell stabilization (Patel, Patel & Gandhi, 2008; Choudhary, 2010) and immunomodulator (Choudhary, 2011)

Further reading

Choudhary GP. Immunomodulatory Activity of Alcoholic Extract of *Onosma bracteatum* in mice. *J Pharm Sci* 2011; 2:1.

Choudhary GP. *In vitro* mast cell stabilization activity of *Onosma bracteatum* Wall *Int J Pharma Bio Sci* 2010; 1: 6.

Dandiya PC, Arora RB. Phytochemical and pharmacological study of *Onosma bracteatum* Wall. *J American Pharm Assoc* 1957; 46: 111-114.

Patel KG, Patel KV, Gandhi TR. Evaluation of the effect of *Onosma bracteatum* Wall (Boraginaceae) on bronchial hyperreactivity in sensitized guinea pigs. *Iranian J Pharmacol Ther* 2008; 7: 35-41.

Tandale NV, Parasar GC, Shilaskar DV. Studies on phytochemistry and tranquilizing effect of *Onosma bracteatum* Wall. *Indian J Med* 1986;4: 1-7.

32. GOKSHURA

Current representative: *Tribulus terrestris and Pedalium murex*

Possible representatives: *Martynia diandra*

Tribulus terrestris L.

It is known as *Chotta gokshura.*

Common name: Puncture-vine.

Ayurvedic names: *Bhaksyatka, candruma, gokantaka, svadukantaka, trikantaka, kshuraka* and *trika.*

Family: Zygophyllaceae.

Habitat: India.

General Description: *T. terrestris* is a taprooted herbaceous perennial plant that grows as a summer annual in colder climates. The stems radiate from the crown to a diameter of about 10 cm to over 1 m, often branching. They are usually prostrate, forming flat patches, though they may grow more upwards in shade or among taller plants. The leaves are pinnately compound with leaflets less than 6 mm (a quarter-inch) long. The flowers are 4–10 mm wide, with five lemon-yellow petals. A week after each flower blooms, it is followed by a fruit that easily falls apart into four or five single-seeded nutlets. The nutlets or "seeds" are hard and bear two to three sharp spines, 10 mm long and 4–6 mm broad point-to-point (Ansari & Prasad, 1976).

Description of the Root

(a) *Macroscopic*

Drug consists of root, 7-18 cm long and 0.3-0.7 cm in diameter, slender, cylindrical, fibrous, frequently branched bearing a number of small rootlets, tough, woody and yellow to light brown in colour, surface becomes rough due to presence of small nodules, fracture fibrous, odour aromatic, taste, sweetish and astringent.

(b) *Microscopic*

Transverse section of primary roots show a layer of epidermis followed by 4-5 layers of thin-walled parenchymatous cortex, endodermis distinct, pericycle enclosing diarch stele, in mature root, cork 4-6 layered, cork cambium single layered followed by 6-14 layers of thin-walled parenchymatous cells with varying number of fibres, distributed throughout, some secondary cortex cells show secondary wall formation and reticulate thickening, fibres found in groups resembling those of phloem, secondary phloem divided into two zones, outer zone characterised by presence of numerous phloem fibres with a few sieve tubes slightly collapsed, inner zone frequently parenchymatous, devoid of fibres often showing sieve tubes and companion cells, phloem rays distinct, few cells get converted into fibres in outer region, cambium 3-5 layered, wood composed of vessels, tracheids, parenchyma and fibres and traversed by medullary rays, vessels scattered, arranged in singles or doubles towards inner side, in groups of three to four on outer side having bordered pits, tracheids long, narrow with simple pits, xylem parenchyma rectangular or slightly elongated with simple pits and reticulate thickening, xylem fibres few, trachieds elongated with simple pits, medullary rays heterogenous, 1-4 cells wide, starch grains and rosette crystals of calcium oxalate present in secondary cortex, phloem and medullary rays cells, few prismatic crystals also present in xylem ray cells (Prakash, Singh & Wahi, 1985).

Chemical constituents: Steroid saponins: disogenin, gitogenin, chlorogenin, ruscogenin, dioscin, neotigogenin, hecogenin, tribulosin and beta-sitosterol-D-glucoside (Shukla & Ranjha, 1969; Purushothaman, Chandrasekharan & Balakrishna, 1976; Deepak *et al.*, 2002), flavonoids, alkaloids: harmine and harmaline.

Specific action as per Ayurveda: *Mutrala* (diuretic).

Therapeutics: Dysuria and urinary tract infections.

Parts used: Fruit and roots.

Bioactivity: Cardio-tonic (Seth & Jagadeesh, 1976), diuretic (Singh *et al.*, 1991), anti urolithic in rats (Anand *et al.*, 1994; Satish, Periasamy & Namasivayam, 1996), immunomodulatory (Mungantiwar *et al.*, 1999), antilipidperoxidative (Ojha *et al.*, 2006) and sex-stimulant (Tyagi *et al.*, 2008).

Clinical Studies: Anti-hypertensive (Murthy, Dubey & Tripathi, 2000).

Further reading

Anand R, Patnaik GK, Kulshreshtha DK, Dhawan BN. Activity of certain fractions of *Tribulus terrestris* fruits against experimentally induced urolithiasis in rats. *Indian J Exp Biol* 1994; 32: 548-552.

Ansari MA, Prasad S. Pharmacognostical studies on roots of *Tribulus terrestris* Linn. (gokhru mool). *J Res Indian Med* 1970;4: 193-200.

Deepak M, Dipankar G, Prasanth D, Asha MK, Amit A, Venkatraman BV. Tribulosin and β-sitosterol-D-glucoside, the anthelmintic principles of *Tribulus terrestris*. *Phytomedicine* 2002;9:753-756.

Mungantiwar AA, Nair AM, Shinde UA, Dikshit VJ, Saraf MN, Thakur VS, Sainis KB. Studies on Immunomodulatory effects of *Boerhaavia diffusa* Alkaloidal fraction. *J Ethnopharmacol* 1999; 65: 125-131.

Murthy AR, Dubey SD, Tripathi K. Anti-hypertensive effect of gokshura (*Tribulus terrestris* Linn.) - a clinical study. *Sach Ayur* 2000;52: 1124-1129.

Ojha SK, Nandave M, Kumari S, Arya DS. Antilipidperoxidative and free radical scavenging activity of *Tribulus terrestris* L. *Indian Drugs* 2006; 43: 136-139.

Prakash D, Singh PN, Wahi SP. An evaluation of *Tribulus terrestris* Linn (Chota Gokharu). *Indian Drugs* 1985; 22: 332-333.

Satish S, Periasamy P, Namasivayam A. Effect of *Tribulus terrestris* on experimental urolithiasis induced by ethylene glycol in albino rats. *Pharm Sci* 1996;2: 437-439.

Seth SD, Jagadeesh G. Cardiac action of *Tribulus terrestris*. *J Res Indian Med* 1976;64: 1821-1825.

Shukla KS, Ranjhan SK. Chemical composition and nutritive value of gokhru (*Tribulus terrestris*). *Indian Vet J* 1969;46: 715-718.

Singh RH, Singh RP, Shukla KP, Singh P. Experimental evaluation of diuretic action of herbal drug (*Tribulus terrestris* Linn.) on albino rats. *J Res Indian Med* 1991;10: 19-21.

Tyagi, RM, Aswar UM, Mohan V, Bodhankar SL, Zambare GN, Prasad A. Thakurdesai Study of furostenol glycoside fraction of *Tribulus terrestris* on male sexual function in rats. *Pharm Biol* 2008;46: 191-198.

Pedalium murex L.

It is known as *Brahat gokshura*.

English name: Land caltrops

Family: Pedialaceae.

Habitat: It is commonly found in Deccan and in some parts of Ceylon and Gujarat.

General Description: A branched succulent herb grows up to 40 cm in height. Leaves simple, opposite, fleshy, serrate or crenate; flowers bright yellow, solitary and axillary; fruits 4 angled with horizontal spines from the angles (Singh, 1963; Mangle & Jolly, 1998).

Chemical constituents: Alkaloid: pedaline (Ansari & Prasad, 1970), heptatriacontan-4-one and tetratriacontanyl octacosanoate (Heptatriacontan-4-one, tetratriacontanyl octacosanoate, 1983). The fruits are rich in flavonoids, sapogenin (diosgenin-0.06%) and soluble proteins (20.14 mg/gm) (Sermakkani & Thangapandian, 2010).

Specific action as per Ayurveda: *Mutrala* (diuretic).

Therapeutics: *P. murex* pacifies vitiated vata, pitta, urinary retention, kidney stone, seminal weakness, amenorrhea, inflammation, flatulence and fever.

Parts used: Fruit.

Bioactivity: Anti hyperlipidemic (Balasubramanian *et al.*, 2008), anti-urolithic (Ananta *et al.*, 2010), nephroprotective (Shelke *et al.*, 2009; Sreedevi *et al.*, 2011), antioxidant (Thakkar *et al.*, 2011) and gastroprotective (Giribabu *et al.*, 2011).

Further reading

Ananta D, Kokilavani KS, Balakrishnan R, Gurusamy A. Effect of ethan olic fruit extract of *Pedalium murex* Linn. in ethylene glycol induced urolithiasis in male Wistar albino rats. *Ancient Sci Life* 2010; 29: 29-34.

Ansari MS, Prasad S. Pharmacognostical studies on gokhru (root of *Pedalium murex* Linn.) *J Res Indian Med* 1970; 5: 50-58.

Balasubramanian, MN, Muralidharan P, Balamurugan G. Anti hyperlipidemic activity of *Pedalium murex* (Linn.) fruits on high fat diet fed rats. *Int J Pharmacol* 2008;4: 310-313.

Giribabu K, Pushpa S, Latha YJP, Ranganayakulu D, Rajasekhar KK, Hari Babu R. Effect of leaves of *Pedalium murex* Linn. against aspirin induced gastric ulcers in rats. *J Pharm Res* 2011; 4: 3.

Mangle MS, Jolly CI.HPTLC studies on *Tribulus terrestris* L. (chota gokhru) and *Pedalium murex* L. (bada gokhru). *Indian Drugs* 1998;35: 189-194.

Sermakkani M, Thangapandian V. Phytochemical Screening for Active Compounds in *Pedalium murex* L. *Recent Res Sci Technol*, 2010; 2:1.

Shelke T, Kotahri T, Adkar PP, Bhaskar VH, Juvale, KC. Nephroprotective activity of ethanolic extract of *Pedalium murex* Linn. *J Cell Tissue Res Vol* 2009; 9: 1687-1690.

Shukla YN, Khanuja SPS. Chemical, Pharmacological and Botanical Studies on *Pedalium murex*. *ChemInform* 2005; 36:27 .

Shukla YN, Thakur RS. Heptatriacontan-4-one, tetratriacontanyl octacosanoate and other constituents from *Pedalium murex. Phytochemistry* 1983; 22: 973-974.

Singh SP. Morphological studies in some members of the family Pedaliaceae II: *Pedalium murex* Linn. *Agra Univ J Res Sci* 1963; 12: 143-161.

Sreedevi AY, Latha JP, Bharathi K. Protective effect of fruits of *Pedalium murex* against gentamicin-induced nephrotoxicity in rats. *Int J Phytopharmacol* 2011, 2:27-36.

Thakkar JH, Solankib AN, Thakkarc MH, Solankia HK, Patel NJ. *In vitro* antioxidant activity of aqueous extract of *Pedalium murex. Int J Preclin Pharm Res* 2011;2:26-29.

Martynia diandra Gloxin nom. illeg.

Syn: *Carpoceras angulata* A. Rich., *Craniolaria annua* L., *Disteira angulosa* (Lam.) Raf., *Martynia angulosa* Lam.

Ayurvedic name: *Vaghnakhi.*

Common name: Devil's claw.

Family: Pedialaceae.

Habitat: India.

General Description: Leaves large, opposite, cordate, sinuately lobed and minutely dentate often covered with a glutinous dew-like substance. Flowers diandrous, handsome, ill-smelling, rose-coloured, in racemes; beetle shaped, hard and woody fruit, with 2 sharp anterior hooks is often an object of curiosity.

Chemical constituents: Fixed oil (epoxy and cyclopropenoid (malvalic acid 7.8% and sterculic acid 4.4%) fatty acids.

Parts used: Leaves.

Bioactivity: Anti-inflammatory (Chatpalliwar *et al.,* 2002), and anti-nociceptive (Chatpalliwar *et al.,* 2003).

Further reading

Chatpalliwar VA, Joharapurkar AA, Wanjari MM, Chakraborty RR. Antinociceptive activity of *Martynia diandra* Gloxin. *Indian J Pharmacol* 2003; 35: 320-321.

Chatpalliwar VA, Joharapurkar AA, Wanjari MM, Chakraborty RR, Kharkar Vt. Anti-inflammatory of *Martynia diandra. Indian Drugs* 2002;39:543-545.

33: GRANTHIPARNI

Current representative: *Angelica glauca*

Possible representatives: *Clerodendron infortunatum*

Angelica glauca Edgew.

Ayurvedic name: *Choraka.*

Family: Apiaceae.

Habitat: E. Asia - Western Himalayas - Kashmir to Simla.

General Description: Plants 1-2 m tall. Stem stout, fistular. Leaves petiolate; compound; 1-2-ternate-pinnate; leaflets oval to ovate, mucronulate-serrate, lower surface glaucous. Peduncle stout, c. 20 cm long. Involucre of 5 long, linear bracts. Rays 20 or more. Involucel of 6-I0 linear bractlets. Fruit 8-12 mm long, 5-6 mm broad, elliptic to oblong; dorsal and intermediate ridges thick, corky, obtuse, approximated; lateral ridges broadly winged; furrows 1-vittate; commissure 2-4-vittate. Inner seed face concave (Shah, Mitra & Kapoor, 1972).

Chemical constituents: Essential oil (Major compounds of the oil of Himalayan *A. glauca* are as α-phellandrene (13.5%), *trans*-carveol (12.0%), β-pinene (11.7%), thujene (7.5%), β-caryophyllene oxide (7.2%), β-caryophyllene (7.0%), γ-terpinene (6.7%), nerolidol (6.5%), β-bisabolene (5.2%) and germacrene D (4.5%) (Kaul, Mallavarapu & Chamoli , 1996; Agnihotri *et al.,* 2004).

Specific action as per Ayurveda: Carminative, diaphoretic and stimulant.

Therapeutics: Dyspepsia and constipation.

Parts used: Roots.

Bioactivity: Hepatoprotective and antioxidant (Joshi *et al.,* 2008)

Further reading

Agnihotri VK, Thappa RK, Meena B, Kapahi BK, Saxena RK, Qazi GN, Agarwal G. Essential oil composition of aerial parts of *Angelica glauca* growing wild in North-West Himalaya (India). *Phytochemistry* 2004; 65: 2411-2413.

Joshi S, Prakash O, Hore SK, Zafar A, Pant AK. Hepatoprotective and antioxidant activity of the aqueous extract of *Angelica glauca* Edgew. root *Asian J Trad Med* 2008; 3: 58-66.

Kaul PN, Mallavarapu GR, Chamoli RP. The essential oil composition of *Angelica glauca* roots. *Planta Med* 1996; 62:80-81.

Shah NC, Mitra R, Kapoor LD. Pharmacognostic studies of *Angelica glauca* Edgew. *Indian J Pharm* 1972; 34:171; *Bull Bot Sur India* 1974; 16: 40-47.

Clerodendron infortunatum Gaertn.

Syn: *Clerodendrum viscosum* Vent. and *Volkameria infortunata* Roxb.

Ayurvedic name: *Bhandira.*

English name: Hill glory bower

Family: Verbenaceae.

Habitat: The species is native to tropical regions of Asia including India, Myanmar, Pakistan, Thailand and Sri Lanka.

General Description: *C. infortunatum* is a flowering shrub or small tree, and is so named because of its rather ugly leaf. The stem is eresct, 0.5–4 m high, with no branches and produce circular leaves with 6 inch diameter. Leaves are simple, opposite; both surfaces sparsely villous-pubescent, elliptic, broadly elliptic, ovate or elongate ovate, 3.5–20 cm wide, 6–25 cm long, dentate, inflorescence in terminal, peduncled, few-flowered cyme; flowers white with purplish pink or dull-purple throat, pubescent. Fruit berry, globose, turned bluish-black or black when ripe, enclosed in the red accrescent fruiting-calyx. The stem is hollow and the leaves are 6-8 inch (15–20 cm) long, borne in whorls of four on very short petioles. The inflorescence is huge, consisting of many tubular snow white flowers in a terminal cluster up to 2 ft (0.6 m) long. The tubes of the flowers are about 4 inch (10 cm) long and droop downward, and the expanded corollas are about 2 inch (5 cm) across. The fruits are attractive dark metallic blue drupes, about a half inch in diameter. Fruit usually with 4 dry nutlets and the seeds may be with or without endosperm. It flowers from April to August.

Chemical constituents: Bitter principle: (clerodin), glycoside: acetoside (Sinha *et al.,* 1980; Sinha *et al.,* 1981), scutellarin and hispidulin-7-O-glucuronide (Subramanian & Nair, 1973)

Specific action as per Ayurveda: Anthelmintic and cholagouge.

Therapeutics: Worm infestation.

Parts used: Roots and leaves.

Bioactivity: Analgesic and anticonvulsant (Analgesic and anticonvulsant, 2009), antioxidant (Gouthamchandra, Mahmood & Manjunatha, 2010) and antihyperglycemic (Das *et al.,* 2011).

Further reading

Das S, Bhattacharya S, Prasanna A, Suresh Kumar RB, Pramanik G, Haldar PK (2011). Preclinical evaluation of antihyperglycemic activity of *Clerodendron infortunatum* leaf against streptozotocin-induced diabetic rats. *Diabetes Ther* 2011;2: 1-9.

Gouthamchandra K, Mahmood R, Manjunatha H. Free radical scavenging, antioxidant enzymes and wound healing activities of leaves extracts from *Clerodendrum infortunatum* L. *Envir Toxicol Pharmacol* 2010; 30:11–18.

Pal D, Sannigrahi S, Mazumder UK Analgesic and anticonvulsant effects of saponin isolated from the leaves of *Clerodendrum infortunatum* Linn. in mice. *Indian J Exp Biol* 2009; 47: 743–747.

Sinha NK, Pandey VB, Shah AH, Dasgupta B. Chemical constituents of the flowers of *Clerodendron [Clerodendrum] infortunatum. Indian J Pharm Sci* 1980; 42: 96-97.

Sinha NK, Seth KK, Pandey VB, Dasgupta B, Shah AH. Flavonoids from the flowers of *Clerodendron infortunatum*". *Planta Med* 1981;42: 296–298.

Subramanian SS, Nair AGR. Scutellarin and hispidulin-7-O-glucuronide from the leaves of *Clerodendrum indicum* and *Clerodendron infortunatum*. *Phytochemistry* 1973; 12: 1195.

34. GRANJANA

Current representative: *Moringa oleifera*

Possible representatives: *Allium cepa, Allium sativum, Brassica campestris, Daucus carota, Eulophia campestris* and *Orchis mascula*.

Moringa oleifera Lam.

Common name: Drumstick plant.

Ayurvedic names: *Shigru and shobanjana.*

Family: Moringaceae.

Habitat: Native to India, Arabia, and possibly Africa and the East Indies; widely cultivated and naturalized in tropical Africa, tropical America, Sri Lanka, India, Mexico, Malabar, Malaysia and the Philippine Islands.

Description: Short, slender, deciduous, perennial tree, to about 10 m tall; rather slender with drooping branches; branches and stems brittle, with corky bark; leaves feathery, pale green, compound, tripinnate, 30–60 cm long, with many small leaflets, 1.3–2 cm long, 0.6–0.3 cm wide, lateral ones somewhat elliptic, terminal one obovate and slightly larger than the lateral ones; flowers fragrant, white or creamy-white, 2.5 cm in diameter, borne in sprays, with 5 at the top of the flower; stamens yellow; pods pendulous, brown, triangular, splitting lengthwise into 3 parts when dry, 30–120 cm long, 1.8 cm wide, containing about 20 seeds embedded in the pith, pod tapering at both ends, 9-ribbed; seeds dark brown, with 3 papery wings. Main root thick (Pal *et al.*, 1996; Padmarao *et al.*, 1996).

Chemical constituents: Essential oil, alkaloids: moringine, moringinine, spirochine and ptergyospermin, resin and mucilage (Jadhav *et al.*, 2000).

Specific action as per Ayurveda: Antispasmodic, expectorant and diuretic.

Therapeutics: Gout.

Parts used: Bark, root, fruit, gum and seeds.

Bioactivity: Antibiotic: ptergyospermin (Raghunandana, Marim & Pandalal, 1946), contraceptive (Shukla, Prakash & Mathur, 1986), anti-implantation (Shukla, Mathur & Prakash, 1987), abortifacient (Sethi *et al.*, 1988), anticonvulsive (Subarnas, Anggadhiraja & Daintini, 1996/97), hepatoprotective (Ruckmani *et al.*, 1998), anti-inflammatory (Rao *et al.*, 1999), analgesic (Panigrahi *et al.*, 1999), radioprotective (Rao, Devi & Kamath, 2001), antioxidant (Kumar & Pari, 2003), nootropic (Mohan *et al.*, 2005), and antiarthritic (Karadi *et al.*, 2006). Moringine is cardiac stimulant and spirochine, uterine tonic.

Further Reading

Jadhav SL, Sharma SR, Pal SC, Kasture SB, Kasture VS. Chemistry and pharmacology of *Moringa oleifera* and *Moringa concanensis*. *Indian Drugs* 2000;37: 139-144.

Karadi RV, Gadge NB, Alagawadi KR, Savadi RV, Mannur VS, Koti BC. Antiarthritic activity of extracts of *Moringa oleifera* Lam. Roots. *Indian Drugs* 2006; 43, 7.

Kumar NA, Pari L. Antioxidant action of *Moringa oleifera* Lam. (Drumstick) against antitubercular drugs induced lipid peroxidation in rats. *J Med Food* 2003;6: 255-259.

Mohan M, Kaul N, Punekar A, Girnar R, Junnare P, Patil L. Nootropic activity of *Moringa oleifera* leaves. *J Nat Rem* 2005;5: 1.

Padmarao P, Acharya BM, Dennis TJ. Pharmacognostic study on stem bark of *Moringa oleifera* Lam. (Sigru). *Bull of Medico-Ethno-Bot Res* 1996;17: 141-151.

Pal SK, Mukherjee PK, Saha K, Pal M, Saha BP. Some pharmacognostical characteristics of *Moringa oleifera* Lam. Leaves. *Ancient Sci Life* 1996;15:288-292.

Panigrahi B, Das MCH, Mishra SS, Mohanty S, Panigrahi M, Kanungo S. Analgesic actions of *Moringa oleifera* bark extract. *Indian J Pharmacol* 1999;31:56 (abstract No. 70).

Raghunandana Rao R, Marim G, Pandalal KM. Pterygospermin: the Antibacterial Principle of *Moringa pterygosperma*, Gaertn. *Nature* 1946; 158:745-746.

Rao AV, Devi PU, Kamath R. *In vivo* radioprotective effect of *Moringa oleifera* leaves. *IJEB* 2001;39: 858-863.

Ruckmani K, Kavimani S, Aanandan R, Jaykar B. Effect of *Moringa oleifera* Lam. on paracetamol induced hepatotoxicity. *Indian J Pharm Sci* 1998;60: 33-35.

Sethi N, Nath D, Shukla SC, Dyal R. Abortifacient activity of a medicinal plant "*Moringa oleifera*" in rats. *Ancient Sci Life* 1988;7: 172-174.

Shukla S, Prakash AO, Mathur R. *Moringa oleifera* Lam., its contraceptive efficacy. *Indian J Pharmacol* 1986;18:19-60.

Shukla, S., R. Mathur and A.O. Prakash Anti-implantation efficacy of *Moringa oleifera* Lam. and *Moringa concanensis* Nimmo in rats. *Int J of Crude Drug Res* 1987;26: 29-32.

Subarnas A, Anggadhiraja EYS, Daintini A. Anticonvulsive activity of alkaloid fraction of *Moringa oleifera* Lam. Roots. *Phytomedicine* 1996/ 1997; 3:260.

Venkateswera Rao KN, Gopalakrishnan V, Loganathan V, Shanmuganathan S. Anti-inflammatory activity of *Moringa oleifera* Lam. *Ancient Sci Life* 1999;18:195-198.

Allium cepa L.

Syn: *Allium ascalonicum* L.

English name: Onion.

Ayurvedic name: Palandu.

Family: Liliaceae.

Habitat: Native to central Asia, and has long been a staple in the Mediterranean region, as well as a frequent seasoning in Asia, Africa, and Europe.

Description: Herb.

Chemical constituents: Volatile oil (containing allyl-prpoyl disulphide), protein, calcium, iron and vitamins A, B_1 and C (Bajaj *et al.,* 1980) and flavonoids: quercetin (Patil *et al.,* 1995).

Specific action as per Ayurveda: *Vedana shamaka* (analgesic).

Therapeutics: Cough, asthma and cholera.

Parts used: Bulbs.

Bioactivity: Hypo and hyperglycaemic (Mathew, & Augusti, 1973), anti-asthmatic (Handa, Singh & Atal, 1983), antiplatelet (Makheja & Bailey, 1990), hypolipidemic (Lata *et al.,* 1991), antioxidant (Helen *et al.,* 2000), antimicrobial (Satya *et al.,* 2005) and neuroprotective (Shri & Bora, 2008).

Clinical Study: Bronchial asthma (Amla *et al.,* 1981).

Further Reading:

Amla V, Verma SL, Sharma TR, Gupta OP, Atal CK. Clinical study of *Allium cepa* Linn. in patients of bronchial asthma. *Indian J Pharmacol* 1981; 13:, 63.

Bajaj KL, Kaur G, Singh J, Gill SPS. Chemical evaluation of some important varieties of onion (*Allium cepa* L.). *Plant Foods Human Nutr* 1980;30:117-122.

Handa G, Singh J, Atal CK. Anti-asthmatic principle of *Allium cepa* Linn. (onion). *Indian Drugs* 1983; 20:239.

Helen A, Krishnakumar K, Vijayammal PL, Augusti KT. Antioxidant effect of onion oil (*Allium cepa* Linn.) on the damages induced by nicotine in rats as compared to alpha-tocopherol. *Toxicol Lett* 2000;116: 61-68.

Lata S, Saxena KK, Bhasin V, Saxena RS, Kumar A, Srivastava VK. Beneficial effects of *Allium sativum, Allium cepa* and *Commiphora mukul* on experimental hyperlipidemia and atherosclerosis - a comparative evaluation. *J Postgrad Med* 1991; 37:132-135.

Makheja AN, Bailey JM. Antiplatelet constituents of garlic and onion. *Inflam Res* 1990;29: 360-363.

Mathew PK, Augusti KT. Isolation of hypo- and hyperglycaemic agents from *Allium cepa* Linn. *IJEB* 1973;11: 573-575.

Patil PS, PikeLM, Hamilton BK. Changes in quercetin concentration in onion (*Allium cepa* L.) owing to location, growth stage and soil type. *New Phytol* 1995;130:349-355.

Satya VK, Radhajeyalakshmi R, Kavitha K, Paranidharan V, Bhaskaran R, Velazhahan R. *In vitro* antimicrobial activity of zimmu (*Allium sativum* L., *Allium cepa* L.) leaf extract. *Arch Phytopathol Plant Prot* 2005;38:185-192.

Shri R, Bora KS. Neuroprotective effect ofmethanolic extracts of *Allium cepa* on ischemia and reperfusion-induced cerebral injury. *Fitoterapia* 2008;79: 86-96.

Eulophia campestris Wall.

Ayurvedic name: *Munjataka.*

Family: Orchidaceae.

Habitat: India.

Description: A tuberous, terestrial herb; tubers irregularly oblong, often lobed. Leaves two, 25-40 cm long, linear, acuminate, plicate, arising from the apex of a slender sheathing pseudostem. Scape 15-30 cm long; raceme laxly many-flowered. Flowers drooping, subsecund, yellowish or green with pink or purple markings, about 2.5 cm across. Capsule 2 cm long, ellipsoid (Trivedi *et al.,* 1974).

Chemical constituents: Mucilage.

Specific action as per Ayurveda: *Balya* (tonic).

Therapeutics: Seminal debility.

Parts used: Rhizome.

Further reading

Dhiman, AK. Medicinal uses of orchid plants. *Sach Ayur* 2005; 58: 56-59.

Trivedi VP, Dixit RS, Lal VK, Joshi P. Clues for identification of a controversial drug (mumjataka) from ancient literature. *J Res Indian Med* 1974; 9: 56-64.

Orchis mascula (L.) L.

Syn: *Graphorchis campestris* (Wall.) Kuntze

English name: Early-purple Orchid.

Family: Orchidaceae.

Habitat: *O. mascula* grows in a variety of habitats, from meadows to mountain pastures and woods, in full sun or shady areas, from 0–2,500 metres (0–8,200 ft) above sea level.

Description: *O. mascula* is a herbaceous plant with stems up to 50–60 centimetres (20–24 in) of height, green at the base and purple on the apex. The root system consists of two tubers, rounded or ellipsoid. The leaves, grouped at the base of the stem, are oblong-lanceolate, pale green, sometimes with brownish-purple speckles. The inflorescence is 7.5–12.5 centimetres (3.0–4.9 in) long and it is composed of 6 to 20 flowers gathered in dense cylindrical spikes. The flower size is about 2.5 centimetres (0.98 in) and the color varies from pink to violet. The lateral sepals are ovate-lanceolate and erect, the median one, together with the petals, is smaller and cover the gynostegium. The labellum is three-lobed and convex, with crenulated margins and the basal part clearer and dotted with purple-brown spots. The spur is cylindrical or clavate, horizontal or ascending. The gynostegium is short, with reddish-green anthers. It blooms from April to June (Rajbir, Kachroo & Agrawal,2009).

Chemical constituents: Mucilage.

Specific action as per Ayurveda: *Nadi balya* (nervine-tonic).

Therapeutics: Diarrhoea and dysentery.

Parts used: Rhizome.

Further reading

Chopra RN, Nayar SL, Chopra IC. *Glossary of Indian Medicinal Plants (Including the Supplement).* Council of Scientific and Industrial Research, New Delhi, 1986.

Rajbir S, Kachroo M, Agrawal SS. Standardization of *Centipeda minima, Mimusops elengi, Orchis mascula* and *Paeonia emodii. Indian J Trad Knowl* 2009;8: 338-345

35. HAMSAPADI

Adiantum lunulatum Burm.

Common name: Walking-maiden hair fern.

Ayurvedic names: *Raktapadi, tripadika, prahladini, kitamari, kitamata* and *madhusrava.*

Family: Adiantaceae.

Habitat: India.

Description

(a) *Macroscopic*

Root - Very thin, fibrous, about 10-15 cm long, reddish-black in colour, soft and branched. Rhizome - Long, upto 2 mm thick, glabrous, prostrate or erect, dark reddish-brown or a black in colour. Frond - Rachis shiny black, simply pinnate, pinna roughly lunulate, subdimidiate, lower edge nearly in line and oblique with its black shiny petiole, upper edge bluntly rounded and more or less lobed, a few sori in a continuous line on the under surface along the edge, with a false indusium.

(b) *Microscopic*

Root mature root shows single layered epidermis consisting of thin-walled, small and irregular cells, followed by 3-4 layers of large thick-walled, polygonal, parenchymatous cells of cortex; endodermis single layered composed of square or somewhat rounded cells; pericycle single layered composed of square shaped sclerenchymatous thick and dark reddish-brown wall; pericycle encloses a diarch stele with a few elements of xylem and phloem.

Rhizome - Mature rhizome consists of thick walled, rectangular, small cells of epidermis, followed by 3-4 layers of sclerenchymatous cells of hypodermis, composed of thickwalled cells; cortex wide, made up of thin-walled, rounded or oval-shaped parenchymatous cells, enclosing an amphiphloic siphonostele; endodermis present; vascular bundle with xylem consisting protoxylem towards both ends and metaxylem in centre; phloem surrounds the xylem externally and also internally; tracheid with scalariform to reticulate thickening present; a central pith consists of thick-walled cells, and fibres, and is sclerenchymatous.

Frond- *Petiole* - Shows concave-convex outline; epidermis single layered; hypodermis consists of 2 or 3 layers, lignified, thick-walled, sclerenchymatous cells; ground tissue composed of oval to polygonal, thin-walled parenchymatous cells; stele single, slightly triangular in shape, located centrally and surrounded by peri cycle and endodermis.

Pinnule - Shows single layered epidermis on either surface; mesophyll round to oval in shape and not differentiated into palisade and spongy parenchyma; a few stomata present only on lower surface; a few sori also seen.

Powder - Dark reddish-brown in colour; shows dark reddish-brown pieces of sclerenchymatous cells and light coloured crushed cells of cortex, a few tracheids having reticulate thickening, fibres and a few spores.

Chemical constituents: Carotenoids, triterpenoids (22,29xi-epoxy-30-norhopane-13beta-ol, fern-9(11)-en-6alpha-ol. fern-9(11)-ene, fern-9(11)-en-25-oic acid, fern-9(11)-en-28-ol, filicenol-B, adiantone and oxidation product of fern-9(11)-en-6alpha-ol obtained as 6-oxofern-9(11)-ene (Reddy *et al.,* 2001).

Specific action as per Ayurveda: Demulcent.

Therapeutics: Fevers and dysentery.

Parts used: Fronds.

Bioactivity: antibacterial (Reddy *et al.,* 2001) and antifungal (Piyali *et al.,* 2005).

Further reading

Reddy VL, Ravikanth V, Rao TP, Diwan PV, Venkateswarlu Y. A new triterpenoid from the fern *Adiantum lunulatum* and evaluation of antibacterial activity. *Phytochemistry* 2001; 56:173-175.

Piyali G, Mukhopadhyay R, Gupta K. Antifungal activity of the crude extracts and extracted phenols from gametophytes and sporophytes of two species of *Adiantum*. *Taiwania* 2005;50: 272-283

36: HASTIKARNA PALASHA

Current representative: *Leea macrophylla*

Leea macrophylla Roxb. ex Hornem.

Syn: *Leea diffusa* M. A. Lawson, *Leea robusta* Roxb.

Ayurvedic name: *Gajkarni, Hastikarnapalasha.*

Family: Leeaceae.

Habitat: India.

Description: Erect shrubs, small trees, or perennial herbs. Branchlets terete with longitudinal ridges, hairy to mealy pubescent. Leaves variable in shape and size, simple, 3-foliolate, or 1-3-pinnate; stipules large and obovate, 4-6 cm long, 2-6 cm wide, caducous; petiole 15-20 cm, pubescent, or glabrescent; simple leaf blade broadly ovate, 40-65 cm long, 35-60 cm wide, base rounded, margin dentate, apex acuminate, pubescent; lateral veins 12-15 pairs, veinlets inconspicuously protruding. Inflorescences opposite to leaves, compound corymbose-dichasial; peduncle 20-25 cm, mealy pubescent; bracts triangular to narrowly so. Pedicel 2-3 mm, pubescent; buds oval-elliptic, 2-3 mm, apex rounded. Calyx tube cupulate, with 5 triangular teeth, pubescent. Petals elliptic, 2.5-4 mm, pubescent. Staminodial tube 2-2.2 mm, with lower part 0.3-0.5 mm, upper 1.4-1.7 mm, lobes 1.3-1.6 mm. Stamens 5; filaments 1.5-2 mm; anthers elliptic, 1.4-1.8 mm long, 0.6-0.8 mm wide. Ovary nearly globose; style 1.2-1.5 mm; stigma slightly capitate. Berry oblate, 0.8-1.3 cm, 6-seeded (Singh & Singh, 1981; Tiwari, 1991)

Therapeutics: *L. macrophylla* is reported to be anti-cancerous properties. Powder of leaves mixed with honey is given to patient of cancer (Swarnkar and Katewa, 2008).

Further reading

Singh RS, Singh AN. On the identity and economic-medicinal uses of *Hastikarnapalasha* (*Leea macrophylla* Roxb.) as evidenced in the ancinet (Sanskrit) texts and traditions. *Indian J Hist Sci* 1981;16:219-22.

Swarnkar S, Katewa SS. Ethnobotanical observation on tuberous plants from tribal area of Rajasthan (India). *Ethnobot Leaflets* 2008; 12:647.

Tiwari, NN. Some crude drugs: source, substitute and adulterant with special reference to KTM crude drug market. *Sach Ayur* 1991;44: 284-290.

37. JINGINI

Current representative: *Odina woider*

Odina woider Roxb.

Syn: *Lannaea grandis* Engl.

Common name: Woider tree.

Family: Anacardiaceae.

Habitat: India.

Habit: Tree.

Chemical constituents: Gum-resin.

Parts used: Gum.

38. JIVANTI

Laptadenia reticulata (Retz.) Wight. & Arn.

Family: Asclepidaceae.

Ayurvedic names: *Jivanti, jivani, jivaniya, jiva, jivya, jivada, jivdatri, shakasrestha, jivabhadra, bhadra, manglya, kshudrjiva, yashaysa, shrngati, jivaprishta, kanjika, shasashimbika* and *supiangla.*

Habitat: Distributed in sub himalayan tracts of India mainly in Punjab, Uttar pradesh and throughout Deccan peninsula upto 900 m.

Description: *L.reticulata* is a much-branched twining shrub. The bark is yellowish brown, corky, deeply cracked; the leaves are ovate-cordate, coriaceous glabrous above, more or less finely pubescent beneath; the flowers are in many-flowered cymes, greenish yellow; the follicles are sub-woody and turgid (Gupta & Kapoor, 1971; Raman *et al.*, 2002).

Phytochemistry: Crystalline principles (Krishna, Rao & Rao, 1975), sterols: stigmasterol, β-sitosterol, β-amyrin acetate, lupanol 3-O–diglucoside, flavonoids: diosmetin and luteolin, triterpenoid: leptadenol, leptidin, n–triacontane and cetyl alcohol (Krishna, Rao & Rao, 1975; Mammen, Daniel & Sane, 2010).

Therapeutics as per Ayurevda: The plant is galactogogue, cooling, nutritive, aphrodisiac, stimulant, diuretic, and eyetonic. It is useful to cure eye-diseases, seminal debility, general weakness, cough, dyspnoea, fever, asthma, constipation, sore throat, and gonorrhea. It promotes health and vigour, improves voice, and alleviates the three dosas *vata, pitta* and *kapha.* It also cures eye diseases, hematemesis, emaciation, cough, dyspnoea, fever and burning sensation.

Parts used: Roots.

Bioactivity: Antimicrobial (Patel & Dantwala, 1958), galacatagouge (Anjaria *et al.,* 1975), antifungal (Mishra *et al.,* 2010), immunomodulator (Girishkumar *et al.,* 2010) and hepatoprotective (Nema, Agarwal & Kashaw, 2011)

Further reading

Anjaria JV, Varia MR, Janakiraman K, Gulati OD. Studies on *Leptadenia reticulata*: lactogenic effects on rats. *Indian J Exp Biol* 1975; 13:448-449.

Girishkumar V, Sreepriya M, Praveenkumar S, Bali G, Jagadeesh MS. Modulating effect of *Leptadenia reticulata* (Retz) Wight & Arn against chromate (VI)-induced immunosuppression and oxidative stress on mouse splenic lymphocytes and bone marrow derived macrophages. *J Ethnopharmacol* 2010 15; 131:505-508.

Gupta RC, Kapoor LD. Pharmacognostical studies on jivanti, part II: *Leptadenia reticulata* Wight & Arn. (syn. *Gymnema aurantiacum* Wall. ex Hook f. and *Aclepias tuberosa* Roxb.). *Bull Bot Sur India* 1971;13: 53-63.

Krishna, P.V.; Rao, E.V. & Rao, D.V. (1975). Crystalline principles from the leaves and twigs of *Leptadenia reticulata. Planta Med* 1975; 27:395-400.

Mammen D, Daniel M, Sane RT. Pharmacognostic and Phytochemical studies on *Leptadenia reticulata* (Retz.) Wight & Arn. and *Ichnocarpus frutescens* R. Br. for identification of distinguishing biomarkers. *Pharmacog J* 2010; 2: 7-12.

Mishra MK, Tiwari P, Dash DK, Jadon RS, Ghosh G, Barik BB. Antifungal activity of *Leptadenia reticulata* Wight and Arn. aerial parts. *Int J Phytomed* 2: 172-176.

Nema AK, Agarwal A, Kashaw V. Hepatoprotective activity of *Leptadenia reticulata* stems against carbon tetrachloride-induced hepatotoxicity in rats. *Indian J Pharmacol* 2011; 43:254-257.

Patel RP, Dantwala AS. Antimicrobial activity of *Leptadenia reticulata. Indian J Pharm* 1958;20: 241-244.

Raman N, Sundaram M, Jayaraman P, Brindha P. Pharmacognostic standardization of *Leptadenia reticulata* (Palaikkodi) - a Siddha drug. J. Joseph Thas (Ed.), 2002; 84-85.

39. KALODAYAM

Current representative: *Ceropegia bulbosa*

Ceropegia bulbosa Roxb.

Syn: *Ceropegia acuminata* Roxb.

Family: Asclepiadeae.

Habitat: Pakistan (Punjab); India (Punjab, Uttar Pradesh, W. Peninsula).

Description: Tuber globose. Stem trailing or twining, usually glabrous. Leaves orbicular, ovate, lanceolate to linear, 3-6 cm long, slightly cordate, fleshy, glabrous, subsessile to petiolate. Cymes pedunculate, 3-5 flowered arising between the petioles, peduncle 1.2-2.6 cm long. Pedicels 3-7 mm long. Sepals 2-2.5 mm long, lanceolate, acute. Corolla 1.2-2.5 cm long, whitish green, purple striate, lobes deep purple, 44.5 mm long, narrow, pilose. Outer corona lobes rounded to shortly acute, inner lobes linear, c. 2 mm long. Follicles 10 x 0.4-0.5 cm, taper towards apex, glabrous. Seeds 8-9 mm long, flattened, ovate-oblong, margin broad, membranous, coma 2.5-3.8 cm long.

Chemical constituents: Gum-resin and an alkaloid, ceropegine, present in tubers.

Parts used: Tubers.

40. KAKAJANGHA

Current representative: *Peristrophe bicalyculata*

Possible representative: *Leea aquatica*

Peristrophe bicalyculata Nees

English name: Panicled peristrophe.

Family: Acanthaceae.

Habit: Throughout India, Afghanistan and Africa.

General Description: Annual herb, up to 35–120 cm tall or more; stems erect, sharply 6-angled, sparsely to densely hairy with up to 2 mm long spreading hairs. Leaf-blades narrowly ovate with prominent veins, at least up to 40 x 14 mm, subglabrous, apex acuminate, base attenuate; petiole up to c. 2–12 mm long. Primary inflorescences arranged in lax axillary cymes spread along the stems, the entire plant looking like a lax inflorescence; leaves of the inflorescence gradually smaller towards the apex of the branches, uppermost ones narrowly triangular-subulate with a thin white margin, c. 4–5 x 0.5 mm; bracts markedly unequal, linear-subulate, up to c. 12 mm, enlarging up to c. 15–20 mm in fruit; bracteoles lanceolate-subulate, up to c. 7 mm long, about as long as calyx. Calyx subglabrous; lobes linear-subulate with a thin white margin, up to c. 6–7 x 1 mm. Flowers pink, c. 12–14 mm long; tube c. 5 mm long; lips up to c. 7 mm long. Capsule 8–9 mm long. Seeds with shortly-stalked small papillae concentrated to the edge (Rashmi *et al.*, 2010).

Detailed Description of the Root (Gaudani *et al.*, 2009)

(a) *Macroscopic*

Root occurs upto 0.7 cm thick, and upto 4 cm long cylindrical with branched lateral roots, dirty brown; fracture, fibrous; odour and taste not characteristic.

(b) *Microscopic*

Shows poorly developed cork, consisting of 2-4 layers of tangentially elonated, thin-walled cells; where cork is not developed, epidermis present, consisting of single layered cells; secondary cortex narrow, consisting of 5-7 layers of elliptical or tangentially elongated, thin-walled, parenchymatous cells; secondary phloem narrow, consisting of sieve elements and parenchyma; phloem rays not distinct; secondary xylem consisting of pitted vessels, fibres, tracheids and parenchyma; vessels occur singly or in groups of 2-4 or more and arranged radially throughout secondary xylem; vessels with simple pits, tracheids thick-walled and lignified.

Powder - Dirty-brown; shows parenchymatous cells, aseptate fibres and pitted vessels.

Therapeutics: Snake bite

Chemical constituents: Aliphatics : 14-methyl-tritriacont-14-en-15-ol and 35-hydroxynonatriacontanal (Singh *et al.*, 2000).

Bioactivity: Antibacterial (Chopra & Chopra, 1986), antipyretic and analgesic (Pandal *et al.*, 2009)

Further reading:

Chopra IC, Chopra CL. Antibacterial activity of *Peristrophe bicalyculata*. *Indian J Med Res* 1986; 47:161-163.

Gaudani R, Saluja AK, Shah UD, Prajapati HM. Pharmacognostical and Phytochemical Investigation on Aerial Parts of *Peristrophe bicalyculata*. *Pharmacog J* 2009; 1:19-22.

Pandal S, Choudhury NSK, Patro JV, Parial KS, Jana GK. Antipyretic and Analgesic Activity of the Dried Aerial Parts of *Peristrophe bicalyculata* Retz. (Acanthaceae) in Albino Rats. *J Pharm Res* 2009; 2:123.

Rashmi G, Jaya P, Hardik P, Bhumi M, Shivani A. *Peristrophe bicalyculata* - A Review. *Pharmacog J* 2010; 2:39-46.

***Leea aequata* L.**

Ayurvedic names: *Nadikanta, kakatikta, sulomaja, paravatapadi, kakamada* and *chardikarini.*

Family: Vitiaceae.

Habitat: India.

Description: A shrub, treelet or less frequently a small tree up to 10 m tall, young branches usually densely hairy. Leaves 1-3-pinnate, leaflets 5-numerous, rachis 7-20(-25) cm long, petiole (5-) 8-14 (-20) cm long, stipules oblong-obovate 1.5-4.5 cm x 3-6 (-10) cm, pubescent to densely hairy, leaflets ovate to ovate-lanceolate or elliptical to elliptical-lanceolate, (3-)10-22 (-30) cm x (1.5-) 4-8 (-12) cm, base cuneate to truncate, sometimes unequal, apex acuminate, margin serrate, pearl-gland rounded to discoidal, large, conspicuous and brown. Cyme 4-14 (-20) cm long, rusty pubescent to hairy, bracts up to 8 mm x 5 mm, conspicuous; flowers greenish white, calyx about 4 mm x 4 mm, glabrous to pubescent, usually with pearl-glands, staminodial tube about 2 mm long, upper free part 1.5-2 mm long, lobes deeply notched, sinuses shallow, ovary 4-7-celled. Berry 8-15 mm in diameter, orange-red, 5-6-seeded; seed 4-6 mm across.

Specific action as per Ayurveda: Astringent.

Therapeutics: Liver diseases.

Parts used: Stem and root.

41. KAKANASA

Current representative: *Pentatropis microphylla*

Possible representative: *Leptadenia reticulata* and *Asclepias curassavica*

***Pentatropis microphylla* (Roth ex Schult.) Wight & Arn.**

Syn: *Asclepias microphylla* Roth ex Schult.

Natural order: Asclepidaceae.

Habitat: It is found in Punjab, Delhi, Upper Gangetic Plain, Rajasthan and Gujarat.

Habit: Herb.

Chemical constituents: An acyclic diterpene ester, cis-phy-tyl-l-palmitate, together with the triter-penes, squalene, n-taraxasterol and taraxasterol, has been isolated from the plant.

Therapeutics: In certain parts of India the plant is used for senile diseases.

Parts used: Roots.

Bioactivity: Antimicrobial (Rama Prabha & Vasantha, 2010)

Further reading

Chandrasekaran K, Gowri Sankar NL, Varadharaju S, Jayaraman P, Vasanthy HR, Manavalan R. Pharmacognostical and preliminary phytochemical studies on *Pentatropis capensis* (Bullock) Wight et Arn. J. Joseph Thas (Ed.), 2003; 74 (abstract).

Rama Prabha M, Vasantha K. Antimicrobial activity of *Pentatropis microphylla* L. leaves. *Int J PharmTech Res* 2010; 2:2022-2024.

Asclepias curassavica L.

Common name: Blood flower, Indian root and wild cotton.

Ayurvedic name: *Kakatundi*.

Family: Asclepidaceae.

Habitat: India.

General Description: Typical plants are evergreen perennial subshrubs that grow up to 1 m (3.3 ft) tall and have pale gray stems. The leaves are arranged oppositely on the stems and are lanceolate or oblong-lanceolate shaped ending in acuminate or acute tips. Like other members of the genus, the sap is milky. The flowers are in cymes with 10-20 flowers each. They have purple or red corollas and corona lobes that are yellow or orange. Flowering occurs nearly year round. The 5–10 cm (2.0–3.9 in) long, fusiform shaped fruits are called follicles. The follicles contain tan to brown seeds that are ovate in shape and 6–7 mm (0.24–0.28 in) long. The flat seeds have silky hairs that allow the seeds to float on air currents when the pod-like follicles dehisce (Mitra *et al.,* 1974).

Chemical constituents: Glucoside:asclepin (Patnaik & Dhawan, 1971).

Action: Anthelmintic and purgative.

Therapeutics: Worm infestation, hemorrhoids, gonorrhea and warts.

Parts used: Roots.

Bioactivity: Abortifacient and cardio tonic (Singh *et al.,* 1977; Patnaik. and Kohler, 1978).

Further reading

Mitra R, Mehrotra S, Mehrotra BN, Kapoor LD. Pharmacognostical study of *Asclepias curassavica* Linn. *Bull Bot Sur India* 1974; 16: 82-88.

Patnaik GK, Dhawan BN. Asclepin: a new cardiactive glycoside from *Asclepias curassavica. Indian J Pharmacol* 1971; 3: 8-9.

Patnaik KG, E. Kohler E. Pharmacological investigation on asclepin - a new cardenolide from *Asclepias curassavica* - part II - Comparative studies on the inotropic and toxic effects of asclepin, g-strophantin, digoxin and digitoxin. *Arzneimittelforschung* 1978; 28: 1368-1372.

Singh N, Gupta ML, Mishra N, Kulshrestha VK, Kohli RP. Elucidation of mechanism of the cardiovascular responses to *Asclepias curassavica* (kakatundi). *J Res Indian Med* 1977; 12: 30-36.

42. KANDIRA

Current representative: *Ranunculus scleretus*

Ranunculus sceleratus L.

Syn: *Hecatonia scelerata* (Linnaeus) Fourreau

Family: Ranunculaceae.

English name: Celery-leaved crowfoot.

Habitat: Western Himalayas.

General Description: Stems erect, glabrous, rooting at base, only very rarely rooting at proximal nodes. Leaves basal and cauline, basal and proximal cauline leaf blades reniform to semicircular in outline, 3-lobed or -parted, 1-5 × 1.6-6.8 cm, base truncate to cordate, segments usually again lobed or parted, sometimes undivided, margins crenate or crenate-lobulate, apex rounded or occasionally obtuse. Flowers: receptacle pubescent or glabrous; sepals 3-5, reflexed from or near base, 2-5 × 1-3 mm, glabrous or sparsely hirsute; petals 3-5, 2-5 × 1-3 mm; nectary on petal surface, scale poorly developed and forming crescent-shaped or circular ridge surrounding but not covering nectary; style absent. Heads of achenes ellipsoid or cylindric heads, 5-13 × 3-7 mm; achenes 1-1.2 × 0.8-1 mm, glabrous; beak deltate, usually straight, 0.1 mm (Pandey, 1966b).

Phytochemistry: Glycoside (protoanemonine), anemonic acid and serotonin.

Therapeutics: Goiter, rheumatism and sciatica.

Parts used: Whole plant.

Bioactivity: Galacatagouge (Anjaria *et al.,* 1975) and antifungal (Mishra & Dixit, 1978; Mishra & Dixit, 1980).

Further reading

Mishra SB, Dixit SN. Antifungal principle of *Ranunculus sceleratus. Econ Bot* 1980;34, 362-367.

Misra SB, Dixit SN. Antifungal properties of leaf extract of *Ranunculus sceleratus* L. *Experientia* 1978;34:1442-1443.

Pandey G. Contribution to the botanical identity and medical efficacy of Ayurvedic drug kandira with special reference to Himachal Pradesh. *Sach Ayur* 1996b;49:216-224.

43. KARAVI

Current representative: *Strobilanthus callosus*

Possible representative: *Carum carvi*

Strobilanthes callosus Nees

Syn: *Carvia callosa* Bremek.

Family: Acanthaceae.

Habitat: It is found mainly in the low hills of the Western Ghats all along the west coast of India.

General Description: A large shrub, sometimes attaining 6-20 ft. in height and 2½ inches in diameter; branches often warted or scabrous-tubercled. Leaves opposite, 7 by 3 in., sometimes much larger, crenate, rough, conspicuously marked with five lines above, nerves 8-16 pair; petiole 2-3 in. Flowers in strobiliform spikes 1-4 in. long, often densely or laxly cymose; bracts 1/2 - 1 in. long, orbicular or elliptic. Calyx 1/2 in., in fruit often exceeding 3/4 in., sub-equally 5-lobed to the base; segments oblong, obtuse, softly hairy. Corolla tubular-ventricose, 1½ in., glabrous without, very hairy within, deep blue; lobes 5, nearly equal, contorted in bud. Stamens 4; filaments hairy downwards; anthers blunt; not spurred at the base. Ovary 4-ovuled; style linear; stigma of one long linear-lanceolate branch the other minute. Capsule 3/4 by 1/3 in., seeds more than 1/3 in. long, thin, obovate, acute, densely shaggy with white inelastic adpressed hair, except on the large oblong areoles.

Chemical constituents: Lupeol and 19 α-H lupeol (Agarwal and Rangari, 2003).

Bioactivity: Antimicrobial (Agarwal and Rangari, 2003), anti-inflammatory (Agarwal and Rangari, 2003) and antiarthritic (Singh, Sahu and Sharma, 2002).

Further readings

Agarwal RB, Rangari VD. Anti-inflammatory and anti-arthritic activities of lupeol and 19α-H lupeol isolated from *Strobilanthus callosus* and *Strobilanthus ixiocephala* roots. *Indian J Pharmacol* 2003; 35:384-387.

Singh B, Sahu PM, Sharma MK. Anti-inflammatory and antimicrobial activities of triterpenoids from *Strobilanthes callosus* Nees. *Phytomedicine* 2002; 9: 355-359.

Behl PN, Captain RM. Skin-Irritant and Sensitizing Plants Found in India, 2nd edn. New Delhi: S Chand, 1979.

Carum carvi L.

English name: Black cumin.

Ayurvedic name: *Krsna jiraka, shyama jiraka, jiraka, jarana, krsna and varsakala-sugandhika.*

Family: Apiaceae.

Habitat: Native to western Asia, Europe and Northern Africa.

General Description: *C. carvi* is a biennial, with smooth, furrowed stems growing 1 1/2 to 2 feet high, hearing finely cut leaves, and umbels of white flowers which blossom in June. The fruitswhich are popularly and incorrectly called seeds - and which correspond in general character to those of the other plants of this large family, are laterally compressed, somewhat horny and translucent, slightly curved, and marked with five distinct, pale ridges. They evolve a pleasant, aromatic odour when bruised, and have an agreeable taste.

Description of the Dried Fruit (Atal & Dutta, 1966)

(a) *Macroscopic*

Fruit, greenish-brown, slightly curved, elongated, mericarps, usually separate, free from the pedicel, carpophores, upto 7 mm long, 2 mm broad almost equally five sided, narrow, tapering to each end, arcuate, glabrous, brown with five very narrow, yellowish primary ridges' endosperm, orthospermous, odour and taste, aromatic and characteristic.

(b) *Microscopic*

Transverse section of fruit shows pericarp with outer epidermis of polygonal tabular cells with a thick outer wall and striated cuticle, trichomes, absent, vittae four dorsal, intercostal and two commissural extending the length of each mericarp, with an epithelium of brown cells and volatile oil in the cavity, mesocarp parenchymatous without reticulate thickening, costae five in each mericarp with vascular strand consisting of an inner group of small vessels and fibres and arched, outer group of pitted sclerenchyma with a small group of phloem on each lateral surface, on the outer margin of each vascular strand a small schizogenous canal extending into both stylopod and pedicel, inner epidermis of thin -walled, subrectangular cells, elongated tangentially each about 8-12 μ wide and 40-100 μ long, arranged parallel with one another, endosperm of thick-walled, cellulosic parenchyma, containing much fixed oil and numerous small aleurone grains upto 10 μ in diameter, each containing one or sometimes two micro-rosette crystals of calcium oxalate, carpophore, when present, passing at the apex to a raphe in each mericarp, and with a small strand of sclerenchyma, the sclereids of which continue into the stylopod.

Powder-Colour fawn to brown, epidermal cells of pericarp with striated cuticle, fragments of brown endothelium of vittae, parenchymatous cells of the mesocarp without reticulate thickening, rectangular, finely pitted sclereids of mesocarp, thickwalled polygonal parenchymatous cells of endosperm containing much fixed oil, numerous small aleurone grains containing micro-rosette crystals of calcium oxalate, trichomes, starch and parquetry layer absent, it contains no less than 2.5 per cent of volatile oil.

Phytochemistry: Volatile oil containing carvene (Atal & Sood, 1967; Patel *et al.,* 1981).

Actions: Carminative and diuretic.

Therapeutics: Loss of appetite.

Parts used: Fruit.

Bioactivity: Antioxidant (Kamaleeswari & Nalini, 2006) and blood-glucose reducing Sushruta *et al.,* 2006).

Further reading

Atal CK, Dutta AK. Pharmacognosy of caraway and its adulterants. *Indian J Pharm* 1966; 28: 80.

Atal CK, Sood NM. Study of Indian caraway and its substitutes, part I: Essential oil of *Carum carvi* Linn. *Indian J Pharm* 1967; 29: 42-44.

Kamaleeswari M, Nalini N. Dose-response efficacy of caraway (*Carum carvi* L.) on tissue lipid peroxidation and antioxidant profile in rat colon carcinogenesis. *J Pharm Pharmacol* 2006; 58: 1121-1130.

Patel KN, Chauhan MG, Bhavsar GC, Shah CS. Studies on commercial caraway. *Indian J Pharm Sci* 1981; 43: 68 (abstract of a paper presented at the XXXII Indian Pharmaceutical Congress, held at Nagpur, 23-25 January 1981).

Sushruta K, Satyanarayana S, Srinivas N, Raja Sekhar J. Evaluation of the blood-glucose reducing effects of aqueous extracts of the selected umbelliferous fruits used in culinary practices. *Tropical J Pharm Res* 2006; 5:613-617.

44. KSHIRA KAKOLI

Current represntative: *Fritillaria royleii*

Possible representative: *Roscoea procera*

Fritillaria roylei Hook.

Syn: *Fritillaria gulielmi-waldemarii* Klotzsch.

Common name: Himalayan fritillary.

Ayurvedic names: *Kshira shukla, paysivani, payasya, ksiram-adhura, ksirvishanika, jivavalli* and *jivshukla.*

Family: Liliaceae.

Habitat: E. Asia - Himalayas.

General Description: Glabrous bulbous herbs; leaves liner-lanceolate; 3-6 in a whorl; flower yellow-green solitary or 2-4 in racemes. Bulb growing to 0.6m. The flowers are hermaphrodite.

Actions: Cooling and spermopiotic.

Therapeutics: Phthisis, pain, blood diseases, burning sensation and fever.

Phytochemistry: Steroidal alkaloids: kashmirine (Chatterjee *et al.,* 1963) and sterol.

Parts used: Corm.

Further reading

Agarwal VS. Economic Plants of India, Kailash Prakashan, Calcutta, 1986;419 pp.

Chatterjee A *et al.*, Kashmirine, a new steroidal alkaloid from *Fritillaria royeli* Hook (Liliaceae). *Tetra Lett* 1963; 33: 2903-2904.

Uniyal MR. Astavarga? Sandigdha Vanaushadhi. *Dhanwantri Partrika.* Sri Jwala Ayurevd Bhawan Aligarh, 1975.

Uniyal MR, Issar RK. Botanical studies of Astavarga. *Nagarjun* 1966; 9:276-284.

Roscoea procera Wall. formerly *Roscoea purpurea* Smith

Synonyms: *Dhawankshśolika, Karnika, Kshira, Madhura, Shukla, Svadumamsi, Vayasoli* and *Vayshashśa.*

English name: Alpine ginger.

Family: Zingiberaceae.

Habitat: Western Himalayas between 1000 m and 2700 m.

Description: *R. purpurea* is a much larger and lusher plant. Stem: 6-20 cm and 30-60 cm, robust. Leaves: 2-6 lanceolate, 15x3 cm. Flowers: Several liliac, purple or white, corolla tube much longer than the calyx, borne in a short stem with or without leaves, upper lobe rounded, hooded and narrower reflexed lateral petals. Lower staminode 5 x 3 cm obscurely 3 lobed spreading, notched at the hip (Tiwari, Yadava & Tiwari, 1975; Tewari, Tewari & Yadava, 1976).

Phytochemistry: No information.

Actions: Cooling and spermopiotic.

Therapeutics: Fever, burning sensation and phthisis.

Part used: Bulb.

Bioactivity: Immunomodulatory (Sahu *et al.*, 2010)

Further reading

Sahu, MS Mali PY, Waikar SB, Rangari VD. Evaluation of immunomodul.atory potential of ethanolic extract of *Roscoea procera rhizomes* in mice. *J Pharm Bioallied Sci* 2010; 2: 346–349.

Tewari VP, Tewari KC, Yadava BBL. Study of *Roscoea procera* Wall. (kakoli, kseerkakoli of astavarga). *Quarterly J Crude Drug Res* 1976; 14: 69-70.

Tiwari KC, Yadava BBL, Tiwari VP. Study on *Roscoea procera* Linn. *J Res Indian Med* 1975; 10: 115-116.

45. KUSHTA

Saussurea lappa (Decne.) C.B.Clarke.

Syn: *Saussurea costus* (Falc.) Lipsch.

Common name: Costus.

Ayurvedic names: *Rogahvaya, vapya, kaubera, pariharya, paribhavya, utpala* and *haribhadraka.*

Family: Asteraceae.

Habitat: E. Asia - Himalayas.

General Description: An erect perennial herb grows up to 1-2 meters in height. Leaves membranous, Basel leaves large, upper ones are small. Leaves irregularly tooted. Flowers bluish purple in axillary or terminal heads. Fruits compressed achenes. The roots are long, stout and with a peculiar odor.

Detailed Description of the Dried Roots (Upadhyay *et al.,* 1993)

(a) *Macroscopic*

Drug greyish to dull brown, thick, stout, fusiform to cylindrical, 7-15 cm long, 1.0-5.5 cm broad, thicker roots with collapsed centre, occasionally ridged, wrinkles longitudinal and anastomosed, rootlets rarely present, cut surface shows two regions, outer periderm ring thin, inner porous woody portion lighter in colour showing fine radial striations and often the central portion collapsed, fracture, short, horny, odour, strong, characteristically aromatic, taste, slightly bitter.

(b) *Microscopic*

Transverse section of thin root shows thin periderm, followed by broad zone of phloem and still broader zone of xylem traversed by wide medullary rays, cork, 3-5 layered wide secondary cortical cells polygonal, mostly elongated, secondary phloem consists of mostly storage parenchyma, small groups of sieve tubes and companion cells and often phloem fibres, bast fibres thick-walled, lignified, upto 350 μ in length, with many simple pits associated with fibre, tracheids and parenchyma, wood fibres smaller than bast fibres, with wider lumen and obtusely tapering ends, meduallary rays multi seriate and wider in phloem region, resin canals found throughout as large cavities, some roots possess a central cylinder of sclerenchyma, while others have parenchymatous centre with scattered xylem elements, in older roots, wood parenchyma collapses and takes a spongy appearance in the centre of root, inulin present in storage parenchyma.

Powder-Deep brown or rusty, under microscope irregular bits of yellow, brown or orange-red fragments of resins and oils associated with thin-walled parenchymatous cells, broken bits of xylem vessels with scalariform, reticulate thickening and horizontal end walls.

Chemical constituents: Alkaloid: saussurine (Dutta, Sastry & Tamhane, 1960), inulin, resin and aromatic oil and kushtin.

Specific action as per Ayurveda: *Sukrashodhaka* (semen purifier).

Therapeutics: Asthma, skin diseases, dysuria and rheumatism.

Parts used: Roots.

Bioactivity: Antiprotozoal, antibacterial, positive iontropic and chronotropic (Dutta, 1960), muscle relaxant activity of saussurine (Chopra, 1982), hypolipidaemic (Upadhya, Ojha & Bajpai, 1993), immunostimulant (Kulkarni and Deasi, 2001), anti-inflammatory and antiproliferative (Damre, Damre and Saraf, 2003), anxiolytic (Ambavade *et al.*, 2006) and antimicrobial (Pandey *et al.*, 2008).

Clinical Studies: Ischaemic heart disease (Upadhyay *et al.*, 1993).

Further readings

Ambavade SD, Mhetre NA, Patil KM, Desai T, Bodhankar SL Anxiolytic activity of root extracts of *Saussurea lappa* in mice. *J Nat Rem* 2006; 6:103-108.

Damre AA, Damre AS, Saraf MN. Evaluation of sesquiterpene lactone fraction of *Saussurea lappa* on transudative, exudative and proliferative phases of inflammation. *Phytother Res* 2003; 17: 722-725.

Dutta NK, Sastry MS, Tamhane RG. Pharmacological actions of an alkaloidal fraction isolated from *Saussurea lappa* (Clarke). *Indian J Pharm* 1960;22: 6-7.

Ghosh S, Chatterjee NR, Dutta A. Chemical examination of the plant and its medicinal uses, with chemical analysis of roots and leaves of *Saussurea lappa* Clarke, part I. *J Indian Chem Soc* 1929;6: 517-522.

Gupta OP, Ghatak BJ. Pharmacological investigations on *Saussurea lappa*. (Clarke). *Indian J Med Res* 1967; 55:1078-1083.

Kulkarni S, Deasi S. Immunostimulant activity of inulin isolated from *Saussurea lappa* roots. *Indian J Pharm Sci* 2001; 63:292-294.

Pandey MM, Singh M, Rastogi S, Rawat AKS. Antimicrobial activity of methanolic extract and oil of *Saussurea costus* roots. *Nigerian J Nat Prod Med* 2008;12.

Upadhya OP, Ojha JK, Bajpai HS. Experimental study on hypolipidaemic activity of kushtha (*Saussurea lappa* C.B. Clarke). *J Res Ayur Siddha* 1993;15: 52-63.

Upadhyay O, Ojha JK, Bajpai HS, Hathwal AK. Study of kustha (*Saussurea lappa* Clarke) in ischaemic heart disease. *Ancient Sci Life* 1993; 13: 11-18.

Upadhyay O, Ojha JK, Datta SK. Pharmacognostic study of the root of *Saussurea lappa* C.B. Clarke. *Sach Ayur* 1993;45: 608-612.

46. LANGALI

Current representative: *Gloriosa superba*

Gloriosa superba L.

English names: Malabar glory lily, flame lily.

Ayurvedic names: *Agnimukhi, agnisikha, garbhapatani, halini* and *indra puspi*. .

Family: Liliaceae.

Habitat: South India.

General Description: Erect perennial, tuberous, scandent or climbing herbs; grasp with tendrils formed at the tip of the leaves. Stem leafy. Leaves sessile, spirally arranged or sub-opposite, 6-7 x 1.5-1.8 cm, lanceolate, acuminate, entire, glabrous; the upper ones with cirrhose tips. Flowers axillary, solitary, large, borne on long, spreading pedicels, actinomorphic, hermaphrodite; perianth segments 6, free, lanceolate, keeled within at base, long-persistent, yellow in lower half, red in upper half; stamens 6, spreading, hypogynous; anthers extrose, medifixed, versatile, opening by longitudinal slits; ovary superior, 3-celled; ovules numerous; style deflected at base, projecting from the flower more or less horizontally. Capsule 2-3 cm long, oblong. Seeds numerous, subglobose, black. The fruit is oblong containing about 20 globose red colored seeds in each valve (Prasad, Khosa & Durve, 1969).

Description of the Root

(a) *Macroscopic*

Tuberous roots thick, almost cylindrical or slightly laterally flattened, occurring in pieces of 15-30 cm long and 2.5 - 3.8 cm thick, often bifurcated with tapering ends, resembling a plough-share, one arm generally more than double the length of the other; brownish externally and yellowish internally; fracture, short; taste, acrid and bitter.

(b) *Microscopic*

Tuberous root shows single layered epidermis, externally cuticularised, consisting of rectangular cells, followed by ground parenchyma, with scattered small vascular bundles; parenchyma cells large, thin-walled, polygonal to circular, having conspicuous intercellular spaces, most of the cells specially of the outer layers filled with starch grains, simple, round to oblong, or polyhedral, measuring 8-33 μ in dia., showing clear hilum and concentric striations, occasionally compound with 2-3 components, measuring 24-36 μ in dia.; vascular bundles collateral, numerous, scattered throughout ground tissue, consisting of xylem and phloem; each vascular bundle enclosed by sclerenchymatous sheath, xylem composed of vessels, tracheids and parenchyma; vessels having mostly reticulate thickening, smaller ones having spiral thickening, tracheids with reticulate thickening; xylem parenchyma cells usually rectangular; phloem consisting of sieve tubes, companion cells and phloem parenchyma; phloem parenchyma cells very small and thin-walled.

Powder - Brown; shows fragments of parenchyma cells, simple starch grains, round to oblong or polyhedral measuring 8-33 μ dia. showing clear hilum and concentric striations, occasionally compound with 2-3 components, measuring 24-36 μ in dia., sclerenchymatous cells, a few xylem vessels and tracheids.

Chemical constituents: Alkaloids:colchicine and superbine (Srivastava & Chandra, 1975; Joshi, Sanmuga Priya & Mathela, 2010) and volatile oil.

Specific action as per *Ayurveda*: *Artavajanana* (emmenagouge).

Therapeutics: Amenorrhea and dysmenorrhea.

Parts used: Tubers.

Bioactivity: Uterine-stimulant (Tewari *et al.*, 1967; Tewari *et al.*, 1972) and anti-spermatogenic (Dixit, Joshi & Kumar, 1983).

Further readings

Dixit VP, Joshi S, Kumar A. Possible anti-spermatogenic activity of *Gloriosa superba* (EtoH extract) in male gerbil (*Meriones hurrianae* Jerdon) - A preliminary study. *Comp Physiol Ecol* 1983; 8: 117-122.

Joshi CS, Sanmuga Priya E, Mathela CS. Isolation and anti-inflammatory activity of colchicinoids from *Gloriosa superba* seeds. *Pharm Biol* 2010;48: 206-209.

Prasad S, Khosa RL, Durve RS. Pharmacognostical studies on kalihari, *Gloriosa superba*, and its adulterant *Costus speciosus* Linn. (Kustha). *J Res Indian Med* 1969;3:162-174.

Srivastava UC, Chandra V. *Gloriosa superba* Linn. (kalihari) - an important colchicine producing plant. *J Res Indian Med* 1975; 10: 92-95.

Tewari P, Prasad DN, Chaturvedi C, Das PK. Preliminary studies on uterine activity of *Gloriosa superba* Linn., and its adulterant *Costus speciosus* Sm. *J Res Indian Med* 1967; 1:196-202.

Tewari PV, Sharma PV, Prasad DN. Experimental studies on the ecbolic properties of *Gloriosa superba* Linn. (kalihari). *J Res Indian Med* 1972;7: 27-38.

47. LINGINI

Diplocyclos palmatus (L.) C. Jeffrey

Family: Cucurbitaceae

Common name: Red-stripped cucumber.

Ayurvedic names: *Bahupatra, ishvari, shevmallika, svayambhu, lingi, citraphala, amrita, pandoli, lingaja, devi, chanda, apstimbhini, shivaja* and *shivavalli.*

Description: *D. palmatus* is a weak stemmed, branched tendril climber. Leaves simple, alternate, 5 lobed, hairy above, pale and smooth beneath; flowers yellow, small, unisexual, male in small fascicles and females solitary. Fruits globose, smooth berry, become bright red when ripe, contain 1-2 small seeds.

Actions and therapeutics: Pungent in taste, hot in potency, foul-smelling, tonic, good for holistic practice and regulation of *parada* (mercury).

48. MATSYAKSI

Current representative: *Alternanthera sessilis*

Possible representative: *Bacopa monniera* and *Centella asiatica*

Alternanthera sessilis R. Br. ex DC.

English name: Sessile joy wood.

Ayurvedic name: *Marisha.*

Family: Amaranthaceae.

Habitat: India.

Description: *A. sessilis* is a perennial herb with prostrate stems, rarely ascending, often rooting at the nodes. Leaves obovate to broadly elliptic, occasionally linear-lanceolate, 1-15 cm long, 0.3-3 cm wide, glabrous to sparsely villous, petioles 1-5 mm long. Flowers in sessile spikes, bract and bracteoles shiny white, 0.7-1.5 mm long, glabrous; sepals equal, 2.5-3 mm long, outer ones 1-nerved or indistinctly 3-nerved toward base; stamens 5, 2 sterile (Sinha *et al.,* 1984; Sinha *et al.,* 1986).

Phytochemistry: Lupeol, β-carotene and Vitamin C.

Actions: Galactagogue, cholagogue, abortifacient and febrifuge.

Therapeutics: Fevers and as antidote to snake poisoning.

Further reading

Sinha P, Arora VK, Ansari MS, Wahi SP. Pharmacognostical studies of the leaf of *Alternanthera sessilis* R.Br. *J Sci Res Plant Med* 1984;5: 41-47.

Sinha P, Arora VK, Ansari MS, Wahi SP. Pharmacognostic studies on the stem of Alternanthera sessilis R.Br. *Int J Crude Drug Res* 1986;24:7-15.

Gupta M, Mazumder UK, Basu SP, Ghosh SK. Effects of extract from *Alternanthera sessilis on* haemopoeietic system and on liver, kidney functions and metabolism. *Indian Drugs* 1985;32: 498-501.

Jalalpure SS, Agrawal N, Patil MB, Chimkode R, Tripathi A. Antimicrobial and wound healing activities of leaves of *Alternanthera sessilis* Linn. *Int J Green Pharm* 2008;2: 3.

See notes on Bacopa monniera and Centella asiatica.

49. MAYURASHIKHA

Current representative: *Adiantum caudatum*

Possible representative: *Celosia argentea* and *Actinopteris dichotoma*

Adiantum caudatum L.

Ayurvedic names: *Mayurahva, sikha, sahasri* and *madhukacchada.*

English name: Tailed maidenhair.

Family: Adiantaceae.

Habitat: Africa to Pacific.

Descrption: Terrestrial fern. Rhizomes short-erect; scales lanceolate, bicolored, centers dark reddish, margins brown, entire. Leaves/Fronds Stipes dark reddish brown, polished but obscured by dense septate hairs. Blades pinnate; pinnae sessile, not articulate to the stalk as in A. tenerum, hairy, deeply incised into narrow segments; lower pinnae reduced; rachis frequently proliferated (Mukherjee, Khosa & Wahi, 1981).

Chemical constituents: Rutin and ß-sitosterol (Khosa, Wahi & Mukherjee, 1978).

Specific action as per Ayurveda: Demulcent.

Therapeutics: Chronic bronchitis, diarrhoea, spleen diseases, hemorrhoids, liver ailments and worm infestation.

Parts used: Fronds.

Bioactivity: Antimicrobial (Singh *et al.,* 2008).

Further reading

Mukherjee AK, Khosa RL, Wahi AK. Pharmacognostical studies on *Adiantum caudatum* Linn. *Bull Medico-Ethno-Bot Res* 1981;1:428-433.

Khosa RL, Wahi AK, Mukherjee AK. Chemical studies on *Adiantum caudatum. Curr Sci* 1978;47: 624.

Singh M, Singh N, Khare PB, Rawat AK. Antimicrobial activity of some important Adiantum species used traditionally in indigenous systems of medicine. *J Ethnopharmacol* 2008; 115:327-329.

Actinopteris dichotoma Kuhn.

Syn: *Actiniopteris radiata* Link., *Actinopteris australis* Link.

English name: Peacock's tail.

Family: Adiantaceae.

Habitat: Throughout the rocky hillsides of Western Ghats.

Descrption: A small fern grows up to 20 cm high. Fronds fan-like with numerous dichotomous segments which are rush-like in texture, having few veins, sub-parallel, segments of fertile frond longer than those of the barren one, sori linear, sub marginal and elongate.

Chemical constituents: Rutin and ß-sitosterol.

Action: Haemostatic.

Therapeutics: *A. dichotoma* pacifies vitiated *kapha, pitta,* diarrhea, dysentery, worm infestations, skin discoloration, skin diseases, diabetes and fever.

Parts used: Whole plant.

Bioactivity: Anti-fertility (Sharma *et al.,* 1999).

Further reading:

Sharma A, Mathur A, Verma P, Joshi SC, Dixit VP. Effects of *Actiniopteris dichotoma* on reproductive function of male rat. *J Endocrinol Reprod* 1999; 2:75-79.

Note: See notes on *Celosia argentea*

50. MARKANDI

Current representative: *Cassia angustifolia*

Possible representative: *Cassia auriculata*

Cassia auriculata L.

Syn: *Cassia densistipulata* Taubert, *Senna auriculata* (L.)Roxb.

Common name: Tanner's cassia/ Mature tea tree.

Family: Fabaceae.

Habitat: Mainly in central part of India, growing in dry stony hills.

Habit: *C. auriculata* is a large shrub growing up to 2 -3 meters in height. Leaves paripinnately compound, leaflets 10-12 pairs, ovate-lanceolate, large stipule at the base of each compound leaf. Flowers yellowish, in axillary and terminal corymbs. Fruits pods, brown when ripe contain 10-15 dark brown seeds.

Chemical constituents: Anthraquinones.

Actions: Astringent, tonic and alterative.

Therapeutics: Diabetes, conjunctivitis, sore throat and menstruation.

Parts used: bark, root, seed and leaf.

Bioactivity: Lipid-lowering (Kumar *et al.,* 2002), antihyperlipidaemic and antidiabetic (Pari and Latha, 2002), renoprotective (Annie, Rajagopal & Malini, 2005) and hepatoprotective (Rajagopalan *et al.,* 2003).

Further reading:

Annie S, Rajagopal PL, Malini S.Effect of *Cassia auriculata* Linn. root extract on cisplatin and gentamicin-induced renal injury. *Phytomedicine* 2005; 12:555-60.

Kumar Rajagopal S, Manickam P, Periyasamy V, Namasivayam N. Activity of *Cassia auriculata* leaf extract in rats with alcoholic liver injury. *J Nutr Biochem.* 2003; 14:452-8.

Pari L. and Lata, M. Effect of *Cassia auriculata* flowers on blood sugar levels, serum and tissue lipids in streptozotocin diabetic rats. *Singapore Med J* 2002;43:617-621.

Rajagopal SK, Ponmozhi M, Viswanathan P, Namasivayam N. Effect of *Cassia auriculata* leaf extract on lipids in rats with alcoholic liver injury *Asia Pacific J Clin Nut* 2002;11:157-163.

51. MAHAMUNDI

***Sphaeranthus africans* Linn.**

Ayurvedic names: *Hapusha, Mundi, Sravani, Alambusha, kadambapushpi.*

Family: Asteraceae.

Habitat: Tropical Africa and Asia through Malaya to Australia.

Description: The plant is more or less hairy or nearly smooth, rather coarse, erect or spreading, branched are prominently winged with three thin, wide, longitudinal structures, which are the extension of the leaf blades. The leaves are obovate to oblong-obovate, 4 to 13 centimeters long, without stalks, and finely toothed at the margins. The heads are very numerous, borne in dense, rounded clusters about 1 centimeters in diameter, and occur singly at the ends of erect, winged stalks. The flowers are greenish-white.

Specific action as per Ayurveda: *Medya.*

Therapeutics: *S. africans* pacifies vitiated vata, pitta epilepsy, migraine, jaundice, fever, cough, hemorrhoids, helminthiasis, skin diseases and as nervine tonic.

Further Reading:

Subramanian PS, Chelladurai V, Selva Gomathy M, Kavitha P. Rare Siddha herbs identified. J. Joseph Thas (Ed.), 2003;65 (abstract)

52. MESHASRINGI

Current representative: *Gymnema sylvestre*

Possible representative: *Dolichandrone falcata*

Gymnema sylvestre R.Br.

Ayurvedic names: *Madhunashini.*

Family: Asclepiadaceae.

Habitat: Native to the tropical forests of southern and central India.

Description: *G. sylvestre* is a large climber, rooting at nodes, leaves elliptic, acuminate, base acute to acuminate, glabrous above sparsely or densely tomentose beneath; Flowers small, in axillary and lateral umbel like cymes, pedicels long; Calyx-lobes long, ovate, obtuse, pubescent; Corolla pale yellow campanulate, valvate, corona single, with 5 fleshy scales. Scales adnate to throat of corolla tube between lobes; Anther connective produced into a membranous tip, pollinia 2, erect, carpels 2,unilocular; locules many ovuled; Follicle long, fusiform (Hooker, 1866; Girija, Agarwal &. Nair, 2005).

Chemical constituents: Dried leaves: Resin, pararabin, triterpene glycoside :gymnemic acid (Pandey, & Yadav, 2010; Srivastava *et al.,* 2010), peptide (gurmarin), alkaloid (gymnamine), bitter principle (having sialagouge activity), lupeol, quercitol, coloring matter and anthraquinones. Bark: Calcium and starch. Alcoholic extract contains saponin. Ash contains alkali, phosphoric acid and manganese.

Specific action as per Ayurveda: Antidiabetic.

Therapeutics: *G. sylvestre* is used in the treatment of diabetes mellitus type-2. The emetic action of *G. sylvestre* is similar to ipecac. Dash (1986) reported the use of *Gymnema sylvestre* in leucoderma.

Parts used: Leaves.

Bioactivity: Insulinotropic (Shanmugasundaram *et al.,* 1981), hypoglycemic (Gupta, 1961; Shanmugasundaram *et al.,* 1981; Khare, Tondon & Tewari, 1983; Shanmugasundaram *et al.,* 1983; Srivastava *et al.,* 1985; Anand *et al.,* 1986), hypolipidaemic and antiatherosclerotic (Bishayee & Chatterjee, 1994), and anti-inflammatory (Diwan, Margaret & Ramakrishna, 1995).

Further reading

Anand P, Mathur S, Mathur R. Effect of feeding *Gymnema sylvestre* leaves on blood glucose in beryllium nitrate treated rats. *J Ethnopharmacol* 1986; 18: 143-146.

Bishayee A, Chatterjee M. Hypolipidaemic and antiatherosclerotic effects of oral *Gymnema sylvestre* R. Br. Leaf extract in albino rats fed on a high fat diet. *Phytother Res* 1994; 8: 118–120.

Diwan PV, Margaret I, Ramakrishna S. Pharmacological and biochemical evaluation of *Gymnema sylvestre*: an anti-inflammatory agent. *Indian J Pharmacol* 1995;27: 50 (abstract no. 25).

Girija PV, Agarwal A, Nair GA. Physico-chemical standardization of *Gymnema sylvestre* R.Br., an Ayurvedic drug. *Bull Medico-Ethno-Bot Res* 2005;26: 1-5.

Gupta SS. Inhibitory effect of *Gymnema sylvestre* (gurmar) on adrenaline-induced hyperglycemia in rats. *Indian J Med Sci* 1961;15:883-887.

Hooker WJ. An examination of the leaves of *Gymnema sylvestre. J Soc Chem India* 1866;6: 380.

Khare AK, Tondon RN, Tewari JP. Hypoglycaemic activity of an indigenous drug (*Gymnema sylvestre*, gurmar) in normal and diabetic persons. *Indian J Physiol Pharmacol* 1983; 27: 257-258.

Pandey AK, Yadav S. Variation in gymnemic acid content and non-destructive harvesting of *Gymnema sylvestre* (Gudmar). *Pharmacog Res* 2010;2: 309-312.

Shanmugasundaram ER, Paneerselvam C, Samudram P, Shanmugasundaram KR. Enzyme changes and glucose utilization in diabetic rabbits: the effect of *Gymnema sylvestre* R.Br. *J Ethnopharmacol* 1983;7:205-234.

Shanmugasundaram KR, Panneerselvam C, Lalitha T, Ranibai AJ. Studies on the hypoglycemic action of *Gymnema sylvestre* R.Br. in diabetes mellitus. Arogya - *J Health Sci* 1981;7:38-60.

Shanmugasundaram KR, Panneerselvam C, Samudram P, Shanmugasundaram ER. The insulinotropic activity of *Gymnema sylvestre* R.Br., an Indian herb. *Pharmacol Res Commun* 1981;13: 475-486.

Srivastava A, Patel PR, Saravanan R, Manivel P. Gymnemic acid content in Gymnema sylvestre accessions collected from different regions of India. *Open Access J Med Arom Plants* 2010;1: 2.

Srivastava Y, Venkatakrishna-Bhatt H, Prem AS, Nigam SK, Verma Y. Hypoglycemic and life prolonging properties of *Gymnema sylvestre* R.Br. lead extract in diabetic rats. *Israel J Med Sci* 1985;21:540-542.

Dolichandrone falcata (Wall. ex DC.) Seem.

Family: Bignoniaceae.

Habitat: Endemic to Peninsular India.

Description: *D. falcata* is a 6-10 mts tall deciduous trees, bark rough, brown, branchlets pubescent; Leaves 5-7 foliate; leaflets 2 x 2.5 cm elliptic, obovate, base acute, pubescent; Flowers 3-4 cm across, white in 4-5 cm long terminal racemes.Capsules 20-30 cm long, compressed, falcately curved, with winged seeds.

Chemical constituents: Phenolic triglycoside: dolichandroside A.

Bioactivity: Glucosidase inhibitor and antioxidant (Apama *et al.*, 2009) and anxiolytic (Saifuddin *et al.*, 2010).

Further reading

Apama P, Tiwari AK, Srinivas PV, Ali AZ, Anuradha V, Madhusudana JR. Dolichandroside A, a new glucosidase inhibitor and DPPH free-radical scavenger from *Dolichandrone falcata* Seem. *Phytother Res* 2009;23, 4:591-596.

Saifuddin K, Dattatraya K, Suresh DK, Shaikh KB, CH Gopalakrishna, Loya PJ. An anxiolytic effect of *Dolichandrone Falcata* Leaves extract In experimental animals. *Res J Pharm Biol Chem Sci* 2010; 1:524-534.

53. MOHANAVALLI

Current representative: *Saxifraga ligulata*

Possible representative: *Acacia catechu*

Note: In our view both the plants taken for *Mohanvalli* seems to be incorrect. *Mohanavalli* as name suggests seems to be a creeper or twinner. *Saxifraga ligulata* Wall. syn. *Bergenia ligulata* Wall. Engl. Family: Saxifragaceae is an herb and taken as *Pashanbheda*. *Acacia catechu* (L.f.) Willd. Family: Fabaceae is a tree and known as *Khadira* in Ayurveda. Morphologically *Saxifraga ligulata* and *Acacia catechu* does not represent *Mohanvalli.* This plant requires further investigation.

54. MORATA

Current representative: *Schrebera swietenioides*

Possible representative: *Elaeondendron glaucum*

Schrebera swietenioides Roxb.

English name: Weaver's Beam Tree.

Family: Oleaceae.

Habitat: It is found in Tropical of Africa and India.

Description: Medium-sized trees with grey, thick, slightly rough bark, exfoliating in woody seal Leaves imparipinnate; leaflets ovate- lanceolate. Flowers yellowish-brown, fragrant, Leaves imparipinnate; leaflets ovate-lanceolate. Flowers dichotomous cymes. Capsule woody, pear shaped, pendulous.

Chemical constituents: Tree exudes a gum containing mannitol, fructose and digalaitoside known as swietenose.

Therapeutics: Leaves used in enlargement of spleen and in urinary discharges. Roots used in leprosy; bark in boils and bums. Fruits used in hydrocele.

Bioactivity: Wound healing (Rasal *et al.,* 2009) and antibacterial (Niranjan *et al.,* 2010).

Further reading

Niranjan IH, Umapathy HK, Sreedharamurthy S, Mysore S, Sudarshana S. Antibacterial activity of *Schrebera swietenioides* Roxb. against some human pathogenic bacteria. *J Pharm Res* 2010;3.

Rasal AS, Nayak PG, Baburao K, Shenoy RR, Mallikarjuna Rao C. Evaluation of the healing potential of *Schrebera swietenioides* in the dexamethasone-suppressed wound healing in rodents. *Int J Low Extrem Wounds* 2009; 8:147-52.

Elaeodendron glaucum (Rottb.) Pers.

Syn: *Cassine glauca* (Rottb.) Kuntze.

English name: Ceylon Tea.

Family: Celastraceae.

Habitat: Distributed throughout India (the hotter part), Australia, America, South Africa and Tropical Asia.

Description: The leaves are 3-7 cm x 5-12 cm, elliptic, acuminate, glabrousb and dark green in color with grey or blakish bark is generally grown for its medicinal importance.

Chemical composition: Nor-triterpenes: elaeodendrol and elaeodendradiol (Anjaneyulu & Narayanarao, 1980).

Bioactivity: Antibacterial (Lanjhiyanal *et al.,* 2011).

Further reading

Lanjhiyanal S, Garabadu D, Ahirwarl D, Bigoniya P, Rana AC, Patra KC, Lanjhiyana SK, Karuppaih M. Antidiabetic activity of methanolic extract of stem bark of *laeodendron glaucum* Pers. in Alloxanized rat model. *Adv Appl Sci Res* 2011; 2:47-62.

Anjaneyulu ASR, Narayanarao M. Elaeodendrol and elaeodendradiol, new nor-triterpenes from *Elaeodendron glaucum*. *Phytochemistry* 1980; 19:1163-1169

55. MURVA

Current representative: *Marsdenia tenacissima*

Possible representative: *Chenomorpha macrophylla, Clematis triloba* and *Senseviera zeylanica.*

Marsdenia tenacissima W.& A.

Syn: *Asclepias tenacissima* (Roxb.)

Common name: Rajmahal hemp.

Family: Asclepiadaceae.

Habitat: Deccan Peninsula and N.W Himalayas.

General Description: Lianas robust, densely pilose to tomentose throughout except for interior of corolla. Petiole 5-6 cm, slender; leaf blade ovate, 8-10 × 6-6.5 cm, base deeply cordate with rounded sinus, apex acuminate; basal veins 5-7, lateral veins 2 or 3 pairs. Inflorescences much branched, broader than long, to 8 × 12 cm, many flowered; peduncle to 2 cm, shorter than first internode. Pedicel 6-8 mm. Sepals elliptic-lanceolate, ca. 3 × 1-1.3 mm, tip rounded. Corolla "yellow," campanulate, with spreading lobes, very densely pilose outside; tube ca. 3.5 × 2.5 mm, interior retrorsely pilose toward base; lobes oblong, ca. 4 × 2.2 mm, apex rounded, minutely velvety. Corona lobes exserted from corolla tube, oblong, apex truncate-emarginate with corners produced into short horns, sometimes toothed between these. Anther appendages oblong, slightly longer than corona lobes; pollinia curved cylindric. Stigma head broadly cylindric, concealed by anther appendages.

Chemical constituents: Alkaloids and pregnane glycosides.

Specific action as per Ayurevda: *Jvarahara* (febrifuge).

Therapeutics: Fever, abdominal colic, constipation, gastric irritation and peptic ulcer.

Parts used: Stem.

Further reading

Lucas, D.S. Studies on murva with special reference to pharmacology and clinical studies of *Marsdenia tenacissima* Wight et Arn. *Bull Indian Med* 1973, 42-43.

Lucas, D.S. Controversial drugs of Ayurveda: identification of murva with special reference to *Marsdenia tenacissima* W and A Publication date: 1997 Published in: in: B.L. Gaur and S.N. Sharma (Eds.), 71-75.

Gupta BL. (1945). *Forest Flora of Chakrata, Dehra Dun and Saharanpur.* Forest Research Institute Press.

Chonemorpha macrophylla G.Don

Family: Asclepiadaceae.

Habitat: E. Asia.

General Description: A large, woody perennial twinner or spreading plant with soft greyish bark. Leaves simple, opposite, large, ovate, acute, orbicular with prominent veins. Flower large, creamy white, with yellow inside, found in terminal or axillary cymes. Fruits long, straight, woody mericarps, containing many seeds crowned with silky coma.

Chemical constituents: Steroidal alkaloids (Banerjee and Chatterjee, 1973).

Therapeutics: *C. macrophylla* pacifies vitiated *vata*, *kapha*, skin diseases, constipation, flatulence, colic, worms, urinary tract diseases, cardiac debility, jaundice, bronchitis and fever.

Bioactivity: Skeletal muscle relaxant (Kumar, Ray and Das, 2005).

Parts used: Roots.

Bioactivity: Muscle relaxant (Roy, Ray & Das, 2005)

Further reading

Roy, R.K., N.M. Ray and A.K. Das Title: Skeletal muscle relaxant effect of *Chonemorpha macrophylla* in experimental animals. *Indian J Pharmacol* 2005; 37: 116-119.

Banerjee J, Chatterjee A. New steroidal alkaloids from *Chonemorpha macrophylla* G.Don. *Indian J Chem* 1973;10:1056.

Clemantis triloba Heyne ex Roth.

Common name: Golden thread root.

Family: Ranunculaceae.

Habitat: Himalayas.

General Description: A woody spreading creeper grows over hedges and bushes. Leaves pinnately compound, 3-5 foliolate; petiole 1-7 cm, sparsely glabrous; leaflet ovate, narrowly ovate, or lanceolate, leathery, both surfaces glabrous, margin dentate, apex attenuate to acuminate; Flower whitish, axillary or terminal cymes.

Chemical constituents: Alkaloids (berberine and coptine).

Specific action as per Ayurveda: *Rakat prasadana* (alterative).

Therapeutics: It pacifies vitiated vata, kapha, flatulence, colic, worms, urinary tract diseases, cardiac debility, jaundice, bronchitis, bleeding disorders, diabetes, thirst, heart disease, itching, skin diseases, fever, vomiting and typhoid fever.

Parts used: Rhizomes.

Bioactivity: Hepatoprotective.

Further reading

Behl PN *et al.*, Skin-Irritant and Sensitizing Plants Found in India, New Delhi. Irwin Hospital, 1966.

Sansevieria roxburghiana Schult.f.

Common name: Indian bowstring hemp.

Ayurvedic name: *Nagadamani.*

Family order: Agavaceae.

Habitat: India and Ceylon.

General Description: Stemless, with a creeping rootstock. Leaves 10-20 x 2-4 cm flat, strap-shaped or narrowly lanceolate, with a subulate point up to 2.5 cm long. Inner or adult leaves ascending and slightly recurving, thick, deeply concave, channelled down the face, rounded or very obtusely keeled on the back, green, transversely marked with darker green lines on both sides. Flowers stem 30-75 cm long. Raceme spike-like, 30-45 cm long. Flowers borne in clusters of 4. Pedicels 6-9 cm long, jointed.

Chemical constituents

Specific action as per Ayurveda: *Balya* (tonic).

Therapeutics: Constipation, tuberculosis, fevers and general debility.

Parts used: Root.

Bioactivity: Antitumor (Haldar *et al.*, 2010).

Further reading

Haldar PK, Kar B, Bala A, Bhattacharya S, Mazumder UK. Antitumor activity of *Sansevieria roxburghiana* rhizome against Ehrlich ascites carcinoma in mice. *Pharm Biol* 2010;48: 1337-1343.

56. MUSTA

Current representative: *Cyperus rotundus*

Possible representative: *Cyperus scariosus, Cyperus esculenta and Cyperus longus.*

Cyperus rotundus L.

Common name: Nut grass.

Ayurvedic names: *Varidhara, musta, meghakhya, kuruvindaka, varaha, abda, ghana, bhadra musta, rajakaseruka, pinda musta, visadhvamshi and nagara.*

Family order: Cyperaceae.

Habitat: India.

General Description: It is a pestiferous perennial weed with dark green glabrous culms, arising from a system of underground tubers. The plant has an elaborate underground system consisting of tubers, rhizomes and roots. The tubers are white and succulent when young, and hard and black when mature.

Detailed Description of the Rhizome

(a) *Macroscopic*

Drug consists of rhizome and stolon having a number of wiry roots, stolon 10-20 cm long having a number of rhizomes, crowded together on the stolons, rhizomes bluntly conical and vary in size and thickness, crowned with the remains of stem and leaves forming a scaly covering, dark brown or black externally, creamish-yellow internally; odour, pleasant.

(b) *Microscopic*

Rhizome shows single layered epidermis, followed by 2-6 layers, suberised sclerenchymatous cells; epidermis and outer sclerenchymatous layers filled with dark brown content; ground tissue of cortex consists of circular to oval, thin-walled, parenchymatous cells with small intercellular spaces; a few fibro-vascular bundles present in this region; endoderm is distinct and surrounding the stele; wide central zone beneath endodermis, composed of circular to oval, thin-walled, parenchymatous cells with intercellular spaces, numerous collateral, closed, vascular bundles surrounded by bundle sheath, scattered in this region; vessels narrow having simple reticulate, and scalariform thickening and oblique pore; simple round to oval starch grains measuring 6-28 μ in dia., a number of pigmented cells filled with reddish-brown content, present throughout the cortex and stele.

Powder - Creamish-brown; shows reddish-brown cells, reticulate and simple pitted vessels; fibre-like, closely packed sclerified cells, narrow vessels with scalariform thickness and oblique pore from the remnants of leaves simple, round to oval, starch grains, measuring 6-28 μ in diameter.

Chemical constituents: Aromatic oil (Radomir S, Dev S, Sirsi, 1956).

Specific action as per Ayurveda: Demulcent, diuretic and digestive.

Therapeutics: Diarrhoea, fevers and urinary infections.

Parts used: Tubers.

Bioactivity: Antibacterial (Radomir S, Dev S, Sirsi, 1956), estrogenic (Indira *et al.,* 1956), anti-inflammatory (Saxena *et al.,* 1971), antioxidant (Nagulendran, Mahesh & Begum, 2004), hepatoprotective (Kumar & Mishra, 2005), antidiarrhoeal (Uddin *et al.,* 2006), antidiabetic (Raut & Gaikwad, 2006), wound healing (Puratchikody *et al.,* 2006) and anticonvulsant (Shivakumar *et al.,* 2009).

Further reading

Indira M, Sirsi M, Radomir S, Sukhdev S. The occurrence of some estrogenic substances in plants, part I: Estrogenic activity of *Cyperus rotundus* (Linn.) *J Sci Ind Res* 1956;15 C:202-204.

Kumar SVS, Mishra SH. Hepatoprotective activity of rhizomes of *Cyperus rotundus* Linn. against carbon tetrachloride-induced hepatotoxicity. *Indian J Pharm Sci* 2005;67: 84-88.

Nagulendran KR, Mahesh R, Begum H. Curative effect of *Cyperus rotundus* rhizome on oxidative stress in male albino ras. J. Joseph Thas Ed., 2004;83-84.

Puratchikody A, Nithya DC, Nagalakshmi G. Wound healing activity of *Cyperus rotundus* Linn. *Indian J Pharm Sci* 2006; 68: 97-101.

Radomir S, Dev S, Sirsi M. Chemistry and antibacterial activity of nut grass. *Curr Sci* 1956;25:118-119.

Raut N, Gaikwad NJ. Antidiabetic activity of hydro-ethanolic extract of *Cyperus rotundus* in alloxan induced diabetes in rats. *Fitoterapia* 2006;77: 585-588.

Saxena R, Punhari KC et al., Preliminary report on the anti-inflammatory effect of *Cyperus rotundus* in conjunctivitis. *Indian J Pharm* 1971;33: 9.

Shivakumar SI, Suresh HM, Hallikeri CS, Hatapakki BC, Handiganur JS, Kuber S, Shivakumar B. Anticonvulsant effect of *Cyperus rotundus* Linn rhizomes in rats. *J Nat Rem* 2009; 9:192-196.

Uddin SJ, Mondal K, Shilpi JA, Rahman MT. Antidiarrhoeal activity of *Cyperus rotundus*. *Fitoterapia* 2006; 77: 134-136.

Cyperus scariosus (R.Br.).

Common name: Nutsedge.

Family: Cyperaceae.

Habitat: India.

General Description: A delicate, slender sedge. The plant produces deep brown tubers with aromatic odor.

Chemical constituents: Steam distillation of the tubers of cyperiol yields 0.075-0.080% of an essential oil, the principal content of which is cyperine.

Actions: Astringent, diaphoretic, diuretic and stomachic.

Therapeutics: Diarrhea and general debility. Decoction of tubers is used in the treatment of gonorrhea and syphilis.

Parts used: Tuber.

Bioactivity: Anti-inflammatory (Gupta, *et al.,* 1972), antimicrobial (Lahariya & Rao, 1979) and immunomodulatory (Bhagwat *et al.,* 2009).

Further reading

Gupta SK, Sharma RC, Aggarwal OP, Arora RB. Anti-inflammatory activity of the oil isolated from *Cyperus scariosus*. *IJEB* 1972;10: 41-42.

Bhagwat D, Kharya MD, Bani S, Pandey A, Chauhan PS, Kour K, Suri KA, Satti NK, Dutt P. *Cyperus scariosus* chloroform fraction inhibits T cell responses in Balb/C mice. *Trop J Pharm Res* 2009;8: 399-408.

Gupta GN, Chandra G. Oil of cyperiol (*Cyperus scariosus*). *Curr Sci* 1951; 20:273.

Lahariya AK, Rao JT. *In vitro* antimicrobial studies of the essential oils of *Cyperus scariosus* and *Ocimum basilicum*. *Indian Drugs* 1979; 16:150

Cyperus longus **Linn.**

Syn: *Chlorocyperus longus* (L.) Palla

Habitat: Northern and central Africa to Kenya and Nigeria, around the Mediterranean, southern and central Europe, western Asia and central Asia to India.

General Description: It is an evergreen, perennial, herbaceous plant reaching 0.7-1.5 m in height.

57. NANDIVRKSA

Current representative: *Cedrela toona*

Possible representatives: *Ficus retusa, Thespesia populnea, Ervatamia coronaria* and *Gmelina arborea.*

Cedrela toona **Roxb. ex Willd.**

Syn: *Toona ciliata* Roemer

Common name: East Indian tree, Indian mahogany, toon tree, tun tree, toona.

Ayurvedic name: *Tuni, ashvattha bheda, prarohi* and *gajapadapa.*

Family: Meliaceae.

General Description: It is a large deciduous tree with a spreading crown, commonly attaining a height of 20-30 m and a girth of 1.8-3 m. Bark dark grey or reddish-brown, smooth up to middle age, afterwards rough, with shallow reticulate cracks exfoliating in irregular woody scales. Blaze 1.3-1.5 m, fibrous throughout, pink or pinkish-brown, sometimes with just a few white bands towards the outside, turning brown on exposure, bitter to the taste, juice turning purple on the blade of a knife. Leaves 30-50 cm long, on young trees up to 90 cm long, usually imparipinnate, sometimes paripinnate by the abortion of the terminal leaflet; leaflets 11-29, opposite or alternate, 5-15 x 2-6 cm, lanceolate or ovate-lanceolate, acuminate, glabrous, pubescent, margin entire or wavy, base oblique; petiolules 0.3-1.3 cm long. Flowers small, honey scented, cream coloured, in drooping or sub-erect terminal panicles, usually shorter than the leaves. Calyx divided nearly to the base. Petals 5 mm long, ovate-oblong, sub-acute, with ciliate margins. Capsule dark brown, 1.8-2.5 x 0.5-0.8 cm, oblong, usually smooth outside, sometimes sparsely lenticellate. Seeds pale brown, very light, winged at both ends, 1.3-1.5 cm long including the wing.

Chemical constituents: Tannins (Poddar & Chakrabartty, 1971).

Actions: Aphrodisiac.

Therapeutics: Ulcer, skin diseases and bleeding diathesis.

Parts used: Bark.

Bioactivity: Antimicrobial (Chowdhury, Hasan & Rashid, 2003),

Further reading

Chowdhury P, Hasan CM, Rashid MA. Antimicrobial activity of *Toona ciliata* and *Amoora rohituka*. *Fitoterapia* 2003;74: 155-158.

Poddar G, Chakrabartty T. A note on the chemical investigation on the bark of *Cedrela toona* Roxb. *Bull Calcutta School Trop Med* 1971;19: 89.

Ficus retusa ssp. *nitida* (Th.) Miq.

Syn: *Ficus microcarpa* var. *nitida*

Common name: Indian Laurel Fig.

Family: Moraceae.

General Description: *F. retusa* is a rapidly-growing, rounded, broad-headed, evergreen tree can reach 15 metres (50 feet) or more in height with an equal spread. The glossy, dark green, leathery leaves are densely clothed on large, somewhat weeping branches and are usually infested with thrips. New growth, produced all year long, is a light rose to chartreuse color, giving the tree a lovely two-toned effect. The smooth, light grey trunk is quite striking, can grow to around a metre (three or four feet) in diameter, and it firmly supports the massively spreading canopy.

Habitat: South America.

Chemical constituents: Calcium carbonate.

Further reading

Subramanian M, Kavitha P. Title: Study of naalmarapattai kuliyal in Kanyakumari district. J. Joseph Thas (Ed.), 2004;75 (abstract).

Apparanantham T, Chelladurai V. Glimpses of folk medicines of Dharmapuri Forest Division, Tamilnadu. *Ancient Sci Life* 1986; 5:182-185

***Ervatamia coronaria* Stapf.**

Syn: *Tabernaemontana coronaria* R.Br.

Common name: Grape jasmine.

Family: Apocynaceae.

General Description: Shrub upto 1 m Branches divaricate, milky latex present. Young shoots green, lenticellate. Leaves opposite, 7-15 x 2.5-6 cm ± elliptic, acuminate or caudate, cuneate at the base, petiole 5-7 mm long, 6-8 pairs of lateral nerves. Inflorescence 1-few flowered cymes, axillary or terminal, peduncle c.5 cm long. Flowers 2.5-5 cm across, white; pedicel 1-1.5 cm long, bract minute or absent. Calyx lobes broad, ovate, acute c. 4 mm long. Corolla tube c. 2-2.5 cm, dialated in the middle, lobes obliquely ovate 2.5-3.75 cm long, double in cultivated form. Ovary glabrous; ovules numerous; style filiform; stigma papillose. Follicles 2.5-4x1 cm, sessile, glabrous, 1-3 ribbed, orange or bright red inside, curved to form a beak. Seeds 3-6, oblong or irregular in shape, striated (Madan & Nayar, 1960).

Habitat: South America.

Chemical constituents: Alkaloids, flavonoid: kaempferol: flowers (Farooq, Rahman & Ilyas, 1959), Cysteine protease: ervatamin C (Thakurta *et al.,* 2004), thiol proteases: ervatamin A-C (Nallamsetty, Kundu & Jagannadham, 2003).

Actions: Aphrodisiac and purgative.

Therapeutics: Diseases caused by *Kapha,* biliousness and diseases of the blood. The milky juice mixed with oil is rubbed onto the head to cure pain in the eye. It is applied to wounds to prevent inflammation (Kirtikar & Basu, 1987).

Parts used: Flowers.

Bioactivity: Anti-inflammatory and antioxidant (Thambi *et al.,* 2006), antimicrobial, antineoplastic and cytotoxic (Singh, Sharma & Vyas, 2011),

Further reading

Gupta M, Mazumdar UK, Gomathi P, Kumar RS. Antioxidant and free radical scavenging activities of *Ervatamia coronaria* Stapf leaves. *Iranian J Pharm Res* 2008;3:119-126.

Kirtikar KR, Basu BD. Indian Medicinal Plants. M/S Periodical Experts, Delhi. 1987; 7:2176.

Madan CL, Nayar SL. Pharmacognostic study of the root of *Ervatamia coronaria* Stapf. *J Sci Ind Res* 1960;19 C:280-282.

Nallamsetty S, Kundu S, Jagannadham MV. Purification and biochemical characterization of a highly active cysteine protease ervatamin A from the latex of *Ervatamia coronaria*. *J Protein Chem* 2003;22:1-13.

Singh B, Sharma RA, Vyas GK. Antimicrobial, antineoplastic and cytotoxic activities of Indole alkaloids from *Tabernaemontana divaricata* (L.) R.Br. *Curr Pharm Anal* 2011; 7: 125-132.

Thakurta PG, Biswas S, Chakrabarti C, Sundd M, Jagannadham MV, Dattagupta JK. Structural basis of the unusual stability and substrate specificity of ervatamin C, a plant cysteine protease from *Ervatamia coronaria*. *Biochemistry* 2004;43:1532-1540.

Thambi PT, Kuzhivelil B, Sabu MC, Jolly CI. Antioxidant and antiinflammatory activities of the flowers of *Tabernaemontana coronaria* (l) R.Br. *Indian J Pharm Sci* 2006;68:352-355.

58. NAGAKESRA

Current representative: *Mesua ferrea*

Possible representative: *Ochrocarpus longifolia, Cinnamomum wightii* and *Dillenia pentagyna*.

Ochrocarpus longifolius Benth. et Hook.f.

Common name: Cobra's saffron.

Ayurvedic name: *Punnaga*.

Family: Clusiaceae.

General Description: A big tree with very pretty & glossy foliage. Tiny flowers are borne in clusters on the tree trunk & mature branches. Flowers have very pleasant scent, which lasts even when flowers wither. Flowers are in great demand for making Gajras (Afaq & Rizwan,1971; Khan *et al.*, 1978).

Habitat: South America.

Actions: Astringent, bitter and stomachic.

Therapeutics: Gastritis and bloody diarrhoea.

Parts used: Dried flower buds (Afaq & Rizwan, 1971).

Bioactivity: Antibacterial, antifungal and anthelmintic (Naqvi, Khan & Vohra, 1976).

Further reading

Afaq SH, Rizwan M. Pharmacognostic study of the flower buds of *Ochrocarpus longifolius* Benth. et Hook.f. *Indian J Pharm* 1971; 33: 20-33.

Khan MSY, Kumar I, Khan NU, Ilyas M. Chemical investigation of Indian medicinal plants used for leprosy I - Constituents of the flowers of *Ochrocarpus longifolia* Benth. & Hook.f. (Guttiferae). *Curr Sci* 1978;47:414-415.

Naqvi SAH, Khan MSY, Vohora SB. Antibacterial, antifungal and anthelmintic studies on *Ochrocarpus longifolius*. *Planta Medica* 1976;29:98-100.

Dillenia pentagyna Roxb.

Syn: *Dillenia floribunda* Hook.f. & Thomson

Common name: Cobra's saffron.

Ayurvedic name: *Bhavya* and *Aksikiphal*.

Family: Dillenaceae.

General Description: *D. pentagyna* is a large deciduous tree grows up to 40 meters in height. Leaves are large, 1-2 ft, alternate, ovate-rhomboid, obtuse or acute. Flowers yellowish, fragrant, 2-3 cm across, arise from the nodes of fallen leaves, on panicles. Fruits 2.5 cm in diameter, globose contain single seed (Ali, 1967).

Habitat: Subtropical Himalaya, India, Ceylon, Burma, Thailand, S. China and Malaysia.

Chemical constituents: Triterpenoids.

Actions: Astringent, bitter and stomachic.

Therapeutics: Gastritis and bloody diarrhea (Dubey *et al.,* 2009). According to Ayurveda, the plant pacifies vitiated *vata, kapha,* anal fistula, wounds, diabetes, diabetic carbuncle, neuritis, pleurisy, pneumonia, and burning sensation.

Parts used: Fruit.

Bioactivity: Antitumor (Rosangkima & Prasad, 2004).

Further reading

Ali US. Nagakeshara as known in South India. *J Res Indian Med* 1967;2: 57-65.

Anandakumar A, Balasubramaniam M, Muralidharan R. Nagakesara - a comparative pharmacognosy. *Ancient Sci Life* 1986;5: 26-268.

Dubey PC, Sikarwar RLS, Khanna KK, Tiwari AP. Ethnobotany of *Dillenia pentagyna* Roxb. in Vindhya region of Madhya Pradesh, India. *Nat Prod Rad* 2009;8: 546-548.

Rosangkima G, Prasad SB. Antitumour activity of some plants from Meghalaya and Mizoram against murine ascites Dalton's lymphoma. *Indian J Exp Biol* 2004;42: 981-988.

59. NALUKA- NALIKA

Current representative for *Naluka*: *Cinnamomum zeylanicum*

Current representative for *Nalika*: *Ipomoea aquatica*

Ipomoea aquatica Forsk.

Syn: *Ipomoea reptans* Poir.

Common name: Water convolvulus, Water Spinach, Water Morning Glory.

Family: Convolvulaceae.

General Description: *I. aquatica* grows in water or on moist soil. Its stems are 2–3 metres (7–10 ft) or more long, rooting at the nodes, and they are hollow and can float. The leaves vary from typically sagittate (arrow-head-shaped) to lanceolate, 5–15 centimetres (2–6 in) long and 2–8 centimetres (0.8–3 in) broad. The flowers are trumpet-shaped, 3–5 centimetres (1–2 in) diameter, usually white in colour with a mauve centre. The flowers can form seed pods which can be used for planting (Satpathy, 1964).

Habitat: India.

Chemical constituents: Phenolics (Rao & Leela , 1992; Leela & Shanmukha, 1994).

Bioactivity: Antioxidant (Mital, Shukla & Saluja, 2008) and hypolipidemic (Dhanasekaran, 2010),

Further reading

Dhanasekaran S. Hypolipidemic activity of *Ipomoea aquatica* Forsk. leaf extracts on lipid profilr in hyperlipidemic rats. *Int J Pharm Biol Arch* 2010;1, 2.

Leela M, Shanmukha RA. Phenolic compounds in the taxonomy of Ipomoea L. (Convolvulaceae). *Feddes Repertorium* 1994;105: 445-448.

Mital B, Shukla S, Saluja AK. *In-vitro* antioxidant activity of the flowers of *Ipomoea aquatica* Forsk. *Pharmacog Mag* 2008;4:16.

Rao S, Leela M. Chemotaxonomy of some Ipomoea L. (Convolvulaceae). *Feddes Repertorium* 1992;103: 351-355.

Satpathy B. Kalami sag (*Ipomoea reptans*) as a new addition to our greens. *Indian Farm.* 1964; 14: 12-16.

60. NICHULA

Current representative: *Barringtonia racemosa*

Barringtonia racemosa Roxb.

Syn: *Barringtonia pallida* (Miers) Koord. & Valeton

Common name: Fish poison wood.

Family: Lecythidaceae.

Habitat: Southeast Asia.

General Description: *B. racemosa* is usually a small tree, 4-8 m in height but occasionally reaching 15 m; bark grey, yellow or brown, mottled, rather smooth to fissured; no aboveground roots but may have spreading surface roots. Leaves alternate, simple, crowded at the ends of the branches, large, obovate-oblong to oblanceolate, 8-35 x 4-13 cm; apex deep green, broadly tapering; base narrowly tapering, running into the petiole; margin entire or very shallowly toothed or scalloped; petiole very short, without hairs. Flowers attractive, white to pale pink, in many-flowered pendulous sprays; up to 60 cm in length or even more; bisexual; all floral parts in 4s; sepals joined at the base, separating in 3-4 lobes, green flushed with pink, about 10 x 7.5 mm; petal elliptic, up to 3 x 1 cm, attached to the staminal tube; stamens many, long, white or pinkish, forming a large central mass 3.5 cm in diameter; ovary 2 to 4 chambered; style red. Fruit conical to ovate, about 3 x 2 cm, crowned by the remains of the persistent calyx; style fleshy at first, later becoming hard, fibrous and yellowish-brown when mature. Seeds aromatic.

Chemical constituents: Bartogenic acid.

Therapeutics: Cancer.

Parts used: Seeds.

Bioactivity: Anti-tumor (Jose *et al.*, 2002), antidiabetic (Mangala *et al.*, 2007), anti-inflammatory and analgesic (Shikha *et al.*, 2010).

Further reading:

Jose TT, Panikkar B, Subramoniam A, Nair MK, Panikkar KR. Antitumour property and toxicity of *Barringtonia racemosa* Roxb. seed extract in mice. *J Ethnopharmacol* 2002; 82: 223-227.

Mangala GP, Tivari AK, Ali AZ, Rao JM. Inhibition of α glucosidase and amylase by bartogenic acid isolated from *Barringtonia racemosa* Roxb. seeds. *Phytother Res* 2007;21: 796-799.

Shikha P, Latha PG, Suja SR, Anuja GI, Shyamal S, Shine VJ, Sini S, Krishnakumar NM, Rajasekharan S. Anti-inflammatory and analgesic activity of *Barringtonia racemosa* Roxb. Fruits. *Indian J Nat Prod Res* 2010;1: 356-361.

61. OLATE KAMBALA

Current representative: *Abroma augusta*

Possible representative: *Crotolaria juncea*

Abroma augusta (L.) L.f.

Syn: *Abroma mollis* DC.

Common name: Devil's cotton.

Family: Sterculiaceae.

Habitat: It is common in many parts of India. It is also found in Pakistan and many Asian countries.

General Description: *A. augusta* is a shrub or small tree 2 to 3 meters in height. The branches and branchlets are downy. The leaves are alternate, ovate, 10 to 15 centimeters in length, and 10 to 12 centimeters in width, with pointed tip, heart-shaped base, and toothed margins. The flowers are bisexual, about 5 centimeters across, and yellowish with purple bases. The fruit (capsule) is obpyramidal, about 3.5 centimeters in diameter, covered with irritating hairs, and ultimately smooth, with five prominent angles or wings, which are 4 by 7 centimeters across. The seeds are subellipsoid, numerous, and finely punctuate (Mitra & Bal, 1947).

Chemical constituents: Gum, wax, non-crystalline extractive matter, water-soluble base and alkaloids: abromine and betaine (Srivastava & Basu 1956, 1959/1960; Dasgupta & Basu, 1970).

Specific action as per Ayurveda: *Artavajanana*.

Therapeutics: According to Kirtikar and Basu, Chopra, and Dymock, the root and the root-bark are used in medicine as an emmenagogue in menstrual disorders. The fresh viscid sap is said to be still more efficacious and is used in dysmenorrhoea. They quote Dr. Thornton, who considers it useful in the congestive and neuralgic varieties of dysmenorrhoea, a regulator of the menstrual flow, and a uterine tonic. According to Kirtikar and Basu, an infusion of the fresh leaves and the stems in cold water is demulcent and very efficacious in gonorrhea.

Parts used: Root bark.

Bioactivity: Androgenic (Venkitaraman & Radhakrishnan, 1975b) and hypoglycemic (Halim, 2003).

Further reading

Dasgupta B, Basu K. Chemical investigation of *Abroma augusta* Linn. - Identity of abromine with betaine. *Experientia* 1970;26: 477-478.

Halim EM. Lowering of blood sugar by water extract of *Azadirachta indica* and *Abroma augusta* in. diabetes rats. *Indian J Exp Biol* 2003; 41:636-640.

Mitra GC, Bal SN. Pharmacognosy of the root bark of *Abroma augusta* Linn. *Indian J Pharm* 1947;9: 120-125.

Srivastava GP, Basu NK. Studies on *Abroma augusta* Linn., I: Chemical examination, II: Pharmacological examination. *Indian J Pharm* 1956; 18:472; *J Sci Res* (Banaras Hindu University) 1959/1960; 9, 1-8.

Venkitaraman S, Radhakrishnan N. Effect of *Abroma augusta* Linn. root on mammary growth and lactation. *Indian J Pharm* 1975b;7:153-154.

Crotolaria juncea L.

Syn: *Crotalaria fenestrata* Sims., *Crotalaria benghalensis* Lam., *Crotalaria tenuifolia* Roxb. ex DC.

Common name: Brown hemp, Deccan hemp, Madras hemp, sunn hemp.

Ayurvedic name: *Jenapa vera.*

Family: Fabaceae.

Habitat: Native to Eurasia. Found throughout plains of India.

General Description: *C. juncea* is a short-day, erect shrubby annual, generally 1 to 4 m in height. The stems are cylindrical and ribbed. Branching in the upper portion is minimized with dense plantings. The simple, elliptic to oblong shaped leaves, are spirally arranged on the stem. The root system is characterized by a long, strong taproot, well developed lateral roots, and much branched and lobed nodules, up to 2.5 cm in length. The inflorescence is a terminal open raceme to 25 cm in length with deep yellow flowers. Flowering is indeterminate.

Chemical constituents: Mucilage, resin and pyrrolizidine alkaloids.

Action: Emollient and emmenagouge.

Therapeutics: Skin diseases and amenorrhea.

Parts used: Seeds and leaves.

Dose: 1-4G.

Bioactivity: Hepatotoxic (Parakash, Dehadrai & Jonathan, 1995).

Further reading

Parakash AO, Dehadrai S, Jonathan S. Toxicological studies on the ethanolic extract of *Crotalaria juncea* seeds in rats. *J Ethnopharmacol* 1995; 45:167-76.

62. PARPATA

Current representative: *Fumaria vaillanti*

Possible representative: *Oldenlandia corymbosa, Rungia repens, Mollugo oppsitifolia* and *Polycarpaea corymbosa*

Fumaria vaillanti Loisel

Syn: *Fumaria indica* Pugsley.

English name: Fumitory.

Family: Fumariaceae.

Habitat: The plant is distributed throughout India, particularly on the banks of the Ganges and in the Himalayas up to an altitude of 2700 m. It is also distributed in the higher elevations of the Mysore plateau and Nilgiris. It is also found in Europe, Africa and many other Asian countries.

General Description: It is an annual, diffuse herb, up to 30 cm high, with grooved branchlets (Plate 29). The leaves are pale green, 2-3 pinnatisect, 5-7 cm long. Flowers are asymmetrical, pale pink or white with purple tips, in terminal or leaf-opposed racemes, with a filiform style and a two-lobed stigma. The calyx consists of two Ian ceo late sepals which are much smaller than the corolla tube. The fruit is a small, indehiscent nutlet, rugose when dry, rounded at the top with two pits and containing one seed (Mehra, Jolly & Puri, 1968).

Chemical constituents: Alkaloids: fumariline, protopine, narceimine, narlumidine, fuyuziphine (Dasgupta *et al.,* 1984; Tripathi, Rathore & Kumar, 1994; Pandey *et al.,* 2008), pentatriacontane, fumaric acid and tannin.

Specific action as per Ayurveda: *Trishna nigrehana* (thirst depressant).

Therapeutics: Liver diseases and fevers.

Parts used: Whole plant.

Bioactivity: central nervous system depressant (Kumar *et al.*, 1986), anti-inflammatory and anti-nociceptive (Rao *et al.*, 2007) and hepatotoxic: (Rao & Mishra, 1998; Rathi *et al.*, 2008).

Further reading

Dasgupta B, Seth KK, Pandey VB, Ray AB. Alkaloids of *Fumaria indica*: Further Studies on Narceimine and Narlumidine. *Planta Med* 1984;50:481-485.

Kumar A, Pandey VB, Seth KK, Dasgupta B, Bhattacharya SK. Pharmacological Actions of Fumariline Isolated from *Fumaria indica* Seeds. *Planta Med* 1986;4:324-325.

Mehra PN, Jolly SS, Puri HS. Pharmacognosy of *Fumaria indica* (Hassk.) Pugsley and its adulterants. *Indian J Pharm* 1968;30: 284.

Pandey MB, Singh AK, Singh JP, Singh VP, Pandey VB. Fuyuziphine, a new alkaloid from *Fumaria indica*. *Nat Prod Res* 2008 15; 22:533-536.

Rao CV, Verma AR, Gupta PK, Vijayakumar M. Anti-inflammatory and anti-nociceptive activities of *Fumaria indica* whole plant extract in experimental animals. *Acta Pharm* 2007; 57:491-498.

Rao KS, Mishra SH. Antihepatotoxic activity of monomethyl fumarate isolated from *Fumaria indica*. *J Ethnopharmacol* 1998; 60:207-213.

Rathi A, Srivastava AK, Shirwaikar A, Rawat AKS, Mehrotra S. Hepatoprotective potential of *Fumaria indica* Pugsley whole plant extracts, fractions and an isolated alkaloid protopine. *Phytomedicine* 2008;15: 470-477.

Tripathi YC, Rathore M, Kumar H. On the variation of alkaloidal contents of *Fumaria indica* at different stages of life span. *Ancient Sci Life* 1994;13: 271-273.

Oldenlandia corymbosa Linn.

Syn: *Hedyotis corymbosa* (L.) Lam., *Oldenlandia affinis Blanco* and *Oldenlandia burmmaniana* G.Don

Common name: Old World diamond-flower.

Family: Rubiaceae.

Habitat: Native to tropical Africa, including Madagascar, and India.

General Description: Annual herb with ascending or erect 4-anged stems; leaves acute at both ends, linear-oblong or narrowly elliptic, 1-3.5 cm long, 1.5-7 mm wide, subsessile or petiole very short, blade pale beneath, midrib prominent; flowers few (2-8) in axillary cymes shorter than the leaves; flowers white or faintly pinkish-purplish, on slender pedicels 4-8 mm long; calyx not exceeding ovary; corolla about 2 mm long; stamens inserted just above the base of the tube; capsule about 2 x 2 mm, exalate, flattened at apex, slightly laterally compressed (Khastgir, Sengupta & Sengupta, 1960; Datta & Sen, 1969; Sivarajan & Biju, 1990).

Actions: Liver tonic and febrifuge.

Therapeutics: Remittent fever, gastric irritability and nervous depression.

Parts used: Whole plant.

Further reading

Datta PC, Sen A. Pharmacognosy of *Oldenlandia corymbosa* Linn. *Quar J Crude Drug Res* 1969;9:1365.

Khastgir HN, Sengupta SK, Sengupta AP. Note on the constituents of the Indian medicinal plant *Oldenlandia corymbosa* Linn. *J Am Pharm Assoc* 1960; 49:562-563.

Sivarajan VV, Biju SD. Taxonomic and nomenclatural notes on the *Hedyotis corymbosa-diffusa* complex. *Taxon* 1990; 39: 665-674.

Rungia repens Nees

Syn: Justicia repens Linn.

Family: Acanthaceae.

Habitat: is a much branched, procumbent or erect herb found as a weed throughout the warmer parts of India.

General Description: Annual herb, 10-30 cm tall. Slender stems prostrate, rooting near the base, and then they become erect. Leaves oppositely arranged, short-stalked, oblong-lancelike Flowers stalkless occur in an erect 1-sided spike, 3-7 cm long, at the end of the stems. Flowers 2-lipped bluish purple, with dark purple and white spots.

Chemical constituents: Chrysoeriol glycoside: isosalipurposide and other flavonoids (Sankar, Joseph & Nair, 1964; Seshadri & Vydeeswaran,1972).

Bioactivity: Antipyretic (Arivukkarasu, Moorthy & Venkatapiah, 2000; Swain SR, Sinha BN, Murthy, 2008), anti-inflammatory, diuretic and antimicrobial (Basu & Ariyuhkarasu, 2006).

Further reading

Arivukkarasu R, Moorthy P, Venkatapiah V. A study on the antipyretic activity of *Rungia repens* (L.) Nees in rats. *Aryavaidyan* 2000;13: 225-227.

Basu SK, Arivukkurasu R. Acute toxicity and diuretic studies of *Rungia repens* aerial parts in rats. *Fitoterapia* 2006;77: 83-85.

Sankar SS, Joseph KJ, Nair AG. Occurrence of isosalipurposide in the flowers of *Rungia repens. Indian J Chem* 1964;8:338–339.

Seshadri TR, Vydeeswaran S. Chrysoeriol glycosides and other flavonoids of *Rungia repens* flowers. *Phytochemistry* 1972; 2:803–806.

Swain SR, Sinha BN, Murthy PN. Antiinflammatory, diuretic and antimicrobial activities of *Rungia pectinata* Linn and *Rungia repens* Nees. *Indian J Pharm Sci* 2008;70: 679-683.

Mollugo oppositifolia L.

Syn: *Mollugo spergula* L., *Mollugo verticillata* Roxb. non L.

Common name: Slender carpetweed.

Family: Molluginaceae.

Habitat: It occurs India to tropical Africa and Australia.

General Description: The plant is a slender, spreading or ascending, smooth branched, annual herb, with the branches 10 to 40 centimeters in length. The leaves are opposite or whorled, spatulate, oblanceolate to oblong-ovate, 1 to 3 centimeters in length, and up to 1 centimeter in width. The flowers are white and fascicled, with the slender stalks up to 1 centimeter in length. The sepals are 3 to 3.5 millimeters long. The capsule is ellipsoid and a little shorter than the sepals. The seeds are numerous and covered with raised tubular points.

Chemical composition: Vitexin 7-glucoside and 2″-*p*-coumaroylvitexin 7-glucoside (Chopind *et al.,* 1984).

Therapeutics: Kirtikar and Basu, and Chopra consider the plant as stomachic, aperient, and antiseptic.

Further reading

Chopind J, Dellamonicad G, Markhamd KR, Ramachandran AG, Gunasegaran R. p-coumaroylvitexin 7-glucoside from *Mollugo oppositifolia. Phytochemistry* 1984; 23: 2106-2108.

Polycarpaea corymbosa (L.) Lamarck

English name: Old man's cap.

Family: Caryllophaceae.

Habit: Tropical Asia, Africa, Australia and America (Gopal, 1984).

General Description: Robust annual. Stems 10-30 cm, erect to ascending, branching from the base, white woolly at the base.

Leaves opposite, or apparently whorled, 10-15 x 0.5-0.75 mm, linear to linear-subulate, 1-nerved, glabrous or nearly so when mature. Stipules c. 3 mm, lanceolate acuminate, ciliate. Flowers in compact, terminal cymes, with corymbose branching. Pedicels as long as or smaller than the sepals, woolly. Sepals 3-3.5 mm, lanceolate, acuminate, glabrous, entirely scarious silvery-white. Petals shorter than the sepals, emarginate. Capsule included in the sepals, dark brown. Valves with a narrow thickened yellow margin. Seeds small subtrigonous, smooth (Pandey *et al.,* 1994; Jyothi *et al.,* 2010).

Further reading

Gopal GV. Is it a weed? *Ancient Scie Life* 1984;4: 114-115.

Jyothi B, Sudarsanam G, Sitaram B, Prasad B. Pharmacognosy of a local market sample of parpataka *Polycarpaea corymbosa* (L.) Lam. *Int J Plant Sci* 2010;5:613-619

Pandey G, Joshi GC, Pandey NK, Tewari KC. Ethnobotanical studies on the medicinal flora of Tarikhet block (Kumaon Himalaya) District Almora, U.P. *Aryavaidyan* 1994; 7: 228-233.

63. PASHANBHEDA

Current representative: *Bergenia ligulata*

Possible representative: *Bryophyllum calycinum, Aerva lanata, Aerva javanica, Ammania baciferra, Bridelia crenulata, Coleus aromaticus, Decalepis arayalpatra* and *Rotula aquatica.*

Bergenia ligulata (Wall.) Engl.

English name: None.

Ayurvedic names: *Ashmarghana, ashmabhedaka, ashmaribheda, drshadbheda, nagabehdana, nagabhida, pashana, pashanbheda* and *shilabheda.*

Family: Saxifragaceae.

Habitat: It grows in the temperate Himalayas between altitudes of 900 and 3,000 m.

General Description: It is a perennial herb with stout rootstock. The leaves are variable, coarsely hairy, sparsely hairy to glabrous, broadly obovate or elliptic, finely or sparsely denticulate or shallowly sinuate-dentate. The flowers are white, pink or purple, in long cymose panicles. Capsules are rounded. The rhizome is buff-brown color outside and pinkish brown inside. The market sample is transversely wrinkled or irregularly shrivelled with exfoliating, thin bark, having rootlets or thin scars, buds and scaly leaves.

Detailed Description of the Rhizome

(a) *Macroscopic*

Rhizome, solid, barrel shaped, cylindrical, 1.5-3 cm long and 1-2 cm in diameter with small roots, ridges, furrows and root scars distinct, tranversely cut surface shows outer ring of brown coloured cork, short middle cortex, vascular bundles and large central pith, odour, aromatic, taste, astringent.

(b) *Microscopic*

Transverse section of rhizome shows cork divided into two zones, outer a few layers of slightly compressed and brown coloured cells, inner zone multilayered consisting of thin-walled tangentially elongatd and colourless cells, followed by a single layered cork cambium and 2-3 layers of secondary cortex composed of thick-walled, tangentially elongated, rectangular cells with intercellular spaces, some cells contain rosette crystals of calcium oxalate and simple starch grains cortex a narrow-zone of parenchymatous cells containing a number of simple starch grains, most of cortical cells also contain large rosette crystals of calcium oxalate, endoderm is and pericycle absent.vascular bundles, arranged in a ring, collateral, conjoint and open, phloem tissues cornposed of sieve elements and parenchyma, in outer region found as compressed masses while in inner region intact. a number of rosette crystals of calcium oxalate also found as crystal fibres, cambium present as continuous ring composed of 2-3 layers of thinwalled, tangentially elongated cells, xylem consist of fibres, tracheids, vessels and parenchyma, with centre occupied by large pith composed of circular to oval, parenchymatous cells, varying in size and containing starch grains with crystals of calcium oxalate similar to those found in cortical region (Mehra & Raina, 1971; Kapoor & Kapoor, 1976).

Chemical constituents: Coumarin (bergenin), gallic acid, tannic acid, (+) –Afzelechin, minerals and wax (Dix & Srivastava, 1989; Umashankar *et al.*, 1999; Chauhan, Singh & Agrawal, 2000).

Specific action as per Ayurevda: *Ashmaribhedaka* (lithinotriptic).

Therapeutics: Renal calculus, leucorrhoea and opium poisoning (Ballabh *et al.*, 2008).

Parts used: Seeds.

Bioactivity: Antilithiatic (Seth *et al.,* 1974; Joshi *et al.,* 2005) and antidiabetic (Anonymous, 2009).

Further reading

Anonymous. Antidiabetic activity of ethanolic extract of root of *Bergenia ligulata* in alloxan diabetic rats. *Indian Drugs* (Pharmaceut. Manufact.) 2009; 46: 3 (article 11).

Ballabh B, Chaurasia OP, Ahmed Z, Singh SB. Traditional medicinal plants of cold desert Ladakh-used against kidney and urinary disorders. *J Ethnopharmacol* 2008;118:331-339.

Chauhan SK, Singh B, Agrawal S. Simultaneous determination of bergenin and gallic acid in Bergenia ligulata wall by high-performance thin-layer chromatography. *J AOAC Int* 2000;83:1480-1483.

Dix BS, Srivastava SN. Tannin constituents of *Bergenia ligulata* roots. *Ind J Nat Prod* 1989; 5:24-25.

Gehlot NK, Sharma VN, Vyas DS. Some pharmacological studies on ethanolic extract of *Bergenia ligulata. Ind J Pharmacol* 1976;8:92.

Joshi VS, Parekh BB, Joshi MJ, Vaidya AD. Inhibition of the growth of urinary calcium hydrogen phosphate dihydrate crystals with aqueous extracts of *Tribulus terrestris* and *Bergenia ligulata. Urol Res* 2005; 33:80-86.

Kapoor SL, Kapoor LD. On the botany and distribution of `pashanbheda'. *Sach Ayur* 1976; 28: 769-791.

Mehra PN, Raina MK. Pharmacognosy of Paashaanbhed. *Quart J Crude Drug Res* 1971;11:1683.

Seth SDS, Prabhakar MC, Bapna BC, Arora RB. Studies on the antilithiatic property of *Berginia ligualata. J Res Indian Med* 1974; 9:1-3.

Umashankar D. Reddy C, Chawla SA, Deepak M, Singh D, Handa SS. High Pressure Liquid Chromatographic Determination of Bergenin and (+) -Afzelechin from Different Parts of Paashaanbhed (*Bergenia ligulata* Yeo). *Phytochem Anal* 1999;10: 44-47.

Bryophyllum calycinum **Salisb.**

Syn: *Bryophyllum pinnatum* (Lam.) Kurz.

Ayurvedic names: Parnabeeja.

English name: Airplant.

Family: Crassulaceae.

Habitat: It is a herb common in India. It is found in USA, Brazil, Nepal, and Pakistan.

General Description: This is a perennial succulent, glabrous, about 1 meter tall herb. It has obtusely qudrangular stem, the younger parts are reddish in color. The leaves are opposite, decussate, the lowers are usually simple and upper usually 4 - 7 foliate, flashy and roundish with dentate margins, long petioled. The lateral veins are furnished with rooting vegetative buds. The flowers are reddish purple, pendulus, in large spreading panicles. The fruits are membranous follicles. The seeds are small, round or oval (Jain, 2008).

Chemical constituents: Citric acid, malic acid and flavonoids (Devbhuti *et al.,* 2008).

Specific action as per Ayurevda: *Raktastambhana* (styptic).

Therapeutics: In Ayurveda, the leaves are used for treatment of vomiting of blood, acute inflammation, and menorrhagia

Parts used: Leaves.

Further reading

Devbhuti D, Gupta JK, Devbhuti P, Bose A. Phytochemical and acute toxicity study on *Bryophyllum calycinum* Salisb. *Acta Pol Pharm* 2008; 65: 501-504.

Jain VC. Microscopical and preliminary phytochemical studies on leaves of *Bryophyllum calycinum* Salisb. *J Pharm Res* 2008; 1.

Aerva lanata Juss.ex Schult.

Ayurvedic names: *Gorakshganja.*

Family: Amaranthaceae.

Habitat: It is a common weed which grows wild everywhere in plains of India.

General Description: A semi erect many branched under shrub grows up to 50 cm in height. Leaves are simple, alternate, short petioled, tomentose, and become smaller in the flowering twigs. Flowers are small sessile, greenish or whitish, often found in spikes. Fruits are greenish round compressed utricle, seeds kidney shaped and small (Prasad, Sankarasubramanian & Guruswamy, 1986).

Chemical constituents: α-amyrin, β-sitosterol and lupeol (Srinivasan *et al.,* 2008).

Specific action as per Ayurveda: *Ashmaribhedaka* (lithontriptic).

Therapeutics: Headache, renal calculus and dysuria.

Parts used: Roots.

Bioactivity: Diuretic (Udupihille, 1986), hypoglycemic and anti-hyperlipidaemic (Vetrichelvan & Jegadeesan, 2002; Appia *et al.*, 2009) and anthelmintic (Rajesh, Chitra & Paarakh, 2010).

Further reading

Appia, KG, Rai VK, Nandy BC, Meena KC, Dey S, Tyagi PK, Tyagi LK. Hypoglycemic and antihyperlipidaemic effect of ethanolic extract of aerial parts of *Aerva lanata* Linn. in normal and alloxan induced diabetic rats. *Int J Pharm Sci Drug Res* 2009;1:193-194.

Prasad KVSRG, Sankarasubramanian S, Guruswamy MN. Pharmacognostical studies on the roots of *Aerva lanata*. *Arogya J Health Sci* 1986; 12:6-13.

Rajesh R, Chitra K, Paarakh PM. *In vitro* anthelmintic activity of aerial parts of *Aerva lanata* Linn Juss. *Int J Pharm Sci Drug Res* 2010;2: 269-271.

Srinivasan GV, Geetha Devi KT, Balachandran I. Comparison of lupeol content in *Aerva lanata* and *Rotula aquatica* by HPLC method. *Aryavaidyan* 2008;21: 200-203.

Udupihille M, Jiffry MTM. Diuretic effect of *Aerva lanata* with water, normal saline and coriander as controls. *Indian J Physiol Pharmacol* 1986; 30: 91-97

Vetrichelvan T, Jegadeesan M. Anti-diabetic activity of alcoholic extract of *Aerva lanata* (L.) Juss. ex Schultes in rats. *J Ethnopharmacol* 2002; 80: 103-107.

Aerva javanica Juss.ex Schult.

Syn: *A. persica* (Burm.f.) Merrill, *Iresine javanica* Burm.f.

English name: Kapok bush, snow bush and aerva.

Family: Amaranthaceae.

Habitat: India.

General Description: Shrubby perennial herb to 1.6 m high. Leaves linear to narrow-ovate, 2–7 cm long, 0.3–1.2 cm wide, bluish-green to whitish-hairy, more hairy on lower surface and leaf narrowing to stem. Fruit an inflated capsule 1–1.5 mm long. Seeds rounded, about 1 mm long and 0.75 mm wide, red-brown, shiny.

Bioactivity: Antimicrobial (1998) and hyperglycaemic (Reddy, 2009).

Further reading

Gehlot D, Bohra A. Antimicrobial activity of various plant part extracts of *Aerva persica*. *Advances in Plant Sci* 1998;11:109-111.

Reddy SK. Hyperglycaemic Activity of Ethanol Extract of *Aerva javanica* Leaves in Alloxan- Induced Diabetic Mice. *J Pharm Res* 2009;2.

Bridelia crenulata Roxb.

Family: Euphorbiaceae.

Habitat: Common on the slopes, inside savanna forests, 500 m - 1400 m. Indian sub-continent.

General Description: A medium sized deciduous tree. Flowers clusters or spikes, greenish to purple. Flowering from May-June. Fuit in crowded drupes tinged red, finally white, pulpy, globose. Seed 1. Fruiting from July-September.

Chemical constituents: Luteoforol (Ramesh *et al.,* 2001)

Specific action as per Ayurveda: Astringent.

Parts used: Stem bark.

Bioactivity: Antimicrobial (Ramesh *et al.,* 2001).

Further reading:

Ramesh N, Viswanathan MB, Saraswathy A, Balakrishna K, Brindha P, Lakshmanaperumalsamy P. Phytochemical and antimicrobial studies of *Bridelia crenulata*. *Pharml Biol* 2001;39: 460-464.

Ramesh N. Viswanathan MB, Saraswathy A, Brindha P, Balakrishna K, Lakshmanaperumalsamy P. Antibacterial activity of luteoforol from *Bridela crenulata*. *Fitoterapia* 2001;72: 409-411.

Coleus amboinicus Lour.

Syn: *Plectranthus amboinicus* (Lour.) Spreng.

English name: Country borage.

Ayurvedic name: *Parnayavani*.

Family: Lamiaceae.

Habitat: *C. amboinicus* is found in India, USA, Carribians and Australia.

General Description: This is a semi spreading herb, it grows to a height of 0.5 meter. The leaves are thick, fleshy, oval to round with small teethed margin. The leaves and stem have characteristic aroma. The flowers are small and pale purple, arranged in panicle (Datta, 1956; Rout *et al.*, 2010; Patel *et al.*, 2010).

Chemical constituents: Essential oil (contains carvacrol).

Specific action as per Ayurveda: *Grahi.*

Therapeutics: Diarrhea associated with cholera, malabsorption, dysuria and abdominal colic.

Parts used: Leaves.

Bioactivity: Antiurolithiatic (Chamachar, 1967; Ghosh *et al.*, 2000), antibacterial (Jeyakumar, Gowri & Rajasekarapandian, 2001), diuretic (Sur *et al.*, 2003), anti-inflammatory (Anonymous, 2010),

Further reading

Anonymous. Evaluation of anti-inflammatory activity of methanolic extract of *Coleus aromaticus* leaves. Indian Drugs (Pharmaceut. Manufact.) 2010;47, 12 (article 9).

Chamachar BR. Use of 'pashanabheda' *Coleus aromaticus* in ashmari (urinary calculi). *Nat Med J Mysore* 1967;9: 475-477.

Datta DD. Pharmacognostic studies of the plant *Coleus aromaticus* Benth. *Indian J Pharm* 1956;18: 163-168.

Ghosh RB, Sur TK, Maity LN, Chakraborty SC. Antiurolithiatic activity of *Coleus aromaticus* Benth. in rats. *Ancient Sci Life* 2000;20: 44-47.

Jeyakumar N, Gowri AR, Rajasekarapandian M. *In vitro* studies on antibacterial activities of *Coleus aromaticus* Linn. J. Foseph Thas (Ed.), 2001;93-94 (abstract).

Patel R, Mahobia N, Waseem N, Upwpwar N, Singh S. Phyto-Physicochemical Investigation of Leaves of *Plectranthus amboinicus* (Lour) Spreng. *Pharmacog J* 2010;2 :536-542.

Rout, OP, Rout KK, Acharya R, Mishra SM. Pharmacognostical and phytochemical evaluation of *Coleus aromaticus* Benth. Leaf. *Int J Pharm World Res* 2010;1: 4 (article 3).

Sur TK, Pandit S, Biswas TK, Ghosh RB, Bhattacharyya D. Diuretic activity of *Coleus aromaticus* Benth. on rats. *Ancient Sci Life* 2003; 22: 146-151.

Decalepis arayalpathra (J. Joseph & V. Chandras.) Venter

This plant is also known as *pashanbheda* (Unnikrishnan, 2004). The Kani tribe of the Trivandrum district of Kerala uses the tuberous roots of this plant to treat peptic ulcers. Ethanolic extract has gastric antisecretory and anti ulcer activity (Shine *et al.*, 2007).

Further reading

Unnikrishnan, E. (2004). Materia Medica of the Local Health Traditions Level Development Centre for Development Studies. Thiruvanthapuram.

Shine VJ, Shyamal S, Latha PG, Rajasekharan S. Gastric Antisecretory and Antiulcer Activities of *Decalepis arayalpathra*. *Pharm Biol* 2007;45:210-216.

Homonoia riporia Lour. syn *Lumanaja fluvatilis* Blanco

Syn: *Lumanaja fluviatilis* Blanco

English name: Willow-Leaved Water Croton.

Family: Euphorbiaceae

Ayurvedic names: *Jalavetasa, kshudra pashanbheda* and *pashanbheda*ka.

Habitat: It is widespread from India to China, Taiwan and throughout Malaysia to New Guinea.

General Description: Trees up to 5 m tall. Trunk & Bark: Bark brownish, flaky; blaze light yellow. Branches and Branchlets: Branchlets terete, pubescent when young. Leaves : Leaves simple, alternate, spiral; stipules in pairs, caducous and leaving scar; petiole to 1 cm; lamina to 16.5 × 2 cm, linear-lanceolate, apex acute, base cuneate, margin entire, subcoriaceous, glandular scales beneath; midrib raised above; secondary nerves 15-19 pairs; tertiary nerves reticulate. Inflorescence / Flower: Inflorescence in axillary spikes; flowers reddish, unisexual, dioecious. Fruit and Seed: Capsule, 3-2 valved cocci; seeds 3, ovoid.

Therapeutics: Chopra explains use of decoction of the root as diuretic and laxative. Isoflavonoids have been reported form the plant (Viswanadh *et al.,* 2005).

Further reading

Viswanadh GS *et al.*, *Homonoia riporia. J Trop Med Plants* 2005; 6:49-54.

Rotula aquatica Lour.

Syn: *Carmona viminea* (Wallich) G.Don; *Ehretia viminea* Wallich; *Rhabdia lycoides* Mart.

Family: Boraginaceae.

Habitat: It is widely distributed in India from kumaun to Assam and western to southern India.

General Description: *R. aquatica* is a shrub or bush reaches a height of 2 to 3 meters. The leaves are alternate, imbricate, ovate-oblong, usually 0.8 to 1.5 centimeters in length, and short-stalked. The flowers are small and crowded, and are borne on terminal inflorescences. There are five sepals which are about 4.5 millimeters long, and narrow, with sharp points. The corolla is purple or pink, 5 to 6 millimeters long, and five-lobed. The fruit is somewhat rounded, fleshy, and about 4 millimeters in diameter, and contains four seeds.

Chemical constituents: Tannins (Patil *et al.,* 2004).

Actions: Anticancer, sudorific, diuretic and laxative.

Therapeutics: Fever, diabetic nephropathy and anuria.

Parts used: Whole plant.

Bioactivity: Psychoactive (Nayar, Kumar & Pushpangadan, 1999) and decoction ahs antilithic property (Christina, Priya Mole & Moorthy, 2002).

Further reading

Christina AJ, Priya Mole M, Moorthy P. Studies on the antilithic effect of Rotula aquatica lour in male Wistar rats. *Methods Find Exp Clin Pharmacol* 2002 24:357-359.

Mamta K, Rohit AB. Pharmacognostic evaluation of the root of *Rotula aquatica. Int J Pharm Bio Sci* 2010;5: 1-4.

Mengi SA, Bakshi VJ. Evaluation of the aqueous extract of *Rotula aquatica* roots in acute and chronic inflammatory conditions in rats. *Int J Drug Delop Res* 2009;47491-495.

Patil S, Narayanan S, Eibl G, Jolly CI. Evaluation of antimitotic activity of *Rotula aquatica* (Lour): a traditional herb used in treatment of cancer. *Indian J Exp Biol* 2004;42:893-899.

Raut AA, Sunder S, Sarkar S, Pandita NS, Vaidya AD. Preliminary study on crystal dissolution activity of *Rotula aquatica, Commiphora wightii* and *Boerhaavia diffusa* extracts. *Fitoterapia* 2008;79:544-547.

Didymocarpus pedicellata R.Br

Syn: *Carmona viminea* (Wallich) G.Don; *Ehretia viminea* Wallich; *Rhabdia lycoides* Mart.

Ayurvedic names: *Shilapushpa, shantapushpi* and sometimes *pasanbheda.*

Family: (Gesneriaceae).

Habitat: India.

General Description: *D. pedicellata* is a small herb with a reduced stem, bearing 2-3 pairs of opposite, roundy ovate, glabrous, glandular-punctate leaves, 3-6 inches in diameter

Chemical constituents: Chalcones (Rathore, Garg & Gupta, 1981): pashanone (Agarwal, Bhaskar & Sheshadri), polyterpenes: didymocarpol and didymacarpenol, flavonoid: didmyocarpin (Bose & Chauadhary, 1978; Garg, Gupta & Sharma, 1979), isodidmyocarpin (Bose & Chauadhary, 1978), pedicin, isopedicin and pedicellin (Sharma & Siddiqui, 1939) and pediflavone (Guha & Bhattacharya, 1992). Dicarboxylic acid: pedicellic acid is considered to be the active principle of *D. pedicellata* extracts (Rao *et al.,* 1966). Essential oil contains didymocarpene.

Actions: Anticancer, sudorific, diuretic and laxative.

Therapeutics: Traditionally *D. pedicellata* is used in the treatment of renal diseases particularly kidney stones (Kapoor & Kapoor, 1976).

Bioactivity: Antioxidant.

Further reading

Agarwal SC, Bhaskar A, Sheshadri TR. Constituents of the roots of *Didymocarpus pedicellata*. Isolation and structure of pashanone, a new chalcone. *Indian J Chem* 1972; 12:2-5.

Bahl CP, Seshadri TR Eds. Pashanbhedi: drugs for urinary calculus, K.N. Udupa 1978, 77-98.

Bose PC, Chauadhary N. Didmyocarpin, a new flavanone from *Didymocarpus pedicellata*. *Phytochem* 1978, 17:587-8.

Bose PC, Chauadhary N. Isodidmyocarpin, a new chalcone from *Didymocarpus pedicellata*. *J Indian Chem* 1978, 25:1198-1200.

Garg SK, Gupta SR & Sharma ND. Synthesis of 7-hydroxy-5, 6, 8-trimethoxyflavone: revision of structure of didmyocarpin. *Indian J Chem* 1979, 17B:394-5.

Guha PK, Bhattacharya A. 5, 8-dihydroxyflavone from the immature leaves of *Didymocarpus pedicellata. Phytochemistry* 1992, 31: 1833-34.

Kapoor SL, Kapoor LD. On the botany and distribution of `pashanbheda', *Sachitra Ayurved* 1976, 28: 769-791.

Kaur G *et al.* Protective effect of *Didymocarpus pedicellata* on ferric nitriloacetate induced renal oxidative stress and hyperprolifer-ative response. *Chem Biol Interact* 2007, 165:33-34.

64. PATHA

Current representative: *Cissampelos pareira*

Possible representative: *Cyclea peltata*

Cissampelos pareira Linn.

Common name: Velvet-leaf.

Ayurvedic names: *Patha, akanadi, vanatikta* and *pilupaha*.

Family: Menispermaceae.

Habitat: India and Sri Lanka.

General Description: Climber.

Detailed Descirption of the Root (Prasad, Gupta & Bhattacharya, 1962)

a) *Macroscopic*

Roots, cylindrical, often tortuous, 1-1.5 cm in diameter, light brown to yellowish in colour, surface rough and at places rugged due to transverse wrinkles, cracks and fissures, fracture short and splintery, odour, faint aromatic, taste, bitter.

b) *Microscopic*

Transvarse section of root shows, 6-10 layers of thin-walled, rectangular cork cells secondary cortex, 1-3 layered of oval to tangentially elongated cells, discontinuous ring consisting of 2-3 rows of stone cells and group of phloem fibres, stone cells variable in shape with simple pits, vascular strands as radiating strips usually 8-12 of xylem and phloem some reaching up to the centre, phloem consists of small strands of sieve elements and parenchyma just below the ring of stone cells, xylem consists of vessels, tracheids, fibres and xylem parenchyma, vessels and tracheids show simple pits on the walls, xylem parenchyma usually thick-walled and lignified but due to delignification patches of thin-walled parenchyma appear in the xylem region., medullary rays 1-3 seriate appear to be very wide at a number of places due to addition of delignified xylem parenchymatous cells, ray cells thin-walled, a few lignified and thick-walled while some show reticulate thickening, plenty of starch grains present in some of ray cells.

Chemical constituents: Protoberberine alkaloids: pelosine or cissampariene, bebeerine, hyatidine, hayatinine and pareirubrines A and B (Bodeker, 1849, Bhattacharji, Sharma & Dhar, 1962;

Bhatnagar *et al.,* 1967; Kupchan, 1967; Bhatnagar & Popli, 1967; Anwer *et al.,* 1968).

Actions: Bitter and galactagouge.

Therapeutics: Fever and urinary tract infections. Expressed juice is given in migraine. Locally the root finds use in snake bite.

Parts used: Roots.

Bioactivity: Antioxidant (Amresh, Rao & Singh, 2007), antifertility (Ganguly *et al.,* 2007), anticancer (Amresh, Rao & Singh, 2007), anti-inflammatory and analgesic (Amresh *et al.,* 2007), antinociceptive and antiarthritic (Amresh, Rao & Singh, 2008).

Further reading

Amresh G, Rao Ch V, Singh PN. Antioxidant activity of *Cissampelos pareira* on benzo (a) pyrene-induced mucosal injury in mice. *Nut Res* 2007; 27:625-32.

Amresh G, Rao Ch V, Singh PN. Evaluation of *Cissampelos pareira* against gastric cancer and enzymes associated with carcinogen metabolism. *Pharm Biol* 2007;45:595-603.

Amresh G, Rao Ch V, Singh PN. Antinociceptive and antiarthritic activity of *Cissampelos pareira* roots. *J. Ethnopharmacol* 2008; 111: 531-536.

Amresh G, Zeashan H, Singh PN, Rao Ch V. Prostaglandin mediated anti-inflammatory and analgesic activity of *Cissampelos pareira. Acta Pharma Sci* 2007;49:153-160.

Anwer F, Popli SP, Srivastva RM, Khare MP. Studies in medicinal plants 3. Protoberberine alkaloids from the roots of *Cissampelos pareira* Linn. *Experientia* 1968; 15.

Bhatnagar AK, Popli SP. Chemical examination of the roots of *Cissampelos pareira* Linn.: III.-Structure and stereochemistry of hayatinine. *Ind J Chem* 1967;5:102-104.

Bhatnagar AK, Bhattacharji S, Roy AC, Popli SP, Dhar ML. Chemical examination of the roots of *Cissampelos pareira* Linn. IV Structure and stereochemistry of hyatidine. *J Org Chem* 1967; 32:819-820.

Bhattacharji S, Sharma VN, Dhar ML. Chemical examination of the roots of *Cissampelos pareira* Linn. *J Sci Indust Res* 1962; 11B: 81-82.

Bodeker C. Constitutions of pelosine and bebeerine. *Ann Chem Pharm* 1849;69:37.

Dandiya PC, Chopra YM. *Cissampelos pareira* Linn. *Indian J Pharm* 1970;2:67.

Ganguly M., Borthakur M, Devi N, Mahanta R. Antifertility activity of the methanolic leaf extract of *Cissampelos pareira in female albino rats. J Ethnopharmacol* 2007; 111:688-61.

Kupchan MS, Patel AC, Fujita E. Tumor inhibitors VI. Cissampareine, new cytotoxic alkaloid from *Cissampelos pareira*. Cytotoxicity of bisbenzylisoquinoline alkaloids. *J Pharm Sci* 1964;54:580- 583.

Mukerji B, Bhandari PR.*Cissampelos pareira* source of a new curariform drug. *Planta Med* 1959;7:250-259.

Pradhan SN, Pandey K, Badola R. A clinical trial of hayatine methiodide as a relaxant in 100 cases. *Brit J Anest* 1964; 36:604-611.

Prasad S, Gupta KC, Bhattacharya LC. Pharmacognostical study of *Cissampelos pareira* Linn. *J Sci Indust Res* 1962; 21:150-154.

Rao V, Mehrotra S, Shirwaikar A. *Standardization and ethnopharmacological evaluation of antidiarrhoeal herbal formulation*. Manipal Academy of Higher Education, Manipal (Dissertation), Karnataka, India, 2003; pp. 24-40.

Roy PK, Dutta AT, Ray GK, Mukerji B. A preliminary note on the pharmacological action of the total alkaloids isolated from *Cissampelos pareira* Linn. (False Pareira brava). *Med Res* 1952;40: 95-100.

Sur RN, Pradhan SN. Studies on Cissampelos alkaloids: L Action of hayatin derivatives on the central nervous system of cat and dogs. *Archiv Int Pharmacodyn Ther* 1963;152:6-14.

Cyclea peltata (Lam.) Hook.f. & Thomas.

Syn: *C.barbata*

English name: Green grass jelly.

Family: Menispermaceae.

General Description: The plant is a slender twining shrub, frequently climbing up on tall trees. The leaves are simple, alternate, heart shaped, 2.5-10 cm long and 2.5-3.75 cm broad, stipule 5-10 cm long and nerves 7-11. The flowers unisexual, pale yellow, in axillary panicles. The fruits are ovoid drupes, brown or scarlet in color. The seeds are covered. The roots are tuberous, cylindrical, irregularly curved, with grayish brown surface. The plant blooms in the rainy season (Prasad, Girija Devi & Hepsibah, 1995).

Habitat: *C. peltata* Lam. grows throughout India and Sri Lanka, up to 800-900 metres elevation.

Chemical constituents: The roots of *C. peltata* are reported to contain alkaloids including *d*-tetrandrine, *dl*-tetrandrine, *d*-isochondrodendrine, fangchinoline, tetrandrine N-2'-oxide, α-cyclanoline, tetrandrine 2'β- N-oxide, (-)-2-norlimacine, (-)-curine, (-)-cycleapeltine, (-)-N-methylcoclaurine, (-)-repandine, (+)-coclaurine, (+)-cycleabarbatine, (+)-cycleanorine, coclaurine and cycleadrine (Kupchan, Yokoyama and Thyagarajan, 1953).

Therapeutics: In tribal medicine, jelly sediment obtained from *C. peltata* is applied to head and washed after 30 minutes in frothy stool in Kerala (Vijayan *et al.*, 2006). The roots are used in coryza, hemorrhoids, diarrhea and burning sensation in Mangalore (Shiddamallayya, Yasmeen & Gopalkumar, 2006).

Parts used: Roots.

Bioactivity: Modulator effect on stone formation (Cristina, *et al.*,2002) and neuroprotective (Vijayan *et al.*, 2007).

Further reading

Christina AJM, Packia Lakshmi M, Nagarajan M, Kurian S. Modulatory effect of *Cyclea peltata* Lam. on stone formation induced by ethylene glycol treatment in rats. *Methods and Findings in Experimental and Clinical Pharmacology* 2002; 24:77.

Kupchan SM, Yokoyama N, Thyagarajan BS. Menispermaceae alkaloids II. The alkaloids of *Cyclea peltata* Diels. *Journal of Pharmaceutical Sciences* 1953; 50:164-167.

Prasad NBR, Girija Devi RS, Hepsibah PTA. *Cyclea peltata* Diels - a possible substitute for *Cissampelos pareira* Linn. *Ancient Science of Life* 1995; 15: 150-152.

Shiddamallayya N, Yasmeen A, Gopalkumar K. Medico-botanical survey of *Kumar parvatta* Kukke Subramanya, Manglore, Karnataka. *Indian Journal of Traditional Knowledge* 2010; 9: 96-99.

Vijayan A, Liju VB, Reena John JV, Parthipan B, Renuka C. Traditional remedies of Kani tribes of Kootor reserve forests, Agasthyavanam, Thiruvananthapuram, Kerala. *Indian Journal of Traditional Knowledge* 2006; 6: 589-594.

Vijayan FP, Rani VKJ, Vineesh VR, Sudha KS, Michael MM, Padikkala J. Protective effect of *Cyclea peltata* Lam on cisplatin-induced nephrotoxicity and oxidative damage. *J Basic Clin Physiol Pharmacol* 2007, 18:101-114.

65. PATLA

Current representative: *Stereospermum suaveolens*

Possible representative: *Stereospermum tetragonum*

Stereospermum suaveolens (Roxb.) DC.

Ayurvedic names: *Kamaduta, kumbhika, kalavrntika, svalpamedha, madhorduti, tamrapuspa, ambuvasini, phaleruha, shveta* and *krsna patala.*

Family: Bignoniaceae.

Habitat: Sub-Himalayan tracts, Rajasthan, Chota Nagpur and Central and Peninsular parts of India.

Habit: A medium sized deciduous tree with grey to dark brown exfoliating bark, pinnately compound leaves, elliptical leaflets, purple, fragrant flowers in large panicles and rough, grey bark (Prasad & Prakash, 1966/67).

Chemical constituents: Flavonoids (Subramanian, Nagarajan & Sulochana, 1972).

Actions: Carminative, diuretic and cardiac tonic.

Therapeutics: General debility and dyspepsia.

Parts used: Bark.

Bioactivity: Anti-inflammatory (Balasubramanian *et al.,* 2010), hepatoprotective (Chandrashekhar *et al.,* 2010), Antiulcer and gastroprotective (Muchandi & Chandrashekhar, 2011)

Further reading

Balasubramanian T, Chatterjee TK, Sarkar M, Meena SL. Anti-inflammatory effect of *Stereospermum suaveolens* ethanol extract in rats. *Pharm Biol* 2010; 48:318-323.

Chandrashekhar VM, Muchandi AA, Sudi SV, Ganapty S. Hepatoprotective activity of *Stereospermum suaveolens* against CCl4-induced liver damage in albino rats. *Pharm Biol* 2010; 48:524-528.

Muchandi AA, Chandrashekhar VM. Antiulcer and gastroprotective potential of *Stereospermum suaveolens* in wistar rats. *J Pharmacol Pharmacother* 2011; 2:117-119.

Prasad S, Prakash A. Pharmacognostical studies on root and stem bark of *Stereospermum suaveolens*. *J Sci Res* 1966/1967; 17:101-111.

Subramanian S. Nagarajan SS, Sulochana N. Flavonoids of the leaves of *Stereospermum suaveolens. Curr Sci* 1972; 41:102-103.

Stereospermum tetragonum DC.

Syn: *S. personatum* (Hassk.) D. Chatterjee

English name: Yellow snake tree.

Family: Bignoniaceae.

Habitat: India.

Description:

Macroscopy: Drug occurs in pieces of variable size and thickness, external creamish brown in colour, rough due to longitudinal fissures and ridges, slightly recurved, when cut across it shows lamellations due to presence of concentric bands of phloem fibres; inner surface tough and short, occasionally peeling off light yellow to cream in colour, fracture, taste and odour not characteristic.

Microscopy: Diagrammatic TS of the bark shows tangentially elongated multilayered layered cork followed by 3 or 4 layers of cork cambium. Cortex wide parenchymatous, consist of groups of stone cells. Phloem wide consists of band of lignified phloem fibres interrupted by phloem rays. Tangential band of cerantenchyma present in the phloem region.

Detailed transverse section of the bark shows cork consisting of about 10 to 20 layers of tangentially elongated, thin-walled, rectangular cells followed by 3 or 4 layers of cork cambium. Cortex very wide composed of tangentially elongated, thick-walled parenchymatous cells having numerous, mostly groups of thick-walled, lignified stone cells of various sizes and cerantenchyma cells and medullary rays. Phloem composed of ceratenchyma, phloem parenchyma, fibres and rays cells. Ceratenchyma present in the form of thick-walled tangential strips between two obliquely running rays. Phloem fibres lignified, mostly in groups arranged in concentric bands interrupted by phloem rays. Rays cells mostly multi seriate, fairly large, 2 to 4 cells wide. Starch grains and prismatic crystals of calcium oxalate embedded in the parenchyma cells of the cortex and phloem.

Powder: Brown; fragments of lignifed, thick-walled, oval to polygonal stone cells with wide lumen and pits upto 120 μ in length, fibres with small tapering and pointed ends upto 1500 μ in length. Cork cells in surface and sectional view, prismatic crystals of calcium oxalate present. Starch grains simple and compound 2 or 3 in groups upto 12 μ in diameter.

Chemical constituents: Lapachol (Purushothaman & Natarajan, 1974).

Parts used: Roots.

Bioactivity: Anticancer and antiviral (Rastogi & Dhawan, 2004).

Further reading

Purushothaman KK, Natarajan RK. Chemical examination of patala (*Stereospermum tetragonum* DC). *J Res Indian Med* 1974; 9:107-108.

66. PITACHANDANA

Current representative: *Jateorrhiza palmata*

Possible representative: *Coscinium fenestartum*

Jateorhiza palmata Miers

Syn: *Jateorhiza columba*

Common name: Calumba.

Ayurvedic name: *Simhalatikta.*

Family: Menispermaceae.

Habitat: East India islands.

General Description: Vine.

Chemical constituents: Bitter principle: calumbin, alkaloids: berberine, palmatine, jatrorrhizine, columbamine, calumbic acid and ash.

Actions: Bitter tonic and febrifuge.

Therapeutics: Diarrhea, poor digestion, dysentery, dyspepsia, nausea and loss of appetite.

Parts used: Roots.

Further reading

Berry A. Title: An account of the male plant which furnishes the medicine generally called Columbo, or Columba root. *Asiatic Res* 1976; 10:385-388.

Mehendale MA. The flora in the Aradnyakaparvan of the Mahabharata. ABORI 1987;67:233-242; reproduced in *Asian Agri-Hist* 2004;8: 103-114.

67. PRIYANGU

Current representative: *Callicarpa macrophylla*

Possible representative: *Aglaia roxburghiana, Prunus mahaleb and Setaria italica.* The source plants of priyangu of the South Indian raw drug market are *Mammea longifolia* Planch. et Triana, *Myristica malabarica* Lam., and *Zanthoxylum budrunga* Wall.

Callicapra macrophylla Vahl.

Synonyms: *C. incana, C. tomentosa*

Ayurvedic names: *Priyangu, phalini, kantha, lata, mahila, cundra, gandhaphala, shyama, viswaksena, kanthapriya* and *angana priya.*

Family: Verbenaceae.

Habitat: Found in Western Ghats and Himalayan peninsula.

General Description: A large shrub or small tree grows up to 6 meters in height. Leaves simple opposite, ovate and acute; flowers rose colored in axillary cymes; fruits are small drupes (Ali, 1975).

Chemical constituents: Diterpenoid: calliterpenone, betulinic acid. Essential oil of the leaves yields 16α, 17-isopropylideno-3-oxo-phyllocladane (isopropylidenocalliterpenone) along with calliterpenone and its monoacetate (Singh *et al.*, 2010).

Actions: Haemostatic.

Therapeutics: Plant pacifies vitiated *vata, pitta*, diarrhea, arthritis, burning sensation, ulcers, poisonous bites and blackish discoloration on face, bleeding plies and menorrhagia.

Parts used: Flowers.

Bioactivity: Anti-inflammatory (Amatya, 1993)

Further reading

Amatya MP. Anti-inflammatory activity of *Callicarpa microphylla*. *J Nepal Chem Soc* 1993;12:39-41.

Ali, US. Contribution to the botanical identity of `priyangu', I, II, III. *J Res Indian Med* 1975;10: 61-64, 65-69, 70-73.

Singh AK, Chanotiya CS, Yadav A, Kalra A. Volatiles of *Callicarpa macrophylla*: a rich source of selinene isomers. *Nat Prod Commun* 2010; 5:269-72.

Singh AK, Agpawal PK. 16oc, 17-isopropylideno-3-oxo-phyllocladane, a diterpenoid from *Callicarpa macrophylla. Phytochemistry* 1994; 37: 587-588

Aglaia roxburghiana (W. & A.) Miq.var beddomei

Syn: *Aglaia odoratissima* Blume, *A. diepenhorstia* Miq., *A. elaeagnoidea* (A. Juss.) Benth.

Family: Meliaceae.

Habitat: West India.

General Description: Large tree, wood bright red; leaves shortly petiolate, pinnate, leaflets 5, rarely 7 or 3, elliptic, glabrescent, obtuse, sometimes acute; panicles rather supra-axillary, pyramidal, elongate; flowers shortly pedicellate, small, 0.2 cm in diameter, yellow; fruits subglobose, small, 0.25 cm in diameter, buff-coloured, orange when ripe; very minutely pilose, rusty, persistent (Sasikala *et al.*, 1999).

Chemical constituents: Tannic acid and triterpenes: roxburghiadiol A and B (Balakrishnan, Kundu & Patra, 1990), a bis-amide of roxburghilin and its two epimers.

Actions: Astringent.

Therapeutics: Liver diseases, diarrhea and leprosy.

Parts used: Seeds.

Bioactivity: Antibacterial and antifungal (Kakrani & Nair, 1982), gastroprotective (Janaki *et al.*, 1998) and anti-inflammatory (Janaki *et al.*, 1999),

Further reading

Balakrishnan A, Kundu AB, Patra A. Roxburghiadiol A and roxburghiadiol B, two 14 alpha-methylsterols from *Aglaia roxburghiana*. *J Nat Prod* 1990; 53:523-536.

Janaki S, Vijayasekaran V, Viswanathan S, Balakrishna K. Anti-inflammatory activity of *Aglaia roxburghiana* var. *beddomei* extract and triterpenes roxburghiadiol A and B. *J Ethnopharmacol* 1999;67: 45-51.

Janaki S, Viswanathan S, Vijayasekaran V, Parvathy SS, Ramanujam TR, Pratibha D, Rao KR. The effect of *Aglaia roxburghiana* (W. et A.) Miq. var. *beddomei* on experimentally-induced ulcers. *Indian J Pharm Sci* 1998; 60: 402-404.

Kakrani HK, Nair G. Antibacterial and antifungal activity of volatile oil from the seeds of *Aglaia odoratissima*. *Fitoterapia* 1982; 53: 107-109.

Sasikala E, Janaki S, Rao B, Vijayasekaran V. Pharmacognostical studies on *Aglaia roxburghiana* Miq. var. *beddomei*. *Ancient Sci Life* 1999; 19: 64-68.

Setaria italica (L.) Beauv.

Common name: Foxtail Millet.

Family: Poaceae.

Habitat: East Asia.

General Description: Foxtail millet is an annual grass with slim, vertical, leafy stems which can reach a height of 120–200 cm (3.9–6.6 ft). The seedhead is a dense, hairy panicle 5–30 cm (2.0–12 in)

long. The small seeds, around 2 mm (less than 1/8 in.) in diameter, are encased in a thin, papery hull which is easily removed in threshing. Seed color varies greatly between varieties.

Chemical constituents: Alpha- and beta-setarins (Naren & Virupaksha, 1990). Flavone glycoside: setaricin (Illyas, 1989).

Further reading

Illyas M 1989. (Other information missing).

Naren AP, Virupaksha TK. Alpha- and beta-setarins: Methionine-rich proteins of Italian millet. *Cereal Chem* 1990; 67: 32-34.

Sharma N, SharmaVK, Seo S-Y. Screening of some medicinal plants for anti-lipase activity. *J Ethnopharmacol* 2005;97: 453-456.

Wankhede DB, Shehnaj A, Raghavendra MR. Carbohydrate composition of finger millet (*Eleusine coracana*) and foxtail millet (*Setaria italica*). *Plant Foods Hum Nut* 1978;28: 293-303.

68. PUNARNAVA

Current representative: *Boerhavia diffusa*

Possible representative: *Trianthema portulacastrum*

Boerhaavia diffusa Linn.

Syn: *Boerhavia repens* L.

Common name: Spreading hogweed.

Ayurvedic names: *Dirgha, dirghapatraka, mandlacchada, prthvika, punarbhu, punarnava, shothaghani, svetamula, varsabhu and visakha.*

Family: Nyctaginaceae.

Habitat: It is native to areas of several continents, including Africa and North and South America, and present as an introduced species in many other areas.

General Description: It is an annual herb forming a prostrate mat or clump with stems up to a meter long. It is green to red in color, hairless except for small lines of hairs near the leaves, and fleshy. The leaves have small round or oval blades up to 4 centimeters long borne on short petioles. Solitary flowers occur in leaf axils. The flower lacks petals but has purple, petallike sepals. The fruit is a curved, cylindrical capsule emerging from the stem.

It is up to half a centimeter long and has two erect, pointed wings on top, where the capsule opens.

Detailed Description of Whole Plant (Sahu *et al.,* 2008)

(a) *Macroscopic*

Stem -greenishpurple, stiff, slender, cylindrical, swollen at nodes, minutely pubescent or n early glabrous, prostrate divericately branched, branches from common stalk, often more than a metre long. Root - wel developed, fairly long, somewhat tortuous, cylindrical, 0.2-1.5 cm in diameter, yellowish brown to brown coloured, surface soft to touch but rough due to minute longitudinal striations and root scars, fracture, short, no distinct odour, taste, slightly bitter.

Leaves -opposite in unequal pairs, larger ones 25-37 mm long and smaller ones 12-18 mm long ovate-oblong or suborbicular, apex rounded or slightly pointed, base subcordate or rounded, green and glabrous above, whitish below, margin entire or subundulate, dorsal side pinkish in certain cases, thick in texture, petioles nearly as long as the blade, slender.

Flowers -very small, pink coloured, nearly sessile or shortly stalked, 10-25 cm, in small umbells, arranged on slender long stalks, 4-10 corymb, axillary and in terminal panicles, bracteoles, small, acute, perianth tube constricted above the ovary, lower part greenish, ovoid, ribbed, upper part pink, funnel-shaped, 3 mm long, tube 5 lobed, stamen 2-3.

Fruit -one seeded nut, 6 mm long clavate, rounded, broadly and bluntly 5 ribbed, viscidly glandular.

(b) *Microscopic*

Stem - Transverse section of stem shows epidermal layer containing multi cellular, uniserite glandular trichome consisting of 9-12 stalked cells and an ellipsoidal head, 150-220 μ long, cortex consists of 1-2 layers of parenchyma, endodermis indistinct, pericycle 1-2 layered, thick-walled often containing scattered isolated fibres, stele consisting of many small vascular bundles often joined together in a ring and many big vascular bundles scattered in the ground tissue, intra fascicular cambium present.

Root-transverse section of mature root shows a cork composed of thin-walled tangentially elongated cells with brown walls in the outer few layers, cork cambium of 1-2 layers of thin walled cells secondary cortex consists of 2-3 layers of parenchymatous cells followed by cortex composed of 5-12 layers of thin-walled, oval to polygonal cells,several concentric bands of xylem tissue alternating with wide zone of parenchymatous tissue present below cortical regions, number of bands vary according to thickness of root and composed of vessels, tracheids and fibres, vessels mostly found in groups of 2-8 in radial rows, having simple pits and reticulate thickening, tracheids, small, thickwalled with simple pits, fibres aseptate, elongated, thick-walled, spindle shaped with pointed ends, phloem occurs as hemispherical or crescentic patches outside each group of xylem vessels and composed of sieve elements and parenchyma, broad zone of parenchymatous tissue, in between two successive rings of xylem elements composed of thin-walled more or less rectangular cells arranged in radial rows, central regions of root occupied by primary vascular bundles, numerous raphides of calcium oxalate, in single or in group present in cortical region and parenchymatous tissue in between xylem tissue, starch grains simple and compound having 2-4 components found in abundance in most of cells of cortex, xylem elements in parenchymatous tissue between xylem elements, simple starch grains mostly rounded in shape and measure 2.75-11 μ in diameter.

Leaves -Transverse section of leaf shows anomocytic stomata on both sides, numerous, a few short hairs, 3-4 celled, present on the margin and on veins, palisade one layered, spongy parenchyma 2-4 layered with small air spaces, idioblasts containing raphides, occasionally cluster crystal of calcium oxalate and orange-red resinous matter present in mesophyll. Palisade ratio 3.5-6.5, stomatal index 11-16, vein islet number 9-15.

Chemical constituents: Alkaloid (punarnavine) and potassium nitrate.

Action: Diuretic.

Therapeutics: Anemia, cardiac edema, ascites and dysuria.

Parts used: Whole plant and seeds.

Bioactivity: Antiinflammatory (Bhalla, Gupta & Bhargava, 1971), diuretic (Gaitonde *et al.,* 1974), anti-convulsant (Adesina, 1979), antifibrinolytic (Jain GK, Khanna, 1989), hepatoprotective (Rawat *et al.,* 1997), hypoglycaemic (Chude *et al.,* 2001) and chemopreventive (Bharali, Azad & Tabassum, 2003).

Further reading

Adesina K Anti-convulsant properties of the roots of *Boerhaavia diffusa. Quarterly J Crude Drug Res* 1979; 17: 84–86.

Bhalla TN, Gupta MB, Bhargava KP. Antiinflammatory activity of *Boerhaavia diffusa* L. *Indian J Med Res* 1971;6: 11–15.

Bharali R, Azad MRH, Tabassum J. Chemopreventive action of *Boerhaavia diffusa* on DMBA-induced skin carcinogenesis in mice. *Indian J Physiol Pharmacol* 2003; 47: 459–464.

Chude MA, Orisakwe OE, Afonne OJ, Gamaniel KS, Vongtau OH, Obi E. Hypoglycaemic effect of the aqueous extract of *Boerhavia diffusa* leaves. *Indian J Pharmacol* 2001; 33: 215–216.

Gaitonde BB, Kulkarni HJ, Nabar SD. Diuretic activity of punarnava (*Boerhaavia diffusa*). *Bull Haffkine Institute* 1974; 2: 24.

Jain GK, Khanna NM. Punarnavoside: a new antifibrinolytic agent from *Boerhaavia diffusa* Linn. *Indian J Chem* 1989; 28: 163–166.

Rawat AKS, Mehrotra S, Tripathi SC, Shome U. Hepatoprotective activity of punaranava—a popular Indian ethnomedicine. *J Ethnopharmacol* 1997; 56: 61–66.

Sahu AN, Damiki L, Nilanjan G, Dubey S. Phytopharmacological review of *Boerhaavia diffusa* Linn.(Punarnava). *Pharmacog Rev* 2008; 2:14–22.

Trianthema portulacastrum L.

Syn: *T. monogyna L.*

Common name: Horse-purslane.

Family: Aizoaceae.

Habitat: It is native to areas of several continents, including Africa and North and South America, and present as an introduced species in many other areas.

General Description: It is an annual herb forming a prostrate mat or clump with stems up to a meter long. It is green to red in color, hairless except for small lines of hairs near the leaves, and fleshy. The leaves have small round or oval blades up to 4 centimeters long borne on short petioles. Solitary flowers occur in leaf axils. The flower lacks petals but has purple, petallike sepals. The fruit is a curved, cylindrical capsule emerging from the stem. It is up to half a centimeter long and has two erect, pointed wings on top, where the capsule opens (Chopra, Chatterjee & Ghosh, 1937;1940).

Chemical constituents: The plant contains water-soluble bases and potassium salts. Punarnavine and a new alkaloid, trianthemine and ecdysterone are present in the aerial parts. They also contain oxalic acid. 5,2'-dihydroxy-7-methoxy-6,8-dimethylflavone and 5,7-dihydroxy-6,8-dimethyl-chromone (leptorumol). Roots contain saponin glycoside.

Therapeutics: *T. portulacastrum* is alexiteric, analgesic, stomachic and laxative; cures bronchitis, piles, ascites and heart diseases. Decoction of the plant is used as vermifuge and in rheumatism. Powdered root is abortifacient and cathartic; used in asthma, amenorrhoea and to remove obstruction of the liver. The leaves are diuretic and used in oedema, dropsy and ascites.

Parts used: Whole plant.

Bioactivity: Anticarcinogenic (Bhattacharya, & Chatterjee, 1998), hepatoprotective: (Kumar *et al.*, 2004; Banu, Kumar & Murugesan, 2009), antioxidant (Kumar, Banu & Paindan, 2005; Shyam Sunder *et al.*, 2010), nephroprotective (Balamurugan *et al.*, 2009), hypoglycemic and hypolipidemic (Balamurugan Banu, Kumar & Murugesan 2010; Shyam Sunder *et al.*, 2010).

Further reading

Anonymous: Anti microbial studies on *Trianthema portulacastrum* Linn extracts. *Indian Drugs* (Pharmaceut. Manufact.) 2009;46: 3(article 12).

Anreddy RNR, Porika M, Yellu NR, Devarakond RK. Hypoglycemic and Hypolipidemic Activities of *Trianthema portulacastrum* Linn. Plant in Normal and Alloxan Induced Diabetic Rats. *Int J Pharmacol* 2010; 6(2): 129.

Balamurugan G, Jagan Mohan CM, Muthusamy P. Protective effect of *Tranthema portulacastrum* Linn leaves on gentamicin induced nephtotoxicity in rats. *J Nat Rem* 2009;9:165-169.

Banu GS, Kumar G, Murugesan AG. Effect of ethanolic leaf extract of *Trianthema portulacastrum* L. On aflatoxin induced hepatic damage in rats. *Indian J Clin Biochem* 2009; 24: 414.

Bhattacharya S, Chatterjee M. Protective role of *Trianthema portulacastrum* against diethylnitrosamine induced experimental hepatocarcinogenesis. *Cancer Lett* 1998; 129: 7–13.

Chopra RN, Chatterjee NR, Ghosh S. A comparative study of *Boerhaavia diffusa* Linn. and the white and red flowered varieties of *Trianthema portulacastrum* Linn. *J Med Sci* 1937;28:475-480; also in *Ind J Med Res* 1940; 28: 475-480.

Kumar G, Banu GS, Pandian MR. Evaluation of the antioxidant activity of *Trianthema portulacastrum* L. *Indian J Pharmacol* 2005;37:331-3

Kumar G, Banu GS, Pappa PV, Sundararajan M, Pandian MR. Hepatoprotective activity of *Trianthema portulacastrum* L. against paracetamol and thioacetamide intoxication in albino rats. *J Ethnopharmacol* 2004;92:37-40.

Shyam Sunder A, Rama Narsimha Reddy A, Kiran G, Thirumurugu S. Antihyperlipidemic and antioxidant activity of methanolic extract of *Trianthema portulacastrum* in rats fed a high-fat diet. *J Herbs, Spices Med Plants* 2010;16: 193-202.

69. RAKATVASA

Current representative: *Adathoda vasica*

Possible representative: *Jacobini tinctoria*

Adhatoda vasica Nees.

Syn: *Justicia adhatoda* L.

Common name: Malabar Nut.

Ayurvedic names: *Vrisha, Atarhusha, Visaka.*

Family: Acanthaceae.

Habitat: It is found throughout the year in plains and sub-Himalayan tracts in India, ascending upto 1200m.

General Description: This is a shrub with lance-shaped leaves 10 to 15 centimeters in length by four wide. They are oppositely arranged, smooth-edged, and borne on short petioles. When dry they are of a dull brownish-green colour. They are bitter-tasting. When a leaf is cleared with chloral hydrate and examined microscopically the oval stomata can be seen. They are surrounded by two crescent-shaped cells at right angles to the ostiole. The epidermis bears simple one- to three-celled warty hairs, and small glandular hairs. Cystoliths occur beneath the epidermis of the underside of the blade.

Detailed Description of the Leaves

(a) *Macroscopic*

Leaves, 10-30 cm long and 3-10 cm broad, lanceolate to ovate-lanceolate, slightly acuminate, base tapering, petiolate, petioles 2-8 cm long, exstipulite, glabrescent, 8-10 pairs of lateral vein bearing few hairs, dried leaves dull brown above, light greyish brown below, odour, characteristic, taste, bitter.

(b) *Microscopic*

Transverse section of leaf shows, dorsiventral surface with 2 layers of palisade cells, in surface view, epidermal cells sinuous with anomocytic stomata on both surfaces, more numerous on the lower, clothing trichomes few, 1-3, rarely upto 5 celled, thinwalled, uniseriate, upto 500 μ and glandular trichomes with nicellular stalk and 4 celled head measuring, 25-36 μ in diameter in surface view, cystoliths in mesophyll layers, elongated and cigar shaped, acicular and prismatic forms of calcium oxalate crystals present in mesophyll , palisade ratio, 5-6, 5-8.5, stomatal index, 10.8-14.2-18.1 for lower surface.

Chemical constituents: Alkaloids: adhatodine, vasicine, vascinone and anisotine, adhatodic acid, volatile oil, vasakin and resin (Atal, 1980; Brantner & Chakraborty, 2001).

Actions: Diuretic.

Therapeutics: Inflammation of liver and spleen and cough.

Parts used: Leaves.

Bioactivity: Anti-anaphylactic, anti-diabetic, antibacterial (Brantner AH & Chakraborty, 1988), anti-inflammatory (Chakraborty & Brantner, 2001), abortifacient (Atal, 1980), anti-tubercular (Gupta & Chopra, 1954), antioxidant (Singh & Padmavathi, 2000), anti-tussive (Dhuley, 1999), bronchodilator (Amih & Mehta, 1959; Johri, & Zutshi, 2000), hepatoprotective (Bhattacharya *et al.*, 2005), hypotensive, thrombopoietic (Sharma & Atal, 1985; Atal *et al.,* 1982), oxytocic (Atal, 1980; Gautam & Sharma, 1981; Sharma & Atal, 1985; Zutshi *et al.,* 1980; Maddapa, Sankaranarayan & Sharma, 1989; Lal & Sharma, 1981), and radio protective (Kumar *et al.,* 2005).

Further reading

Atal CK. Chemistry and Pharmacology of Vasicine: A New Oxytoxic and Abortifacient. RRL, Jammu, India, 1980.

Brantner AH, Chakraborty A. *In vitro* antibacterial activity of alkaloids isolated from *Adhatoda vasica* Nees. *Pharm Pharmacol Lett* 1988;8: 137-39.

Chakraborty A, Brantner AH. Study of alkaloids from *Adhatoda vasica* Nees on their anti-inflammatory activity. *Phytother Res* 2001; 15:532-534.

Chandoke N. *Indian Drugs*. 1987, 24: 425-429.

Chauhan SK, Kimothi GP *et al.* Development of HPLC method for vasicine and vasicinone in *Adhatoda vasica* Nees. *Indian J Nat Prod* 1999; 15: 21-24.

Chowdhury BK, Bhattacharyya P. A further quinazoline alkaloid from *Adhatoda vasica. Phytochemistry* 1985; 24: 3080-3082.

Dhar KL. *et al.* Vasicol, A new alkaloid from *Adathoda vasica, Phytochemistry* 1981; 20: 319-321.

Dhuley JN. Antitussive effect of *Adhatoda vasica* extract on mechanical or chemical stimulation-induced coughing in animals. *J Ethnopharmacol* 1999; 67:361-365.

Mathew AS, Patel KN, *et al.* Investigation on antifeedant and anthelmintic potential of *Adhatoda vasica Nees. Indian J Nat Prod* 1988;14: 11-16.

Nath D, Sethi N *et al.* 1992. Commonly used Indian abortifacient plants with special reference to their teratologic effects in rats. *J Ethnopharmacol* 1992; 36: 147-154.

Pundarikakshudu K, Bhavsar GC. Variations in pharmacognostic characters and alkaloid content in two types of *Adhatoda vasica* Nees. *Int J Crude Drug Res* 1988; 26: 88-96.

Rajani M, Pundarikakshudu K. A note on the seasonal variation of alkaloids in *Adhatoda vasica* Nees. *Int J Pharmacog* 1996;34:308-09.

70. RATANJOTA

Current representative: *Alkanna tinctoria*

Possible representative: *Onosma echioides, Geranium wallichianum, Clausena pentaphylla* and *Potentilla nepalensis.*

Alkanna tinctoria (L.) Taush.

Syn: *Alkanna tuberculata* (Forssk.) Meikle., *Anchusa tinctoria* (L.) L.

English name: Alkanna.

Family: Boraginaceae.

Habitat: S. Europe.

General Description: The species are hispid or pubescent herbs, with oblong, entire leaves, and bracteated racemes, rolled up before the flowers expand. The corolla is rather small, between funnel and salver-shaped; usually purplish-blue, but in some species yellow or whitish; the calyx enlarges in fruit. The root, which is often very large in proportion to the size of the plant, yields in many of the species a red dye from the rind.

Chemical constituents: Napthoquinone: Alkannin (Leo, *et al.,* 1992; Mita, *et al.,* 1994; Gerardi, *et al.,* 1998) and pyrrolizidine alkaloids: 7-angeloylretronecine, triangularine and dihydroxytriangularine.

Actions: Alterative and purgative.

Therapeutics: Varicose veins.

Parts used: Roots.

Bioactivity: Antimicrobial antioxidant (Assimopoulou, Papageorgiou & Boskou, 2004)

Further reading

Assimopoulou AN, Papageorgiou VP, Boskou D. *Food Chemistry*. 2004, 87: 433-438.

Leo P De. *et al. Agricoltura Mediterranea*. 1992, 122: 334-339.

Mita G *et al. Plant Cell Reports*. 1994, 13: 406-410.

Gerardi C *et al.* 1998. 10:14-27 (Ed. Y.P.S. Bajai) Springer-Verlag.

Note: The description has been compiled from PDR for Herbal Remeides, American Herbal Pharmacopoeia and Pharmacographia Indica.

Onosma echioides L.

Family: Boraginaceae.

Habitat: S. Europe to N. Africa.

General Description: A perennial, erect or ascending herb. Root tapering, purplish,red,stem simple or rarely branched, upto 75 cm long and 3-6 mm thick towards the base; leaves usually without visible veins, hispid with coarse hair having discoid bases, radical leaves narrowly obianceoiate and cauline leaves linear-oblong; flowers odorous, whitish or pale yellow, in cymes; nutlets 6 mm long, smooth or obscurely roughened, somewhat lustrous

Detailed description of the leaf and stem (Bisht, Kundu & Mukerji, 1961)

(a) *Macroscopic*

Stem - Cut pieces available in 5-9 cm long and 3.2 to 4.7 cm in dia., flattened, erect, stout; rough due to white, hard, hispid hairs and cicatrices, and longitudinal wrinkles; colour greenish-yellow; fracture, short; odour and taste not characteristic. Leaf - Lanceolate to ovate-lanceolate, 12-30 cm long, 1.5-3.5 cm broad, acuminate tubercle-based hispid hairs present on both surfaces; greenish to light yellow on top and white beneath.

(b) *Microscopic*

Stem - shows single-layered epidermis, covered with thick cuticle, some epidermal cells elongate to form long, warty, tubercle-based unicellular hairs, cortex differentiated in two zones, 5-7 layered outer collenchyrna, 3-4 layered inner parenchymatous cells, consisting of thin-walled, round to oval cells; phloem composed of usual elements; phloem fibres absent; xylem consisting of usual elements, vessels mostly solitary or rarely 2-3 in groups having spiral thickening, and fibres and tracheids having blunt tips and simple pits; xylem ray not distinct: pith consisting of round, thin-walled, parenchymatous cells.

Leaf –

Midrib -single layered epidermis with thick cuticle and long warty, tubercle-based unicellular hairs present on both surfaces followed by 5-7 layers of collenchymatous and 3-4 layers parenchymatous cortical cells; vascular bundle situated centrally. Lamina - isobilateral, single layered epidermis on either surface covered with thick cuticle, long warty, tubercle-based, simple, unicellular hairs present on both surfaces; palisade 2 layered, spongy parenchyma 8-10 layered, stomata paracytic.

Powder - Greenish-brown; shows groups of oval to polygonal, thin-walled straight epidermal cells; spiral vessels; a few fibres entire or in pieces, elongated with blunt tips; long warty, tubercle-based unicellular hairs and a few paracytic stomata.

Chemical constituents: Alkannin and shikonin (Sharma, Khan & Sultana, 2004).

Actions: Alterative and purgative.

Therapeutics: The leaves are alterative. They are made into a powder and given to children as a purgative. The flowers are used as a cordial and stimulant in the treatment of rheumatism and palpitations of the heart. The root is bruised and used as an external application to skin eruptions.

Parts used: Roots.

Bioactivity: Anticarcinogenic (Sharma, Khan & Sultana, 2004) and anti-inflammatory (Bisht, Kundu & Mukerji, 2006)

Further reading

Bisht BS, Kundu BC, Mukerji B. Pharmacognostical study of the root of *Onosma echioides* Linn. *J Sci Ind Res* 1961; 20C: 218-222.

Omkar A, Thurackal J, Gadgoli C. Evaluation of anti-inflammatory activity of *Nyctanthes arbor-tristis* and *Onosma echioides*. *Pharmacog Mag* 2006;2: 8.

Sharma S, Khan N, Sultana S. Effect of *Onosma echioides* on DMBA/ croton oil mediated carcinogenic response, hyperproliferation and oxidative damage in murine skin. *Life Sci* 2004; 75:2391-410.

Geranium wallichianum D.Don

Family: Geraniaceae.

Habitat: E. Asia - Himalayas from Afghanistan to Bhutan.

General Description: *G. wallichianum* is an herbaceous perennial with a low, spreading habit. It has trailing stems and five-lobed green leaves. In summer and autumn it bears saucer-shaped, light violet-blue flowers with large white centres and dark veins.

Chemical constituents: Tannin and gallic acid.

Actions: Astringent.

Therapeutics: The plant has astringent properties. A decoction of the root, mixed with the bark of Prunus cerasoides, is used in the treatment of peptic ulcers. It has been used as a cure for toothache and has been applied externally to the eyes. The juice of the plant is applied to fresh cuts to stem the bleeding. A paste of the plant is used as a poultice to relieve joint pains.

Parts used: Rhizome.

Kumar V, Rao RR. Some plant beverages used in traditional medicine. *Ethnobotany* 2001;13, 36--39

Clausena pentaphylla (Roxb.) DC.

Syn: *C. kanpurensis* Molino.

Family: Rutaceae.

Habitat: Asia.

General Description: Tree.

Chemical constituents: Terpenoid: O-Methylclausenol (Manandhar *et al.,* 1977) and coumarins: clausarin (Anwer, 1977) and clausmarins (Shoeb *et al.,* 1978; Patnaik & Dhawan, 1982; Raj *et al.,* 2009).

Parts used: Bark.

Bioactivity: Spasmolytic: clausmarin (Patnaik & Dhawan, 1982).

Further reading

Anwer F, Shoeb A, Kapil RS, Popli SP. Clausarin - a novel coumarin from *Clausena pentaphylla* (Roxb.) DC. *Experientia* 1977; 33:412-413.

Manandhar, MD, Shoeb A, Kapil RS, Popli SP. O-methylclausenol - A C_{33} terpenoid from *Clausena pentaphylla. Experientia.* 1977; 33:153.

Patnaik GK, Dhawan BN. Evaluation of spasmolytic activity of Clausmarin A, a novel coumarin from *Clausena pentaphylla* (Roxb.). *J Ethnopharmacol* 1982;5:127-137.

Raj K, Sharma M, Misra N. Clausmarin-C: a novel terpenoid coumarin from *Clausena pentaphylla. Nat Prod Res* 2009;23:1671-1676.

Shoeb A, Manandhar MD, Kapil RS, Popi SP. Clausmarins A and B: two novel spasmolytic terpenoid coumarins from *Clausena pentaphylla* (Roxb.) DC. *J Chem Soc* 1978; 7:281-282.

***Potentilla nepalensis* Hook.**

Common name: Nepal Cinquefoil

Family: Rosaceae.

Habitat: E. Asia - W. Himalayas.

General Description: Perennial, erect or ascending, rarely prostrate herb. Rootstock woody. Stem 30-50 cm long, angular, covered with thick pilose indumentum. Leaves 5-foliate at the base, upper 3-foliate, petioles 5-15 cm long, densely and thickly pilose. Basal stipules membranous, auriculate, entire, cauline stipules entire to divided, leafy. Leaflets sessile, variable in shape obovate to elliptic, serrate dentate, upper and lower surfaces sparsely adpressed pilose. Flowers 1.5-2.5 cm diam, pedicels 2-4 cm long, densely sericeous. Sepals densely pilose, outer elliptic ovate, obtuse, inner ovate, acute. Petals 1.0-1.2 cm long and broad, bright red, obcordate. Stamens 20-25 dark red. Carpels numerous, styles subterminal, c. 2 mm long, filiform, a little thickened at the base, stigma not dilated. Achenes wrinkled (Chopra, Nayar & Chopra, 1986).

Chemical composition: Polyphenolics.

Action: Depurative (Pandey *et al.,* 1994).

Parts used: Root.

Therapeutics: The ashes are mixed with oil and applied to burns (Pandey *et al.,* 1995; Ahmed & Borthakur, 2005).

Further reading

Ahmed, AA, Borthakur SK. Ethnobotanical wisdom of Khasis (Hynniew Treps) of Meghalaya, 2005.

Chopra RN, Nayar SL, Chopra IC. *Glossary of Indian Medicinal Plants (Including the Supplement).* Council of Scientific and Industrial Research, New Delhi,1986.

Pandey G, Joshi GC, Pandey NK, Tewari KC. Ethnobotanical studies on the medicinal flora of Tarikhet block (Kumaon Himalaya) District Almora, U.P. *Aryavaidyan* Parts I, II, III, 1994; 1994; 1995 ;7, 4, 228-233, 8, 1, 40-46, 8, 2, 74-79 and 8, 3, 159-164.

71. RASNA

Current representative: *Pluchea lanceolata*

Possible representative: *Alpinia galanga, Aristolochia indica and Vanda roxburghii.*

Pluchea lanceolata C.B.Clarke.

Ayurvedic Names: *Ramya, yuktarasa, gandhanakuli, suganda-mula, atirasa, shreyasi, suvara* and *sara.*

Common Name: Pluchea, Indian Camphorweed, Indian Fleabane

Family: Asteraceae.

Habitat: Grows in upper gangetic plain and several other parts of India.

General Description: The plant is a stout herb growing 0.33-2 meters in height. The stem is grooved, rough and very hairy. The leaves are elliptical, large, 3-6 cm long and 2-3 cm broad, and have long petioles. The fruits, slender achene's, 0.4 cm long, bearded with 0.75 cm long pappus hairs. The flowers are yellow, many in heads, 0.5-1 cm in diameter. The fresh root is brown and becomes grayish on drying. The fresh roots resemble in aroma of camphor (Prasad *et al.,* 1970).

Chemical constituents: Bases: pluchine and betaine, dihydro-isoalan tolactone, isoalantolactone, alantolactone, sitosterol, octadecanoic acid, D-mannitol, sesquiterpene lactones inunal, alantolactone, isoalantolactone, a germacranolide, inunolide (from root oil) and flavonoids including quercetin and quercitrin (Dasgupta, Basu & Dasgupta, 1968; Dixit & Tewari , 1991; Inderjit & Dakshini, 1994).

Specific action as per Ayurveda: Anti-inflammatory.

Therapeutics: The plant is used for the inflammations and bronchitis, psoriasis, cough and piles. It is also used as antipyretic, analgesic, laxative and nervine tonic. The decoction of plant is used to prevent the swellings of joint in arthritis, rheumatism and neurological diseases. The roots are antipyretic, bitter, laxative and thermogenic and are used for allaying the pain caused by the sting of scorpions. Plant extract is used as a cooling agent in summer. The leaves are aperient and used as a laxative, analgesic and atypyretic.

Bioactivity: Anti-inflammatory (Prasad, Bhattacharya & Das, 1966; Chawla *et al.,* 1991) and immunosuppressive (Bhagwat *et al.,* 2010).

Further reading

Bhagwat DP, Kharya MD, Bani S, Kaul A, Kour K, Chauhan PS, Suri KA, Satti NK. Immunosuppressive properties of *Pluchea lanceolata* leaves. *Indian J Pharmacol* 2010; 42: 21–26.

Chawla AS, Kaith BS, Handa SS, Kulshreshtha DK, Srimal RC.Chemical investigation and anti-inflammatory activity of *Pluchea lanceolata. Fitoterapia* 1991;62:441-444.

Dasgupta B, Basu K, Dasgupta DS. Chemical investigation of *Pluchea lanceolata* II. Identity of pluchine with betaine hydrochloride. *Cel Mol Life Sci* 1968; 24: 882.

Dixit GS, Tewari RP. Chemical constituents of *Pluchea lanceolata. Sach Ayur* 1991;43: 841.

Inderjit KM, Dakshini M. Quercetin and Quercitrin from *Pluchea lanceolata* and Their Effect on Growth of Asparagus Bean. *Allelopathy* 1994; 6: 86–93.

Pathak S, Anand OP. *Pluchea lanceolata* DC. as an antifertility plant. B.K. Gupta (Ed.), 1991;257-261.

Prasad DN, Bhattacharya SK, Das PK. A study of anti-inflammatory activity of some indigenous drugs in albino rats. *Indian J Med Res* 1966;54: 582-590.

Prasad S, Shah CS, Shah NS, Ansari MS, Mitra R. Pharmacognostic study of leaves of *Pluchea lanceolata* {C}.{B}. Clarke and Olive (rasna). *J Res Indian Med* 1970;4: 203-207.

Sharma GP, Sharma PV. Experimental study of anti-inflammatory activity of some rasna drugs. *J Res Indian Med* 1977;12: 18-21.

Alpinia galanga Willd.

Common name: Java galangal.

Ayurvedic names: *Mahabhari vacha, sthulagranthi* and *sugandhi.*

Family: Zingiberaceae.

Habitat: Native to Java and Sumatra. It is found in Eastern Himalayas.

General Description: A perennial, aromatic, rhizomatous herb,1.8-2.1 m in height, leaves large compared to Alpinia calcarata, oblong, lanceolate, glabrous, acuminate, very shortly petioled, ligule short, rounded, ciliate, flowers fragrant, greenish white in dense panicles; fruits capsules, orange red when ripe (Inamdar, 1962; Chitra & Thoppil, 2008).

Detailed Description of the Rhizome

(a) *Macroscopic*

Root - The roots are adventitious, in groups, fibrous, persistent in dried rhizomes, about 0.5 to 2 cm long and 0.1 to 0.2 cm in diameter and yellowish-brown in colour.

Rhizome - Rhizome cylindrical, branched, 2 to 8 cm in diameter, longitudinally ridged with prominent rounded warts (remnants of roots) marked with fine annulations; scaly leaves arranged circularly; externally reddish-brown, internally orange yellow in colour; fracture, hard and fibrous; fracture, surface rough; odour, pleasant and aromatic; spicy and sweet in taste.

(b) *Microscopic*

Root - T.S. of root circular in outline, single layered epidermis with barrel shaped cells having unicellular root hairs, hypodermis 3 or 4 cells deep and sclerenchymatous, cortex parenchymatous, many cells deep, with well developed intercellular spaces; endodermis showing prominent casparian strips and 'v' shaped thickening, followed by many celled sclerenchymatous pericycle; xylem and phloem in separate radial strands; centre occupied with a parenchymatous pith.

Rhizome - T.S. of young rhizome circular in outline; epidermal cells small and angular, thick cuticle present, rhizome differentiated into a wide cortex and a central cylinder, both regions having irregularly scattered vascular bundles, each vascular bundle with a prominent fibrous sheath; inner limit of cortex marked by rectangular parenchymatous cells; stele with irregular, closely placed vascular bundles towards periphery, root traces present, schizogenous canals and oil cells with suberized walls found in cortex and in central region; most of the parenchymatous cells filled with starch grains which are ellipsoidal to ovoid, sometimes beaked, simple, 10 to 64 μm, hilum eccentric, circular or crescent shaped at the broad end, the narrow beak-like end become black when stained with dil. iodine water and chlorzinc iodide but the remaining part become light blue or brown. Macerated prepration shows vessels 95 to 710 μm long and 19 to 190 μm broad, tracheidal fibres 68 to 920 μm long and 19 to 30 μm broad.

Powder - Powder is orange brown in colour, spicy and sweet in taste, shows parenchymatous cells containing starch (as described under microscopy of rhizome), oil cells, schizogenous canals, vessels with scalariform and reticulate thickenings and tracheidal fibres.

Chemical constituents: Flavones (alpinin, galangin and kaempferide), essential oil containing methyl cinnamate, cineole, camphor and d-pinene (Indrayan, Agrawal & Tyagi, 2002; Raina , Srivastava & Syamasunder, 2002) and linoleic acid (Jaju *et al.*, 2009).

Specific action as per Ayurveda: *Kanthya* (beneficial for throat disease).

Therapeutics: Laryngeal palsy, bronchitis, bronchial asthma, rheumatism, loss of appetite, polyuria and impotency (Agnihotri, Wakode & Agnihotri, 2010).

Parts used: Rhizome.

Bioactivity: Antibacterial (Chopra, Khajuria & Chopra, 1957).

Further reading

Agnihotri S, Wakode S, Agnihotri A.: *Alpinia galanga* Willd. - An overview on phyto-pharmacological properties. *Indian J Nat Prod Res* 2010;1: 143-149.

Chitra M, Thoppil JE. A pharmacognostical rapport on the rhizome of *Alpinia galanga* Linn. (Willd). *Ancient Sci Life* 2008;27: 9-21.

Chopra IC, Khajuria BN, Chopra CL. Antibacterial properties of volatile principles from *Alpinia galanga* and *Acorus calamus*. *Antibio Chemother* 1957;7:378-383.

Inamdar MC. Studies on indigenous drugs (*Adhatoda vasica* Nees, *Terminalia chebula* Retz. and *Alpinia galanga* Willd.). *Bombay Technol* 1962; 12: 126-127.

Indrayan AK, Agrawal P, Tyagi PK. Properties of essential oil and the oils isolated by extracting with petroleum ether and ethyl alcohol from the rhizomes of *Alpinia galanga* Willd. *Indian Drugs* 2003;40: 542.

Jaju SB, Indurwase NH, Sakarkar DM, Fuloria NK, Ali MD, Basu SP. Linoleic acid isolated from rhizomes of *Alpinia galangal*. *Nigerian J Nat Prod Med* 2009:13.

Raina VK, Srivastava SK, Syamasunder KV. The essential oil of 'greater galangal' (*Alpinia galanga* (L.) Willd.) from the lower Himalayan region of India. *Flav Frag J* 2002;17: 358-360.

Aristolochia indica L.

Common name: Java galangal.

Family: Zingiberaceae.

Habitat: The plant is distributed throughout all the provinces of India and in Sri Lanka, Nepal and Bangladesh. It is usually found scrambling over hedges and bushes.

General Description: A perennial climber shrub with woody base stocks. Leaves simple, alternate, entire, with undulate margins, acute; flowers greenish white, in axillary cymes; fruits rounded oblong, 6 chambered contain numerous winged compressed seeds.

Detailed Description of the Root (Bal & Gupta, 1956)

(a) *Macroscopic*

Root considerably long, cylindrical, a few irregularly bent; 2-10 mm in dia; surface almost smooth with fine longitudinal wrinkles and transverse cracks; external surface, light greyish-brown; inner whitish; fracture, short and splintery; odour, camphoraceous; taste, strongly bitter.

(b) *Microscopic*

Cork 8-10 layers, composed of tabular, thin-walled cells excepting the outer most layer, having thick-walled cells externally and filled with brownish content; cork cambium single layered; secondary cortex 15 to 17 layers of thin-walled, somewhat rounded and isodiametric cells in the outer region but tangentially elongated in the inner region; plenty of simple, round to oval starch grains measuring 5-18 μ in dia. and compound starch grains having 2-4 components measuring 10-15 μ in dia. and oil globules present in a few cells; in the middle region stone cells round, rectangular, oval or elongated present in small irregular patches having simple pits and radiating canals; centre occupied by xylem, split into strips of radiating arms by wedgeshaped masses of parenchyma; each xylem arm is capped by thin patches of phloem consisting of sieve elements and phloem parenchyma, phloem fibres, and occasionally stone cells also found in this region; a ring of cambium present between phloem and xylem; xylem consists of large vessels, tracheids, fibres tracheids and parenchyma, all being lignified; in older roots, tyloses formation takes place in vessels; medullary rays 8 to 10 in number, multiseriate and dilating towards periphery and alternating with radiating arms of wood; scattered group of stone cells present in a few wider rays; micro-crystals with a few appearing as elongated small prisms and unaffected by acids, are present in a few cortical and ray cells.

Powder - Brownish-yellow; fragments of cork cells, very few, oval to rectangular, lignified, thick-walled stone cells having distinct striations with narrow lumen, vessels with spiral thickenings, non-lignified, thick-walled tracheids, numerous simple, round to oval, starch grains measuring 5-18 μ in dia., and compound grains having 2 to 4 components, measuring 10 - 15 μ in dia., a few crystals and oil globules

Chemical constituents: Flavones (alpinin, galangin and kaempferide), essential oil (contains methyl cinnamate, cineole, camphor and d-pinene).

Specific action as per Ayurveda: Ecbolic.

Therapeutics: Laryngeal palsy, bronchitis, bronchial asthma, rheumatism, loss of appetite, polyuria and impotency.

Parts used: Rhizome.

Bioactivity: Antifertility (Pakrashi, Chakrabarty & Dasgupta, 1976; Pakrashi & Chakrabarty, 1978), antispermatogenic (Pakrashi & Pakrashi, 1977), Anti-implantation and anti-oestrogenic (Pakrashi & Shaha, 1977) and antibacterial (Shafi *et al.,* 2002).

Further reading

Bal SN, Gupta B. Morphology of roots of *Aristolochia indica* L. Sci Cul 1956; 22:276-277.

Pakrashi A, Chakrabarty B, Dasgupta A. Effect of the extracts from *Aristolochia indica* Linn. on interception in female mice. *Cel Mol Life Sci* 1976; 32: 394-395.

Pakrashi A, Chakrabarty B. Antifertility effect of aristolic acid from *Aristolochia indica* (Linn.) in female albino rabbits. *Cel Mol Life Sci* 1978; 34: 1377.

Pakrashi A, Pakrashi PL. Antispermatogenic effect of the extract of *Aristolochia indica* Linn. on male mice. *Indian J Exp Biol* 1977;15: 256-259.

Pakrashi A, Shaha C. Anti-implantation and anti-oestrogenic activity of a sesquiterpene from the roots of *Aristolochia indica* Linn. *Indian J Exp Biol* 1977;15:1197-1198.

Shafi PM, Rosamma MK, Jamil K, Reddy PS. Antibacterial activity of the essential oil from *Aristolochia indica*. *Fitoterapia* 2002;73: 439-441.

Vanda roxburghii R.Br.

Ayurvedic names: *Vrikshakadali, Vrikshadani, Gandhanakuli, Vandaka* and *Rasna*.

Family: Orchidaceae.

Habitat: Himalayas.

General Description: A glabrous epiphytic orchid. Leaves simple, opposite, long and fleshy; flowers in long panicled spikes, rose colored, fragrant, and last for one month without falling off. Flowering season is only once in a year. Fruits are small capsules with numerous spores containing. Plant sticks on the host tree with its fleshy roots. Plant possesses hanging velemen roots to absorb the atmospheric moisture.

Chemical constituents: Starch.

Specific action as per Ayurveda: Hypno-sedative.

Therapeutics: *V. roxburghii* pacifies vitiated *pitta, vata*, pain, arthritis, burning sensation, skin diseases, otitis media, nervine weakness and in poison bites.

Parts used: Rhizome.

Bioactivity: Anti-arthritic (Prasad & Achari,1966) and anti-inflammatory (Basu *et al.,* 1971).

Further reading

Basu K, Dasgupta B, Bhattacharya SK, Lal R, Das PK. Anti-inflammatory principles of *Vanda roxburghii. Curr Sci* 1971;40:86-87.

Gupta JC, Roy PK, Sen Gupta KK. Pharmacological action of an active constituent isolated from *Vanda roxburghii* R.Br. *Indian J Med Res* 1946;34:253-255.

Prasad DN, Achari G. Study of anti-arthritic action of *Vanda roxburghii* in albino rats. *J Indian Med Assoc* 1966;45: 234-237.

72. RENUKA

Current representative: *Vitex agnus castus*

Possible representative: *Vitex negundo*

Vitex agnus castus L.

Ayurvedic names: *Kapila, kaunti, pandu patri* and *harenuka.*

Family: Verbenaceae.

Habitat: The chaste tree is native to the Mediterranean region but is regularly cultivated in warm, temperate and sub-tropical climates.

General Description: The chaste tree is a flowering shrub that grows to a height of 1 to 5 meters. The tree produces a spray of beautiful flowers ranging from lilac to blue and white. The flowers bloom late summer or early fall. The plant has a pleasing aromatic quality. It has slender, finger-like leaves, purple-black berries

Chemical constituents: Bitter principle (castine).

Action: Hormone like activity.

Therapeutics: Postmenopausal syndrome.

Further reading

Vasudevan Nair K, Yoganarasimhan K. Gopakumar, Shantha TR, Keshavamurthy KR. Studies on some South Indian market samples of Ayurvedic drugs – IV. *Ancient Sci Life* 1985;4:211-216.

Vitex negundo L.

Common name: Five leaved chastle.

Ayurvedic names: *Sveta kusuma, sinduka* and *sinduvaraka.*

Family: Verbenaceae.

Habitat: India.

General Description: *V. negundo* is an erect shrub or small tree growing from 2 to 8 m (6.6 to 26 ft) in height. The bark is reddish-brown. It leaves are digitate, with five lanceolate leaflets, sometimes three. Each leaflet is around 4 to 10 cm (1.6 to 3.9 in) in length, with the central leaflet being the largest and possessing a stalk. The leaf edges are toothed or serrated and the bottom surface is covered in hair.The numerous flowers are borne in panicles 10 to 20 cm (3.9 to 7.9 in) in length. Each is around 6 to 7 cm (2.4 to 2.8 in) long and are white to blue in color. The petals are of different lengths, with the middle lower lobe being the longest. Both the corolla and calyx are covered in dense hairs.

Detailed Description of the Leaf

(a) *Macroscopic*

Leaves palmately compound, petiole 2.5 - 3.8 cm long; mostly trifoliate, occasionally pentafoliate; in trifoliate leaf, leaflet lanceolate or narrowly lanceolate, middle leaflet 5- 10 cm long and 1.6 -3.2 cm broad, with 1- 1.3 cm long petiolule, remaining two sub-sessile; in pentafoliate leaf inner three leaflets have petiolule and remaining two sub-sessile; surface glabrous above and tomentose beneath; texture,leathery.

(b) *Microscopic*

Petiole - shows single layered epidermis having a number of unicellular, bicellular and uniseriate multicellular covering trichomes and also glandular trichomes with uni to tricellular stalk and uni to bicellular head; cortex composed of outer collenchymatous tissue and inner 6 - 8 layers of parenchymatous tissue; collenchyma well developed in basal region and gradually decreases in middle and apical regions; pericyclic fibres absent in basal region of petiole and present in the form of a discontinuous ring in apical region surrounding central horse shoe-shaped vascular bundle; a few smaller vascular bundles present ventrally between arms of central vascular bundle and two, or rarely three, bundles situated outside the arms.

Lamina - shows single layered epidermis having mostly unicellular hairs, bi and multicellular and glandular trichomes being rare; hypodermis 1 - 3 layered interrupted at places by 4- 8 palisade layers containing chlorophyll; a large number of veins enclosed by bundle sheath traverse mesophyll; stomata present only on the ventral surface, covered densely with trichomes; vein-islet and vein termination number of leaf are 23-25 and 5-7 respectively.

Powder - shows number of pieces or whole, uni-bi and multicellular covering trichomes, glandular trichomes, palisade tissues with hypodermis, and upper and lower epidermis, xylem vessels with pitted walls.

Detailed Description of the Root

(a) *Macroscopic*

Roots cylindrical, hard, tough with irregular fractures; external surface rough due to longitudinal, narrow, cracks and small rootlets; cut surface shows cork region greyishbrown, middle region greyish-white, and xylem region cream coloured; bark thin, easily separates from wood; wood hard, forming major part of root.

(b) *Microscopic*

Root shows 10 to 18 or more tangential rows of rectangular to cubicular, moderately thick-walled cork cells with a few rows of radially arranged cork cells also being present, inner 3 to 5 rows of cork cells thin-walled; cork cambium consists of single row of squarish to transversely elongated cells; secondarycortex composed of 4 to 12 rows of rectangular to elongated cells, some contain starch grains; numerous, small groups of stone cells found scattered in this zone; stone cells vary in shape and size; secondary phloem consists of sieve tubes with companion cells, fibres and phloem parenchyma traversed by phloem rays; distal portion of phloem conical, due to dilating phloem rays; each band of phloem composed of thin-walled, phloem tissues alternating with transverse strips of thick-walled phloem fibres; a few tangential strips of obliterated phloem tissues also present in outer-phloem region; each fibre group composed of 6 to 60 or more thick-walled, long and short fibres, short fibres comparatively thick-walled, a few fibres show forked tips; inner zone of phloem omposed of intact, thin-walled, phloem tissues mainly sieve tubes, companion cells and phloem parenchyma; cambium composed of one, or sometimes two, rows of cells; central major part of root consists of xylem; vessels varying in size, scattered throughout xylem region, either in small groups of 2 to 4 or singly; a few xylem vessels show tail on one or both the ends; xylem fibres long, having thick-walls and pointed tips; xylem parenchyma contains starch grains similar to those found in cortical region; medullary rays are uni-to triseriate, almost straight, extend from pith to cork, medullary rays in xylem region radial while in phloem region they dilate; cells contain starch grain, simple and compound, oval to circular, having 4 components and measuring 8 to 12 μ in dia.

Powder - Pale yellow; shows parenchymatous cells containing simple oval to round and compound starch grains with 4 components, measuring 8 to 12 μ in dia; stone cells elongated, rectangular and squarish in shape with wide and narrow lumen, radiating canals and conspicuous striations; xylem vessels with pitted thickening, xylem and phloem fibres with thick walls.

Chemical constituents: Oil (containing β-sitosterol), vanillin, alkaloids, glucosides (acubin, agnuside and orientin), resin (Chawla *et al.,* 1992), triterpenoids (Chawla *et al.,* 1992).

Specific action as per Ayurveda: *Vedna shamaka* (analgesic).

Therapeutics: Arthritis and rheumatism.

Parts used: Leaves and seeds.

Bioactivity: Anti-inflammatory (Jain & Pande, 1976; Sharma & Singh, 1980; Chawla *et al.,* 1992; Chattopadhyay, & Shaw, 1999; Dharmasiri *et al.,* 2003), analgesic (Sharma & Singh, 1980; Gupta *et al.,* 1997; Dharmasiri *et al.,* 2003), antihistaminic (Sharma & Singh, 1980), anti-stress (Tandon, & Gupta, 2005) and anti-convulsant (Gupta *et al.,* 1997).

Further reading

Chawla AS, Sharma AK, Handa SS, Dhar KL. Chemical investigation and anti-inflammatory activity of *Vitex negundo* seeds. *J Nat Prod* 1992;55:163-167.

Chawla AS, Sharma AK, Handa SS, Dhar KL. Chemical investigation and anti-inflammatory activity of *Vitex negundo* seeds, part I. *Indian J Chem* 1991; 30 B, 7, 773--776; also in: *J Nat Prod* 55, 2, 163-167.

Datta SC, Datta DD. Pharmacognostic studies of *Vitex negundo* Linn. *Indian J Pharm* 1950; 2:36-39.

Dharmasiri MG, Jayakody JR, Galhena G, Liyanage SS, Ratnasooriya WD. Anti-inflammatory and analgesic activities of mature fresh leaves of *Vitex negundo*. *J Ethnopharmacol* 2003; 87:199-206.

Gupta M, Mazumder UK, Bhawal AR, Swamy SMK. CNS activity of petroleum ether extract of *Vitex negundo* linn in mice. *Indian J Pharm Sci* 1997; 59: 240-245

Gupta OP. Pharmacognostic features of nirgundi (*Vitex negundo* Linn.) *Nagarjun* 1980;16: 21-24.

Jana U, Chattopadhyay RN, Shaw BP. Preliminary studies on anti-inflammatory activity of *Zingiber officinale Rosc., Vitex negundo Linn* and *Tinospora cordifolia* (Willid) Miers in Albino Rats. *Indian J Pharmacol* 1999;31:232-233.

Kaushik RD, Kumar A, Arora C. Fractionation, Characterization and Evaluation of biocidal potential of active principles of leaves of *Vitex negundo*. *Asian J Chem* 2003; 15:1659-1664.

Lal B, Udupa KN, Tripathi VK. Study on the antifertility effect of nirgundi, *Vitex negundo* - A preliminary trial. *J Res Ayur Siddha* 1992; 13: 89-93.

Mohiddin SG. The role of shambalu (*Vitex negundo*) in rheumatoid arthritis. *Rheumatism* 1979; 14: 97-112.

Nag BN, Datta MK, Das PC, Patra BB, Sikdar S. Some observations on the anti-inflammatory properties of *Vitex negundo* in rats. *Indian J Pharmacol* 1978;10: 57.

Sadekar SS, Chauhan VB, Shirole RL,. Salve BU, Kasture VS, Kasture SB. Anti-dopamimergic activity of *Vitex negundo* Linn. Leaves. *J Nat Rem* 2006;6: 165-169.

Sharma AK, Singh RH. Screening of the anti-inflammatory activity of certain indigenous drugs on carageenin induced hind paw oedema. *Bull Med Ethnobot Res* 1980; 1: 262.

Singh RH, Chaturvedi GN. Studies on the anti-arthritic effect of an indigenous drug *Vitex negundo* - a clinical and experimental study. *Curr Med Prac* 1966;11:406-414.

Singh V, Dayal R, Bartley JP. Volatile constituents of *Vitex negundo* leaves. *Planta Med* 1999;65: 580-582.

Srinivas K, Rao SS, Rao MEB, Raju MBV. Chemical constituents of the roots of *Vitex negundo*. *Indian J Pharm Sci* 2001;63: 422-424.

Tamhankar CP, Saraf MN. Anti-arthritic activity of *Vitex negundo* Linn. *Indian J Pharm Sci* 1993;56: 158.

Tandon V, Gupta RK. Effect of *Vitex negundo* on oxidative stress. *Indian J Pharmacol* 2005;37: 38-40.

Tandon VR, Khajuria V, Kapoor B, Kour D, Gupta S. Hepatoprotective activity of *Vitex negundo* leaf extract against anti-tubercular drugs induced hepatotoxicity. *Fitoterapia* 2008;79: 533-538.

Telang RS, Chatterjee S, Varshneya C. Studies on analgesic and anti-inflammatory activities of *Vitex negundo* Linn. *Indian J Pharmacol* 1999; 31: 363-366.

Tripathi YB, Tiwari OP, Nagwani S, Mishra B. Pharmacokinetic interaction of *Vitex negundo* Linn. and paracetamol. *Indian J Med Res* 2009; 130: 479-483.

Uppalapati L, Rao JT. Antimicrobial properties of the essential oil of *Vitex negundo* Linn. *Indian Drugs* (Pharmaceutical Industry) 1979;14: 31-33.

73. ROHITAKA

Current representative: *Aphanamixis polystachya / Amoora rohitaka*

Possible representative: *Tecoma undulata, Maba nigrescens, Ventilago madrasptana, Polygonum glabrum, Myristica attenuate* and *Rhamnus wightii.*

Aphanamixis polystachya (Wall.) R. N. Parker

Syn: *Amoora rohituka*

Family: Meliaceae.

Habitat: Asia.

General Descirption: A medium or large sized tree, 15 to 20 meters in height. The leaves are large, imparipinnate. The leaflets are opposite, elliptic-oblong, acuminate, glabrous on both the surface. The male flowers are numerous in axillary panicles; female or bisexual flowers are larger than male axillary or supra-axillary solitary spikes. The fruits are globular, smooth, yellow when ripe. The seeds are with scarlet aril (Sanyal & Datta, 1981; Vikramaditya, Varma & Datta, 1991)

Chemical constituents: Glycoside (Srivastava & Agnihotri, 1984), limonoid: amooranin (Agnihotri, Srivastava & Srivastava, 1987)

Bioactivity: Antitumor (Rabi , Gupta & Gulati, 1994), hepatoprotective (Gole & Dasgupta, 2002), cytotoxic (Rabi *et al.,* 2002), and antimicrobial (Chowdhury, Hasan & Rashid, 2003).

Clinical study: Leukaemia (Prasad GC, Deshpande, 1968)

Further reading

Agnihotri VK, Srivastava SD, Srivastava SK. A new limonoid, amoorinin, from the stem bark of *Amoora rohituka*. *Planta Med* 1987; 53: 298-299.

Chowdhury P, Hasan CM, Rashid MA. Antimicrobial activity of *Toona ciliata* and *Amoora rohituka*. *Fitoterapia* 2003;74: 155-158.

Prasad GC, Deshpande PJ.Effect of rohitaka (*Amoora rohitaka*) on leukaemia. *J Res Indian Med* 1968;3: 36-42.

Rabi T, Gupta RC, Gulati AK. Influence of amoora rohituka on mcf-7 human mammary adenocarcinoma cells *in vitro*. *Indian J Pharma Sci.* 1994;56(4): 136-137.

Rabi T, Gupta RC. Antitumor and cytotoxic investigation of *Amoora rohituka*. *Int J Pharmacog* 1995;33: 359-361.

Rabi T, Karunagaran D, Nair MK, Bhattathiri VN. Cytotoxic activity of amooranin and its derivatives. *Phytother Res* 2002; 16: 84-86.

Sanyal M, Datta PC. Pharmacognosy of *Aphanamixis polystachya* stem bark. *Quart J Crude Drug Res* 1981;19:113-126

Srivastava SK, Agnihotri VG. Chemical investigation of the stem bark of *Aphanamixis polystachya*. *Curr Sci* 1984;53: 1288-1289.

Vikramaditya SM, Varma PN, Datta PC. Pharmacognostic evaluation of *Aphanamixis polystachya* seed drug. *J Econ Taxon Bot* 1991;15: 121-127.

***Tecoma undulata* (Sm.) G. Don**

Syn: *Tecomella undulata* (Sm.) G. Don; *Bignonia undulata* Sm.

English names: Desert teak, marwar teak and rohida tree.

Family: Bignoniaceae.

Habitat: It is found in Thar Desert regions of northwest and western India.

General Descirption: mall tree, 2.5-5 m tall. Leaves opposite, entire; lamina elliptic-oblong to elliptic-lanceolate or linear-oblong, 35-95 x (8-) 10-20 mm, margin undulate, petiole 6-18 mm long. Racemes condensed, 2-7-flowered. Bracts linear, 1.5-2 mm long, stellate pubescent; bracteoles 2, similar, c. 1.5 mm long. Pedicel 13-14 mm long. Flowers large, attractive, orange-red. Calyx campanulate, 12-13 mm long, unequally 5-lobed; obtuse-apiculate, minutely and sparsely glandular-pubescent. Corolla tube broad funnel-shaped, dilated at base; lobes sub-orbicular, 20-25 x 28-32 mm. Filaments of longer stamens 37-38 mm long; staminode 16-17 mm long. Disk cupular-compressed, 5-lobed. Ovary broad linear, c. 4.5 mm long; style c. 40 mm long, stigmas 2, oblong-ovate, compressed, 4.5-5 mm long. Capsule 17-34 x 0.9-1.2 cm, linear-oblong, laterally compressed, slightly curved, apex beaked with wing, 20 mm broad.

Chemical constituents: Quinones (Joshi *et al.*, 1977; Hoshi, Singh & Pardasani, 1977), ester glycoside: bark (Pandey & Dasgupta, 1970; Bhattacharya *et al.,* 1971), a new iridoid glucoside: undulatin (Verma *et al.,* 1986), triacontanol and triterpenes (Joshi & Singh , 1977), flavone glycoside: 5,7-Dihydroxy-6-methoxyflavone 7-O-alpha-l-rhamnopyranoside (Rao & Prakash, 1999).

Specific action as per Ayurveda: Hepatoprotective.

Therapeutics: Liver and spleen diseases.

Parts used: Bark.

Bioactivity: Antribacterial (Patel & Trivedi, 1959), and hepatoprotective (Khatri, Garg & Agrawal, 2009).

Further reading:

Bhattacharya SK, Lal R, Pandey VB, Das PK. Preliminary pharmacological investigation on the glycosides of *Tecomella undulata* (rugtrora). *J Res Indian Med* 1971;6: 226-227.

Joshi KC, Singh LB. Chemical examination of *Tecomella undulata* (G.Don) Seem. *Curr Sci* 1977;46:145-146.

Joshi KC, Singh P, Pardasani RT, Quinones and other constituents from the roots of *Tecomella undulata. Planta Med* 1977; 31:14-16.

Khatri A, Garg A, Agrawal SS. Evaluation of hepatoprotective activity of aerial parts of *Tephrosia purpurea* L. and stem bark of *Tecomella undulata. J Ethnopharmacol* 2009, 122:1-5.

Pandey VB, Dasgupta B. A new ester glucoside from the bark of *Tecomella undulata Experientia..* 1970, 26(11), 1187-1188.

Patel RP, Trivedi BM. The *in vitro* antibacterial activity of *Tecoma undulata. Ind J Med Sci* 1959;13:188-90.

Rao JT, Prakash EO. A new flavone glycoside from the seeds of *Tecoma undulata Fitoterapia* 1999;70: 287-289.

Verma KS, Jain AK, Gupta SR, Structure of Undulatin: A New Iridoid Glucoside from *Tecomella undulata. Planta Med* 1986; 5:359-362.

Ventilago madersapatana Gaertn.

Syn: *Funis viminalis*

English name: Red creeper.

Ayurvedic name: *Rakatvalii.*

Family: Rhamnaceae.

Habitat: South India.

General Descirption: A large, perennial woody climber thick foliage and hanging branches. Leaves simple, alternate, oblong-lanceolate or elliptic-ovate, obtuse or acute, entire or crenate. base rounded or acute, main nerves 4-8 pairs; flowers greenish with bad odor, found terminal panicles; fruits yellowish globular nuts, supported by the persistent calyx, wing linear-oblong, 1-nerved, rounded at the apex (Sanobar *et al.*, 2010).

Chemical constituents: Quinones.

Therapeutics: *V. madersapatana* pacifies vitiated *kapha*, dyspepsia, colic, flatulence, erysipelas, scabies, general debility fever and other skin diseases.

Parts used: Root bark.

Bioactivity: Antifeedent (Krishnakumari, Bhuvaneswari & Raja, 2001) and anti-oxidant (Duganath *et al.*, 2010).

Further reading:

Duganath N, Kumari SR, Kumanani R, Jayaveera KN. Evaluation of anti-denaturation and anti-oxidant activity of traditionally used medicinal plants. *Int J Pharma Bio Sci* 2010;6:1-4.

Krishnakumari GN, Bhuvaneswari B, Raja S. Antifeedant activity of quinones from *Ventilago maderaspatana*. *Fitoterapia* 2001;72:671-675.

Sanobar S, Sandhya S, Vinod KR, David Banji, Rao KNV, Narender Prasad D, Nema RK. Pharmacognostic Studies on The Leaf of *Ventilago maderaspatana* Graetn. *IJPCR* 2010; 2:51-53.

Rhamnus wightii W.& A.

Syn: *Rhamnus catharticus* L.

English name: Buckthorn.

Family: Rhamnaceae.

Habitat: Asia.

General Descirption: *R. cathartica* is a deciduous shrub or small tree growing up to 10 m tall, with grey-brown bark and spiny branches. The leaves are elliptic to oval, 2.5–9 cm long and 1.2–3.5 cm broad; they are green, turning yellow in autumn, and are arranged somewhat variably in opposite to subopposite pairs or alternately. The flowers are yellowish-green, with four petals; they are dioecious and insect pollinated. The fruit is a globose black drupe 6–10 mm diameter containing two to four seeds.

Chemical constituents: Resin, tannins, ash, bitter principle and napthelene glucoside.

Parts used: Bark and fruit.

Further reading

Pandey G. Joshi GC, Pandey NK, Tewari KC. Ethnobotanical studies on the medicinal flora of Tarikhet block (Kumaon Himalaya) District Almora, U.P. Publication date: 1994; 1994; 1995 Published in: Parts I, II, III, *Aryavaidyan* 7, 4, 228-233, 8, 1, 40-46, 8, 2, 74-79 and 8, 3, 159-164.

73. RUDANTI

Current representative: *Cressa cretica*

Possible representative: *Capparis mooni, Crotalaria laburnifolia*

Cressa cretica Linn.

Ayurvedic name: *Amritsrava.*

Family: Convolvulaceae.

Habitat: Throughout India.

General Description: An erect small dwarf shrub grows up to 40 cm in height. Leaves simple, numerous, subsessile, ovate acute, with hairs; flowers white in small clusters; fruits ovoid capsules, pointed at the apex; seeds usually solitary (Suganthi, Sripathy & Manian, 2007; Suganthi, Sripathy & Manian, 2008).

Chemical constituents: Flavonoids (Parekh & Sitakara Rao, 1965; Tiwari & Kakkar, 1990).

Actions: Alterative, diuretic and tonic.

Therapeutics: Cough and general debility.

Parts used: Whole plant.

Bioactivity: Spermopiotic (Gupta *et al.,* 2006) antibacterial (Suganthi, Sripathy & Manian, 2008), and antitussive (Sunita, Jha & Pattanayak, 2009).

Further reading

Gupta RS, Kachhawa JBS, Khushalani V, Tanwar K, Joshi YC. Effect of *Cressa cretica* methanol extract on testicular function of albino rats. *Pharm Biol* 2006;44: 382-388.

Parekh AV, Sitakara Rao V. Chemical composition of *Cressa cretica* Linn. and *Salicornia brachiata* Roxb. *Curr Sci* 1965;34:247-249.

Satakopan S, Karandikar GK.Rudanti: a pharmacognostic study of *Cressa cretica* Linn. *J Sci Ind Res* 1961;20C:156-160.

Sen DN. *Cressa cretica* Linn. - in search of Sanjeevani. *Curr Sci* 2009;97:1523

Suganthi G, Sripathy SK, Manian K. Anatomical features and histological studies on *Cressa cretica* Linn. *Ancient Sci Life* 2007;27: 31-37.

Suganthi G, Sripathy SK, Manian K. HPTLC and antibacterial analysis of extracts of *Cressa cretica* Linn. *Ancient Sci Life* 2008;27: 1-14.

Sunita P, Jha S, Pattanayak SP. *In-vitro* antitussive activity of *Cressa cretica* Linn. using cough model in rodents. *Pharmacog Res* 2009;1: 152-156.

Tiwari HP, Kakkar A. Phytochemical examination of *Cressa cretica* Linn. (rudanti). *J Indian Chem Soc* 1990;67: 785.

Capparis moonii Hook. f. Thoms.

Ayurvedic name: *Rudanti*.

English name: Indian Caper.

Family: Capparidaceae.

Habitat: Western Ghats.

General Descirption: *C. moonii* is a spiny climbing shrub.

Chemical constituents: Gallotannins.

Bioactivity: Bactericidal (Shah, 1962), tuberculostatic (Shah, 1962) and insulinomimetic (Kanaujia *et al.*, 2010).

Further reading

Kanaujia A, Duggar R, Pannakal ST, Yadav SS, Katiyar CK, Bansal V, Anand S, Sujatha S, Lakshmi BS. Insulinomimetic activity of two new gallotannins from the fruits of *Capparis moonii*. *Bioorg Med Chem* 2010; 1:3940-3945.

Shah HH. Study on tuberculostatic activity of Rudanti (*Capparis moonii*). *Indian J Med Sci* 1962;16:343-346.

Shah HH. *In-vitro* study on bactericidal activity of Rudanti (*Capparis moonii*). *Indian J Med Sci* 1962; 16:347-349.

Note: It is endemic and threatened.

Crotalaria laburnifolia L.

Syn: *Crotalaria australis* Bak. f.

English name: Rattlepod.

Family: Fabaceae.

Habitat: Southern Africa.

Habit: Flowers yellow. Shrub to 2 m tall. Stipules absent, leaves with 3 leaflets. Flowers in a loose raceme, standard petal yellow to pale yellow, sometimes with red-brown markings. Pod pendulous, mottled light brown 4.5-15 cm long.

Chemical constituents: Pyrrozolidine alkaloid: crotalaburnine (Subramanian & Nagarajan, 1968; Snehalata & Ghosh, 1968).

Further reading

Subramanian SS, Nagarajan S. Phytochemical studies on *Crotalaria laburnifolia. Planta Med* 1968; 16:432-435.

Snehalata S, Ghosh MN. Pharmacological investigations of crotalaburnine: an alkaloid from the seeds of *Crotalaria laburnifolia* Linn. *Ind J Med Res* 1968;56:1386-1390.

75. RUDRAJATA

***Aristolochia longa* Linn.**

Ayurvedic names: *Raudri, jata, rudra, saumya, sugandha, suhata, ghana, eshvari, rudralata, supatra, sugandhpatra, surbhi, shiva, patravalli, rudrani, netrapushkara* and *mahajata.*

Description: The root is spindle-shaped from 5 cm. to 3 dm. in length, about 2 cm. in thickness, fleshy, very brittle, greyish externally, brownish-yellow inside, bitter and of a strong disagreeable odour when fresh.

Actions: Pungent in taste, appetizer and laxative.

Therapeutics: Cough, heart ailments and bacterial infections.

76. SAPTALA

Current representative: *Acacia concinna*

Possible representative: *Euphorbia nivulia, Euphorbia dracunuloides and Andrographis paniculata.*

***Acacia concinna* DC.**

Syn: *Acacia sinuata* (Lour.) Merill.

Ayurvedic name: *Satala, virala, sari, bahuphenaka, carmasahva, carmakasa, phena, dipta* and *nalika.*

Family: Fabaceae.

Habitat: Throughout India.

General Description: A common, prickly, scandent shrub, occuring in tropical jungles throughout India, especially in the Deccan. Leaves bipinnate; flowers in yellow, globose, anxillary heads; pods brown, wrinkled and notched when dry; seeds 6-10 in a pod.

Chemical constituents: The bark contains saponin, which on hydrolysis yields lupeol, (-spinasterol and acacic acid lactone. The sugars identified are glucose, arabinose and rhamnose. It also contains hexacosanol and (-) spinasterone. The leaves contain oxalic, tartaric, citric, succinic and ascorbic acids. They also contain two alkaloids calyctomine and nicotine, besides rutin and an enzyme tartaric racimase. A new triterpenoid saponin having the basic skeleton of oleanolic acid. Other constituents present in the leaves are tannins, amino acids and proteins (Gupta & Nigam, 1970).

Actions: Bitter, Alterative, diuretic and tonic.

Therapeutics: An infusion of the leaves is used in malarial fever. A decoction of the pods relieves biliousness and acts as a purgative. It is used to remove dandruff. An ointment, prepared from the ground pods, is good for skin diseases. The pods, known as Shikai or Shikakai, are extensively used as an detergent, and the dry ones are powdered and perfumed, and sold in the market as soapnut powder.

Parts used: Bark.

Bioactivity: Hepatoprotective (Setti *et al.*, 2011)

Further reading

Setti KT, Ramakrishnan. G, Jaykar KR, Dadi VD. Hepatoprotective activity of ethanolic extract of pods of *Acacia concinna. Int J Pharm Res Develop* 2011; 3: 1-4.

Andrographis paniculata Nees.

Common name: Kalmegh.

Ayurvedic name: *Bhunimba.*

Family: Acanthaceae.

Habitat: Brazil.

General Description: *A. paniculata* is an erect branched annual, 0.3-0.9 meters high, branches sharply quadrangular winged in the upper part; leave - lanceolate, acute, undulate, pale beneath; Flowers small, solitary distant, in axillary or terminal racemes or panicles, bracts lanceolate; Corolla - 2 lipped, upper lip 2-toothed, lower 2 lobed, rose coloured; Flowers - Capsule, linear - oblong, acute at both ends; Seeds many, rugosely pitted, yellowish brown (Singh, Kapoor & Chandra , 1972; Tomar, Tiwari & Chaturvedi, 1982; Padma Rao , Ramesh & Rajagopal,1987).

Chemical constituents: Diterpenoids (andrographolide, neoandrographolide and kalmeghin) and sodium chloride.

Actions: According to Ayurveda the plant is bitter, acrid, cooling, laxative, vulnerary, antipyretic, antiperiodic, anti-inflammatory, expectorant, depurative, soporific, anthelmintic and digestive.

Therapeutics: *A. paniculata* is useful in hyperdispsia, buring sensation, wounds, ulcers, chronic fever, malarial and intermittent fevers, inflammations, cough, bronchitis, skin diseases, leprosy, colic, flatulence, diarrhoea, dysentery and haemorrhoids.

Parts used: Leaves and flowers.

Bioactivity: Hepatoprotective (Dwivedi *et al.,* 1987; Handa & Sharma, 1990; Rana & Avadhoot, 1991; Kapil *et al.,* 1993; Trivedi & Rawal, 2000), antipyretic (Kanniappan, Mathuram & Natarajan, 1991), antimalarial (Misra *et al.,* 1992), immunostimulant (Puri *et al.,* 1993; Puri *et al.,* 1997), antioxidant (Koul & Kapil, 1994), cholretic (Chaudhuri, 1978; Tripathi & Tripathi, 1991; Shukla *et al.,* 1992), antiallergic (Gupta, Tandon & Patnaik, 1998), antimicrobial (Singh, Roy & Dey, 2003), anticancer (Rajagopal *et al.,* 2003), immunomodulatory (Naik & Hule, 2009)

Clinical study: Infective hepatitis (Chaturvedi *et al.,* 1983; Gupta & Rai, 1999)

Further reading

Chaturvedi GN, Tomar GS, Tiwari SK, Singh KP.Clinical studies on kalmegh (*Andrographis paniculata* Nees) in infective hepatitis. *Ancient Sci Life* 1983;2: 208-215.

Chaudhuri SK. Influence of *Andrographis paniculata* (kalmegh) on bile flow and hexobarbitone sleeping in experimental animals. *Indian J Exp Biol* 1978;16: 830-832.

Dwivedi SK, Sharma MC, Mukherjee SC, Lal J, Pandey NN. Comparative efficacy of Liv. 52 and *Andrographis paniculata* Nees in experimental liver damage in rabbits. *Indian Drugs* 1987;25: 1-4.

Gupta PP, Tandon JS, Patnaik GK. Antiallergic activity of andrographolides isolated from *Andrographis paniculata* (Burm.f.) Nees. *Pharm Biol* 1998;36: 72-74.

Gupta SK, Rai NP. Kalmegha (*Andrographis paniculata*) and kamala roga (infective hepatitis) - a clinical trial. *Sach Ayur* 1999;51: 544-546.

Handa SS, Sharma A. Hepatoprotective activity of andrographolide from *Andrographis paniculata* against carbontetrachloride. *Indian J Med Res* 1990; 92:276-283.

Kanniappan M, Mathuram LN, Natarajan R. Study of the antipyretic effect of chiretta (*Andrographis paniculata*). *Indian Vet J* 1991;68: 314-316.

Kapil A, Koul IB, Banerjee SK, Gupta BD. Antihepatotoxic effects of major diterpenoid constituents of *Andrographis paniculata*. *Biochem Pharmacol* 1993; 46:182-185.

Koul IB, Kapil A. Effect of diterpenes from *Andrographis paniculata* on antioxidant defense system and lipid peroxidation. *Indian J Pharmacol* 1994;26:296-300.

Misra P, Pal NL, Guru PY, Katiyar JC, Srivastava V, Tandon JC. Antimalarial activity of *Andrographis paniculata* (kalmegh) against Plasmodium berghei NK 65 in Mastomys natalensis. *Int J Pharmacog* 1992;30, 4, 263—274.

Mitra GC, Bal SN. Pharmacognostic studies of the leaves and stems of *Andrographis paniculata* Nees. *Indian Pharm* 1950;5: 168.

Naik SR, Hule AK. Evaluation of immunomodulatory activity of an extract of andrographolides from *Andrographis paniculata*. *Planta Med* 2009;75: 785-791.

Nalamolu KR. Anti-hyperglycemic and renal protective activities of *Androgarphis paniculata* chloroform extract. *Iranian J Pharmacol* Ther 2006;5: 47-50.

Padma Rao P, Ramesh D, Rajagopal T. Pharmacognostic studies on kalmegh (*Andrographis paniculata* Nees). International Seminar on Unani Medicine, New Delhi, 1987;13-15.

Puri A, Saxena R, Saxena RP, Saxena KC, Srivastava V, Tandon JS. Immunostimulant agents from *Andrographis paniculata*. *J Nat Prod* 1993; 56: 995-999.

Rajagopal S, Kumar RA, Deevi DS, Satyanarayana C, Rajagopalan R. Andrographolide, a potential cancer therapeutic agent, isolated from *Andrographis paniculata*. *J Exp Ther Oncol* 2003; 3: 147-158.

Rana AC, Avadhoot Y. Hepatoprotective effects of *Andrographis paniculata* against carbon tetrachloride induced liver damage. *Archives Pharm Res* 1991;14: 93-95.

Shukla B, Visen PK, Patnaik GK, Dhawan BN. Choleretic effect of andrographolide in rats and guinea pigs. *Planta Med* 1992;58:146-149.

Singh A, Kapoor LD, Chandra V. Pharmaco-botanic studies of kalmegh (*Andrographis paniculata* Nees). *J Res Indian Med* 1972; 7: 93-95.

Singha PK, Roy S, Dey S. Antimicrobial activity of *Andrographis paniculata Fitoterapia* 2003;74: 692-694.

Tomar GS, Tiwari SK, Chaturvedi GN. Kalmegh (*Andrographis paniculata* Nees) and its medicinal status. *Nagarjun* 1982; 26: 76-78.

Tripathi GS, Tripathi YB. Choleretic action of andrographolide obtained from *Andrographis paniculata* in rats. *Phytother Res* 1991;5: 176-178.

Trivedi N, Rawal VM. Hepatoprotective and toxicological evaluation of *Andrographis paniculata* on severe liver damage. *Indian J Pharmacol* 2000;32:288-293.

77. SARIVA

Current representatives: *Hemidesmus indicus and Cryptolepis buchanana.*

Possible representative: *Clitoria ternatea, Phyllanthus disichus, Smilax ornate* and *Vallaris solanacea.*

Hemidesmus indicus R. Br. (White variety of *sārivā*)

Syn: *Periploca indica* L.

English name: Indian sarsaparilla.

Ayurvedic names: *Asphota, gopakanya, gopangana, gopavalli, kashta sariva, latahava, pratanika, sarada,* and *utpala sariva.*

Family: Asclepidaceae.

Habitat: *H. indicus* occurrs over the greater part of India, from the upper Gangetic plain eastwards to Assam and throughout central, western and southern India.

General Description: *H. indicus* is a slender, laticiferous, twining, sometimes prostrate or semi-erect shrub. Roots are woody and aromatic. The stem is numerous, slender, terete, thickened at the nodes. The leaves are opposite, short-petioled, very variable, elliptic-oblong to linear-lanceolate. The flowers are greenish outside, purplish inside, crowded in sub-sessile axillary cymes (Nair *et al.,* 1978; Togunashi, Yoganarasimhan & Venkataram,1978).

Detailed Description of the Root (Prasad & Wahi, 1965; Austin & Jegadeesan, 2003)

(a) *Macroscopic*

Roots occur in pieces, about 30 cm long and 3-8 mm in diameter, cylindrical, thick, hard, somewhat tortuous, sparcely branched, provided with few thick rootlets and secondary roots, external appearance dark brown, sometimes with violet grey tinge, centre yellow, woody, surrounded by a mealy white cortical layer, bark brownish, corky, marked with transverse cracks and longitudinal fissures and easily detachable from the hard central core, odour, characteristic, taste, sweetish, slightly acrid and aromatic.

(b) *Microscopic*

Transverse section of root shows periderm consisting of three layers of tissues, cork, cork cambium and secondary cortex, cork cells radially flattened and rectangular in appearance filled with dark brown contents giving reactions of tannins, cork cambium, 2 or 3 layered, compressed, and filled with deep brown contents, secondary cortex, 3-4 layers of cells, similar to cork cells, with very little or no dark brown contents, secondary phloem consists of sieve elements, parenchyma, phloem ray cells along with several laticiferous ducts, parenchyma cells filled with starch grains, diameter 7-10 μ, occasional prismatic crystals of calcium oxalate, laticiferous ducts scattered in parenchymatous tissue, cambium very narrow: xylem traversed by narrow medullary rays, vessels and tracheids characterised by the presence of pitted markings, pith absent and centra region occupied by woody tissues.

Chemical composition: Coumarin and volatile oil (containing p-methoxy salicylic aldehyde), saponin, resin, tannins, sterols and glucosides. Roots contain ß-sitosterol, 2-hydroxy-4-methoxyb-enzaldehyde, α- amyrin, ß-amyrin and its acetate, hexatriacontane, lupeol octacosonoate, lupeol and its acetate. 2-hydroxy-4 methoxy benzoic acid has been reported.

Specific action as per Ayurveda: *Rakta prasādana* (Alterative).

Therapeutics: It is employed in nutritional disorders, syphilis, gonorrhea, chronic rheumatism, gravel and other urinary diseases and skin affections (eczema, acne vulgaris).

Parts used: Root.

Bioactivity: Antileprotic (Gupta,1981), antimicrobial (Prasad *et al.,* 1983), anti-ulcer (Austin & Jegadeesan, 2003), antioxidant and antithrombotic (Mary *et al.,* 2003; Ravishankara *et al.,* 2009), radiatioprotective (Shetty *et al.,* 2005), antinociceptive (Verma *et al.,* 2005), anti-inflammatory and antipyretic (Lakshman *et al.,* 2006), hypoglycaemic (Dash *et al.,* 2008), tranquillizer (Chidrawar *et al.,* 2009), nephroprotective (Chidrawar *et al.,* 2009), anti-diabetic (Gayathri & Kannabiran, 2009).

Further reading

Austin A, Jegadeesan M. Anti-ulcer potential of *Hemidesmus indicus* var. *pubescens* R.Br. Hamdard Medicus 2003; 46: 3.

Austin A, Jegadeesan M. Preliminary phyto-chemical screening of *Hemidesmus indicus* (L.) R.Br. var. *indicus* and *Hemidesmus indicus*

(L.) R.Br. var. *pubescens* (Wight et Arn.) Hook.f. J. Joseph Thas (Ed.), 2001; 76 (abstract).

Chidrawar VR, Patel NJ, Ushir YV, Patel KN. Pre-clinical evolutionary study of *Hemidesmus indicus* Linn. as minor tranquillizer. *J Res Edu Indian Med* 2009; 15:1.

Chidrawar VR, Ushir YV, Sudarshan S, Patel KN, Patel N, Vadalia KR. Possible role of natural nephroprotective: *Hemidesmus indicus* in congestive heart failure. *Pharmacog Res* 2009; 1: 367-374.

Das S, Niranjali Devaraj S. Glycosides derived from *Hemidesmus indicus* R.Br. root inhibit adherence of *Salmonella typhimurium* to host cells: receptor mimicry. *Phytother Res* 2006; 20: 784-793.

Dash GK, Bal SK, Annapurna MM, Suresh P: Studies on the hypoglycaemic activity of *Hemidesmus indicus* R.Br. roots. *Pharmacog Mag* 2008; 4:16.

Gayathri M, Kannabiran K. Effect of 2-hydroxy-4 methoxy benzoic acid from the roots of *Hemidesmus indicus* on streptozotocin-induced diabetic rats. *Indian J Pharm Sci* 2009; 71:581-584.

Gupta PN. Antileprotic action of an extract from anantamul (*Hemidesmus indicus* R.Br.) *Lep India* 1981; 53:354-359.

Lakshman K, Shivaprasad HN, Jaiprakash B, Mohan S. Anti-inflammatory and antipyretic activities of *Hemidesmus indicus* root extract. *African J Trad Comp Alt Med* 2006; 3: 90-94.

Mary NK, Achuthan CR, Babu BH, Padikkala J. *In vitro* antioxidant and antithrombotic activity of *Hemidesmus indicus* (L.) R.Br. *J Ethnopharmacol* 2003; 87:187-191.

Nair RC, Pattanshetty JK, Mary Z, Yoganarasimhan SN. Pharmacognostical studies on the root of *Decalepis hamiltonii* Wt. et Arn. and its comparison with *Hemidesmus indicus* R.Br. *Proc Indian Acad Sci* 1978;87B:37-48.

Prasad S, Wahi SP. Pharmacognostical investigation on Indian sarsaparilla, part I: root and root-stock of *Hemidesmus indicus* R.Br. *Indian J Pharm* 1965;27:2, 35-39.

Prasad YR, Alankararao GSJG, Baby P. Antimicrobial studies on essential oils of *Hemidesmus indicus* R.Br. *Indian Perfumer* 1983;27:197.

Ravishankara MN, Shrivastava N, Padh H, Rajani M. Evaluation of antioxidant properties of root bark of *Hemidesmus indicus* (anantmul). *Phytomedicine* 2002;9:153-160.

Shetty TK, Satav JG, Nair CKK. Radiation protection of DNA and membrane in vitro by extract of *Hemidesmus indicus*. *Phytother Res* 2005; 19:387-390.

Togunashi VS, Yoganarasimhan SN, Venkataram BS. A new source for sariva and identification of shweta sariva. *J Res Indian Med* 1978; 13:75-80.

Verma PR, Joharapurkar AK, Chatpalliwar VA, Asnani AJ: Antinociceptive activity of alcoholic extract of *Hemidesmus indicus* R.Br. in mice. *J Ethnopharmacology* 2005; 102: 298-301.

Cryptolepis buchanani Roem et Schult. (Black variety of *sariva*)

Ayurvedic names: *Bhadra, bhadracandana sariva, candana-gopa, candana, krsnamula* and *krshnavalli.*

Family: Asclepidaceae.

Habitat: Southeast Asia.

General Description: Twining, evergreen, glabrous shrub; leaves simple, opposite, petiole 1.25 cm, lamina 7.5-15.0 em by 2.5-6.2 cm, leathery, shining above, acuminate; cymes very shortly peduncted, paniculate; flowers white, tubular, 1.25 cm in diameter; follicles 5-10 cm long, straight, rigid; seeds compressed.

Detailed Description of the Root (Wahi, Ansari & Prasad, 1971 b)

(a) *Macroscopic*

Roots vary in length and are 1 to 1.5 cm thick; slender, cylindrical, dark brown orblackish; rough due to fine longitudinal ridges and wrinkles running sinuously lengthwise; thicker roots show a few transverse cracks, fissures and longitudinal wrinkles with remnants of rootlets and a few lenticels; cork easily peelable; fracture, short and fibrous; odour, slightly aromatic; taste, sweet and astringent.

(b) *Microscopic*

Shows thin cork consisting of 4 to 14 layers of thin-walled, rectangular to tangentially elongated cells, arranged radially; cork cambium single layered, followed by a wide zone of secondary cortex composed of polyhedral, oval to tangentially elongated cells having fibres in single or in groups of two to ten; fibres long, thick-walled but very occasionally appear also as elongated stone cells; secondary phloem wide consisting of sieve elements, phloem parenchyma, fibres and a few crystal fibres, and traversed by phloem rays; phloem fibres occur in small groups or rarely in singles, somewhat similar in shape to those of secondary cortex with comparatively thicker walls; crystal fibres elongated, thick-walled and divided into chambers, usually 7 to 17 in number, each chamber containing a prismatic crystal of calcium oxalate; medullary rays urn-to triseriate; cambium 2 to 4 layered; secondary xylem composed of vessels, tracheids, fibre-tracheids, fibres and parenchyma and traversed by xylem rays; vessels with bordered pits, and filled with tyloses; tracheids long and narrow having bordered pits, and moderately thick-walls; xylem parenchyma usually rectangular in shape with pitted walls but some of the pits become T or Y shaped with reticulate thickening;

xylem elements thick-walled and lignified; simple and compound starch grains found in abundance in all parenchymatous cells simple being elliptical to oval, measuring 3 to 19 μ in dia., with central hilum and compound with 2 or 3 components.

Powder - Light grey; shows fragments of cork cells, vessels having bordered pits, tracheids, fibres, prismatic crystals of calcium oxalate, starch grains numerous, simple and compound, elliptical to oval, measuring 3 to 19 μ in dia., with central hilum and compound with 2 or 3 components.

Chemical composition: *C. buchanani* contains pyridine alkaloid buchananine. Roots contain cardiac glycoside buchanin. Stems contain n-triacontanol, n-triacontanoic acid, (alpha-amyrin, Beta-sitosterol glucoside, cryptolepine, quindoline, nicotinoyl glucoside (Tewari, Lal & Joshi, 1979).

Specific action as per Ayurveda: *Rakta prasadana* (Alterative).

Therapeutics: Indigestion, asthma, menorrhagia and diarrhea.

Parts used: Root.

Bioactivity: Anti-dermatophyte, immunomodulator and immuno-restorative (Kaul, Zutshi & Sharma, 1996; Kaul *et al.,* 2003).

Further reading

Kaul A, Bani S, Zutshi U, Suri KA, Satti NK, Suri OP. Immunopotentiating properties of *Cryptolepis buchanani* root extract. *Phytother Res* 2003; 17:1140-1144.

Kaul A, Zutshi U, Sharma ML. Immunostimulatory and immunorestorative properties of *Cryptolepis buchanani. Indian J Pharmacol* 1996; 28: 46 (abstract No. 57).

Tayung K, Saikia N. *Cryptolepis buchanani* - A less-known medicinal plant used in bone fracture. *Indian J Trad Knowl* 2003;2; 371-374.

Tewari LC, Lal VK, Joshi P. Chemical studies on *Cryptolepis buchanani* Roem. et Schult. (sariva). *J Res Indian Med* 1979:14;2.

Wahi SP, Ansari MS, Prasad S. Pharmacognostical investigations on Indian sarsaparilla, part III: root of *Cryptolepis buchanani* Roem. and Schult. *J Res Indian Med* 1971b;6: 159-166.

Phyllanthus distichus (L.) Müll. Arg.

Syn: *Cicca acida* Merr., *Cicca disticha* L. and *Phyllanthus acidus* (L.) Skeels

Common name: Star Gooseberry.

Family: Euphorbiaceae.

Habitat: Brazil.

General Description: The plant is a curious intermediary between shrubs and tree, reaching 2 to 9 m (6½ to 30 ft) high. The tree's dense and bushy crown is composed of thickish, tough main branches, at the ends of which are clusters of deciduous, greenish, 15-to-30-cm long branchlets. The branchlets bear alternate leaves that are ovate or lanceolate in form, with short petioles and pointed ends. The leaves are 2-7.5 cm long and thin, they are green and smooth on the upperside and blue-green on the underside.

Chemical constituents: Gallic acid, tannic acid, saponin and a crystalline substance (possibly lupeol).

Actions: Liver-tonic, haemanitic, stomachic, demulcent, sudorific and drastic purgative.

Therapeutics: Syrup of fruits is used in loss of appetite. The leaves along with pepper are used as cataplasm in sciatica, lumbago and rheumatism. A decoction of the leaves is given in fevers. Leaves are used in gonorrhea due to presence of mucilage.

Parts used: Whole plant.

Further reading

http://www.hort.purdue.edu/newcrop/morton/otaheite_gooseberry.html.

Vallaris solanacea (Roth) Kuntze

Habitat: Cambodia, India, Indonesia, Laos, Myanmar, Pakistan, Sri Lanka, Thailand, Vietnam.

Family: Apocynaceae.

General Description: Shrubs climbing, often twining, to 10 m. Bark dirty whitish gray; flowering branchlets, slender, grayish pubescent. Petiole 0.2-2 cm; leaf blade elliptic to narrowly so, 2-15 X 0.8-6 cm, densely pubescent on both surfaces, base cuneate or rounded; lateral veins 5-12 pairs. Cymes di- or trichasial; peduncle 0.5-3 cm. Flowers fragrant; pedicel 0.5-2.5 cm. Sepals ovate or narrowly elliptic, 2-7 mm. Corolla white or pale yellow, limb 1.4-2.5 cm in diam., tube 5-10 mm, lobes rounded at apex. Staminal glands yellow, globose; disc shorter than ovary, apex pilose. Follicles oblong, 8-14 X 1.5-3.5 cm. Seeds ellipsoid, 9-10 mm, coma 3-4 cm (Jamwal, Anand & Chopra, 1961; Madan & Kundu, 1962; Wahi, Khosa & Mukherjee, 1980).

Chemical constituents: glycosides (Jamwal, Anand & Chopra, 1961; Dhawan BN, Patnaik, 1985).

Action: Diuretic.

Bioactivity: Cardiac tonic (Vohra & De, 1961)

Further reading

Desai VB. Antimicrobial activity of *Vallaris heynei* Spreng. *Indian J Pharm* 1971; 33: 53.

Dhawan BN, Patnaik GK. Investigation in some new cardio-active glycosides from *Vallaris solanacea. Indian Drugs* 1985; 22: 288-290.

Jamwal KS, Anand KK, Chopra IC. Pharmacological activity of glycosides derived from the leaves of *Vallaris solanacea* Kuntze. *J Sci Indust Res* 1961; 20 C: 21-24.

Madan CL, Kundu BC. Pharmacognosy of the leaf and petiole of *Vallaris solanacea* Kuntze. *J Sci Indust Res* 1962; 21 C: 128-130.

Vohra MM, Patnaik GK, Kapil RS, Anand N. Chemistry and pharmacology of a glycoside of *Vallaris solanacea J Pharm Sci* 1966; 55: 1425-1428.

Vohra MM, De NN. Digitalis activity of glycosides from *Vallaris solanacea. J Sci Indust Res* 1961; 20 C:161-162.

Wahi AK, Khosa RL, Mukherjee AK. Pharmacognostical studies on *Vallaris solanacea* O.Kuntze. *Bull Med Ethnobot Res* 1980; 1: 506-524.

77. SOMA

Current representatives: *Ephedra gerardiana.*

Possible representatives: *Sarcostemma brevistigma* and *Periploca aphylla*

Ephedra gerardiana Wall.

Common name: Ephedra.

Family: Pinaceae.

Habitat: Temperate and Alpine Himalayas.

General Description: An evergreen shrub, it grows to a height of 1m with a spread of up to3m; the stem is slender, erect, grooved, green and branching; the leaves are smalll, scale-like, articulate, and sheathing at each joint; a dioecious plant, the male flowers have prominent, yellow pollen sacks; the cones are succulent and contain two carpels, with a single seed in each.

Chemical constituents: Alkaloids (ephedrine and pesudoephedrine).

Specific action as per Ayurveda: *Shavshara* (bronchodilator).

Therapeutics: Hysteria, nocturnal enuresis, bronchial asthma, narcolepsy, dysmenorrhoea and common-cold.

Parts used: Branches.

Further reading

Nadkarni AK. *The Indian Materia Medica*, Vol. I, 1982; pp: 486

Sharma PV. *Dravya Guna Vigyana*, Vol II, 1964; pp: 302.

Periploca aphylla Decne. var. *laxiflora* Bornm.

English name: Milk broom.

Habitat: Southwestern Asia.

General Description: A plant with leafless erect stems.

Family: Asclepiadaceae.

Therapeutics: Milky juice is sued for the treatment of indolent swellings.

Further reading

Nadkarni KM. Indian Materia Medica. 3rd ed., Vol. 1 & 2, 1976. Bombay: Popular Prakashan.

78. SHANKHAPUSHPI

Current representative: *Convolvulus plauricaulis*

Possible representative: *Evolvulus alsinoides*

Convolvulus plauricaulis Choisy

Common name: Bindweed.

Ayurvedic names: *Ksheerpushpi* and *mangalyakusuma.*

Family: Convolvulaceae.

Habitat: India.

General Description: *C. plauricaulis* is a perennial herb having stoloniferous rhizome. Leaves are flat, liner green with prominent nerves and scabrous. Flowers are 10-20 flowered pale or purplish with three stamens. Fruit is a nut; oblong, trigonous, stramineous and stiptate (Rakhit & Basu, 1958; Aulakh, Narayanan & Mahadevan, 1988).

Chemical constituents: An active alkaloid, sankhpuspine, has been isolated from *C. pluricaulis*. Also a volatile oil has been obtained by steam distillation of the fresh plant. Additional alkaloids have been identified, including convolvine, convolamine, phyllabine, convolidine, confoline, convoline, subhirsine, convosine, convolvidine and scopoline. In addition, flavonoids and coumarins have been reported. β-sitosterol glycoside, hydroxy cinnamic acid and pentatriacontane, triacontane, octacosanol tetracosane along with glucose and sucrose have also been isolated from the plant (Karandikar & Satakapan, 1959; Bisht & Singh, 1978; Aulakh, Narayanan & Mahadevan, 1988; Basu & Dandiya, 2006; Madhavan, Yoganarasimhan & Gurudeva, 2008).

Specific action as per Ayurveda: *Medhya* (Brain tonic), alterative and febrifuge.

Therapeutics: Therapeutically it is used in fever, nervous debility and loss of memory. The whole herb is used medicinally in the form of decoction with cumin and milk in fever, nervous debility, loss of memory, also in syphilis and scrofula.

Parts used: Whole plant.

Bioactivity: Hypotensive (Chaturvedi, Sharma & Sen, 1964), hypolipidemic (Sharma *et al.,* 1965; Sinha *et al.,* 1989), anti anxiety (Dandiya & Chopra, 1970; Pawar, Dhuley & Naik, 2001; Nahata, Patil & Dixit, 2009; Sharma *et al.,* 2009), antidepressant (Dandiya & Chopra, 1970; Dhingra & Valecha, 2007), antifungal (Gupta & Mudgal, 1974), antiepileptic (Mulchandani *et al.,* 1995), spasmolytic, gastro protective (Sairam, Rao & Goel, 2001), immunomodulatory (Ganju *et al.,* 2003), thyrotropic (Panda & Kar, 2001) and nootropic.

Clinical studies: Thyrotoxicosis (Gupta *et al.,* 1981).

Drug interaction: Phenytoin (Dandekar *et al.,* 1992).

Further reading

Aulakh GS, Narayanan S, Mahadevan G. Phytochemistry and pharmacology of shankapushpi - four varieties. *Ancient Sci Life* 1988; 7:149-156.

Basu NK, Dandiya PC. Chemical investigation of *Convolvulus pluricaulis* Choisy. *J American Pharm Assoc* 2006; 37:27-28.

Bisht NPS, Singh R. Chemical Studies of *Convolvulus microphyllus* Sieb. *Planta Med* 1978; 34:222-223.

Barar FSK, Sharma VN. 1966. Preliminary pharmacological studies on *Convolvulus pluricaulis* Chois: an Indian indigenous herb. *Indian J Physiol Pharmacol* 9:2;99-102.

Chaturvedi GN, Sharma RK, Sen SP. Hypotensive effect of certain indigenous drugs with special reference to shankhapuspi (*C. pluricaulis*) in anaesthetized dogs. *J Res Indian Med* 1964; 2:57-67.

Dandekar UP, Chandra RS, Dalvi SS, Joshi MV, Gokhale PC, Sharma AV, Shah PU, Kshirsagar NA. Analysis of a clinically important interaction between phenytoin and Shankhapushpi, an Ayurvedic preparation. *J Ethnopharmacol* 1992; 35:285-288.

Dandiya PC, Chopra YM. CNS-active drugs from plants indigenous to India. *J Ethnopharmacol* 1970; 2: 67-90.

Dhingra D, Valecha R. Evaluation of the antidepressant-like activity of *Convolvulus pluricaulis* choisy in the mouse forced swim and tail suspension tests. *Int Med J Exp Clin Res* 2007; 13:155-161.

Ganju L, Karan D, Chanda S, Srivastava KK, Sawhney RC, Selvamurthy W. Immunomodulatory effects of agents of plant origin. *Biomed Pharmacother* 2003; 57:296-300.

Gupta RC, Mudgal V. Antifungal effect of *Convolvulus pluricaulis* (shankhpushpi). *J Res Indian Med.* 1974; 9:67-68.

Gupta RC, Singh PM, Prasad GC, Udupu KN. Probable mode of action of shankhapuspi in the management of thyrotoxicosis. *Ancient Sci Life* 1981; 1:46-54.

Karandikar GK, Satakapan S. Shankhapushpi-{A} pharmacognostic study--{II}. *Convolvulus microphyllus* Sieb. *Indian J Pharmacol* 1959:204-207.

Madhavan V. Yoganarasimhan N, Gurudeva MR Pharmacognostical studies on Sankhapushpi (*Convolvulus microphyllus* Sieb. ex Spreng and *Evolvulus alsinoides* (L.) *Indian J Trad Knowl* 2008; 7:529-541.

Mulchandani MA, Barve A, Gokhale PC *et al.*, Mechanism of anti-epileptic effect of *Convolvulus pluricaulis*: effect on veratrine-induced amino acid release. *Indian J Pharmacol.* 1995; 27:48 (abstract no.19).

Nahata A, Patil UK, Dixit VK. Anxiolytic activity of *Evolvulus alsinoides* and *Convolvulus pluricaulis* in rodents. *Pharma Biol* 2009; 47:444-451.

Panda S, Kar A. Inhibition of T3 production in levothyroxine-treated female mice by the root extract of *Convolvulus pluricaulis. Horm Metabol Res.* 2001; 1:16.

Pawar SA, Dhuley JN, Naik SR. Neuropharmacology of an Extract derived from *Convolvulus microphyllus. Pharma Biol* 2001; 39: 253-258.

Rakhit S, Basu NK. *Convolvulus pluricaulis. Indian J Pharm* 1958; 20:357-359.

Sairam K, Rao CV, Goel RK. Effect of *Convolvulus pluricaulis* Choisy on gastric ulceration and secretion in rats. *Indian J Exp Biol* 2001; 39:350-354.

Sharma K, Arora V, Rana AC *et al.*, Anxiolytic Effect of *Convolvulus pluricaulis* Choisy Petals on Elevated Plus Maze model of anxiety in mice. *J Herbal Med Toxicol* 2009; 3:23-26.

Sharma VN, Barar FSK, Khanna NK *et al.*, Some pharmacological actions of *Convolvulus pluricaulis* Choisy. An Indian indigenous herb, part II. *Indian J Med Res* 1965; 53:871-876.

Shukla, SP. A comparative study on the barbiturate hypnosis potentiation effect of medhya rasayana drugs - shankhapushpi (*Convolvulus pluricaulis*) *Bull Med Ethnobot Res* 1980; 1:554.

Sinha SN, Dixit VP, Madnawat AVS, Sharma OP. Hypolidemic effects of the ethanolic extract of *C. pluricaulis*. *Indian Med* 1989; 1:1-2.

Evolvulus alsinoides Linn.

Common name: Slender dwarf morning-glory.

Ayurvedic names: *Kambumalini, medhya, sankhanamani, kriti, sankhuli, smritihita, varnavilasini* and *vishnukrandi*.

Family: Convolvulaceae.

Habitat: India.

General Description: *E. alsinoides* is a perennial herb with a small woody and branched rootstock. Its branches are annual, numerous, more than 30 cm long, often prostrate, slender and wiry with long hairs. Leaves are small, entire, elliptic to oblong, obtuse, apiculate, base acute and densely hairy. Petiole is minute or nearly absent. Bracts are linear and persistent. Flowers mostly solitary in upper axils. Corolla blue rotate and broad funnel shaped, Calyx 4 is lobed, lanceolate and the tip acute. Peduncle is long and axillary. Capsule is globose and 4 valved. Seeds are 4 and glabrous (Goyal & Singh, 2005).

Chemical constituents: Alkaloids (evoline, shankhapushpine and betaine) and an unidentified compound.

Specific action as per Ayurveda: *Medhya* (Brain tonic).

Therapeutics: The whole herb is used medicinally in the form of decoction with cumin and milk in fever, nervous debility, loss of memory and syphilis. Decoction of the drug, with *Ocimum sanctum* is administered in fevers accompanied by indigestion or diarrhea. Decoction was given in cases of malarial fever. The oil promotes the growth of hair.

Parts used: Whole plant.

Bioactivity: Anthelmintic and antimicrobial (Dash *et al.*, 2002), antioxidant (Auddy *et al.*, 2003), antibacterial (Tharan *et al.*, 2003), adaptogenic and anti-amnesic (Sripurapu *et al.*, 2005; Gupta *et al.*, 2007), antiulcer and anticatatonic (Siripurapu *et al.*, 2005), gastroprotective and immmunomodulator. Evolvine has sympathomimetic activity. Fraction 'd' base was shown to decrease the rate and force of contraction in frog heart, in low doses (0.02mg) while high doses (0.1 mg) causes cardiac arrest. Experiments on rabbit intestinal muscle showed that, in doses of 1: 500,000 to 1:2,50,000 it decreased the amplitude of contraction and increased the tone of intestinal muscle (Guruswami, Krishnamurthy & Vaidyanathan, 1956).

Further reading

Agarwal N, Dey CD. Behavioural and lethal effects of alcoholic extracts of *Evolvulus alsinoides* in albino mice. *J Physio Allied Sci* 1997; 31:81.

Asolkar LV, Kakkar KK, Chakre OI. Glossary of Indian medicinal plants with active principles. Publication and Information Directorate, Council of Scientific and Industrial Research, New Delhi, India: 1992, pp. 27

Auddy B *et al.* Screening of antioxidant activity of three Indian medicinal plants, traditionally used for the management of neurodegenerative diseases. *J Ethnopharmacol* 2003; 84:131-138.

Dandekar UP *et al.* Analysis of a clinically important interaction between phenytoin and Shankhapushpi, an Ayurvedic preparation. *J Ethnopharmacol* 1992; 35:285-288.

Dash GK, Suresh P, Sahu SK, Kar DM, Ganapaty S, Panda SB. Evaluation of *Evolvulus alsinoides* Linn. for anthelmintic and antimicrobial activities. *J Nat Red* 2002; 2:19-22.

Goyal PR, Singh KP. Shankhpuspi (*Evolvulus alsinoides* Linn.): a medicinal herb. *Int J Mendel* 2005; 22:124.

Gupta P, Siripurapu KB, Ahmad A, Palit G, Arora A, Maurya R. Anti-stress Constituents of *Evolvulus alsinoides*: An Ayurvedic Crude Drug. *Chem Pharma Bull* 2007; 55:771.

Guruswami MN, Krishnamurthy TR, Vaidyanathan TS. A Preliminary Note on the Pharmacology of Evolvine. *Curr Sci* 1956;1:19-120.

Purohit MG, Shanthaveerappa BK, Badami S, Swamy HKS, Shrishailappa B, Antiulcer and anticatatonic activity of alcoholic extract of *Evolvulus alsinoides* (Convolvulaceae). *Indian J Pharma Sci* 1996; 58:110-112.

Shah V, Bole PV. Botanical identity of *Shankapushpi*. Indian J Pharm 1961; 23:223-224.

Siripurapu KB, Gupta P, Bhatia G, Maurya R, Nath C, Palit G. Adaptogenic and anti-amnesic properties of *Evolvulus alsinoides* in rodents. *Pharmacol Biochem Behav* 2005; 81:424-432.

Tharan NT, Vadivu R, Palanisamy M, Justin V. Antibacterial Activity of *Evolvulus alsinoides. Indian Drugs* 2003, 40:585-586.

80. SHATAVARI

Current representative: *Asparagus racemosus*

Possible representative: *Asparagus sarmentosus*

Asparagus racemosus Willd.

Common name: Common asparagus.

Ayurvedic names: *Dvipishatru, dvipaka, varakantaka, narayani, shatapadi, shatahva* and *bahupatrika.*

Family: Liliaceae.

Habitat: Found through out tropical Africa, Java, Australia, India, Sri Lanka and southern parts of China. In India it is found in plains to 4,000 ft high, in tropical, sub-tropical dry and deciduous forests and in the Himalayas.

General Description: It is an under-shrub, climbs up to 1-3 m high, with stout and creeping root stock. The root occurs in clusters or fascicle at the base of the stem with succulent and tuberous rootlets. The stem is scandent, woody, triquetrous, striate, terete and climbing. The young stem is delicate, brittle and smooth. The spines are long, sub-recurved or straight. Cladodes are in tufts of 2-6 in a node, slender, finely acuminate, falcate divaricate. The flowers, solitary or fascicles, simple or branched racemes of 3 cm long. The pedicel is slender and jointed in the middle. Perianth lobes white, fragrant and 3 mm in length. The anthers minute and purple. The berry globular or obscurely 3 lobbed, purple-reddish, seeds hard with brittle testa (Mehra & Bhatnagar, 1958b).

Detailed Description of the Root (Kanitkar,1970)

(a) *Macroscopic*

Root tuberous, 10 to 30 cm in length and 0.1 to 0.5 cm thick, tapering at both ends with longitudinal wrinkles; colour cream; taste, sweetish.

(b) *Microscopic*

Shows an outer layer of piliferous cells, ruptured at places, composed of small, thin-walled, rectangular asymmetrical cells, a number of cells elongated to form unicellular root hairs; cortex comprises of 25 to 29 layers, distinct in two zones, outer and inner cortex; outer cortex consists of 6 or 7 layers, compactly arranged, irregular to polygonal, thick walled, lignified cells; inner cortex comprise of 21 to 23 layers, oval to polygonal, thin-walled, tangentially elongated cells with intercellular spaces; stone cells, either singly or in groups, form a discontinuous to continuous ring in the upper part of this region; raphides of calcium oxalate also present in this region; 2 or 3 layers of stone cells encirle the endodermis; endodermis composed of thin-walled parenchymatous cells; pericycle present below endodermis; stele ex arch and radial in position; xylem consist of vessels, tracheids and parenchyma; xylem vessels have pitted thickening; phloem patches consists of usual element; pith composed of circular to oval parenchymatous cells, a few cells slightly lignified.

Powder - Yellowish-cream; fragments of lignified, thick-walled cells; vessels with simple pits, pieces of raphides, numerous, lignified, rectangular elongated' stone cells having clear striations with wide as well as narrow lumen and groups of parenchyma.

Chemical constituents: Diogenin (Nair, & Subramanian, 1969), sarsapogenin (Rao , 1952; Kar & Sen, 1985), saponins: shatavarin 1 to 4 (Joshi, 1988; Joshi, & Dev, 1988), isoflavones; 8-methoxy-5, 6, 4'- trihydroxyisoflavone 7-O-beta-D-glucopyranoside (Saxena & Chourasia, 2001).

Specific action as per Ayurveda: Demulcent and spermopiotic.

Therapeutics: Chronic bronchitis, urinary tract infection and oligozoospermia.

Parts used: Tubers.

Bioactivity: Galactogouge (Joglekar *et al.,* 1967), antioxytocic and anti-ADH (Gaitonde & Jetmalani, 1969), anticancer (Rao, 1981), cytoprotective (Dhankur *et al.,* 1983), antibacterial (Grover & Tirumala, 1976), immunomodulator (Rege *et al.,* 1989), antiinflammatory (Imandai *et al.,* 1988) prokinetic (Dalvi *et al.,* 1990), immunomodulator (Rege, & Dahanukar, 1993), antihepatotoxic (Muruganadan *et al.,* 2000), antitussive (Mandal *et al.,* 2000), and antioxidant (Kamat *et al.,* 2000).

Clinical studies: Dueodenal ulcer (Kishore, 1980; Maheshwari & Tewari, 1996)

Further reading

Ahmad S, Ahmad S, Jam PC. Chemical exammation of Shatavari *Asparagus racemosus. Bull Med Ethnobot Res* 1991;12:157-160.

Maheshwari CM, Tewari SK. A clinical study of parinamasula and its treatment with satavari *Asparagus racemosus* Willd. *Ancient Sci Life* 1996; 15: 162-165

Dahanukar SA, Thatte U, Pai N, Mose PB, Karandikar SM. Protective effect of *Asparagus racemosus* against induced abdominal sepsis. *Indian Drugs* 1986;24:125-128.

Dahanukar SA, Date SG, Karandikar SM. Cytoprotective effect of *Terminalia chebula* and *Asparagus racemosus* on gastric mucosa. *Indian Drugs* 1983;21:442-445.

Dalvi SS, Nadkarni PM, Gupta KC. Effect of *Asparagus racemosus* shatavari on gastric emptying time in normal healthy volunteers. *J Postgrad Med* 1990;36: 91-94.

Gaitonde BB, Jetmalani MH. Antioxytocic action of saponin isolated from Asparagus racemosus Willd (Shatavari) on uterine muscle. *Arch Int Pharmacodyn Ther* 1969; 179:121-129.

Gaitonde BB, Jetmalani MH. 1969a, Antioxytocic action of saponin isolated from *Asparagus racemosus* Willd. shatavari on uterine muscle. *Arch Int Pharmacodyn Ther* 1969a; 179: 121-129.

Gaitonde BB, Jetmalani MH. Antioxytocic and anti-ADH activity of saponin fraction isolation from *Asparagus racemosus* Willd. *Indian J Pharm* 1969b; 31:175-176.

Grover GS, Tirumala Rao J. *In vitro* antimicrobial studies of the saponin obtained from *Asparagus adscendens*. *Indian Drugs* 1976; 14:103.

Grover GS, Tirumala Rao J. Chemical and antimicrobial studies of a saponin of *Asparagus adscendens*. *Fitoterapia* 1988; 59: 415-416.

Imandai SC, Maiti BC, Maity TK, Pal M, Saba BP. Evaluation of antiinflammatory activity *Asparagus racemosus* Willd. root extract. *Nat Prod Sci* 1988; 4:230-233.

Joshi J,. Dev S. Chemistry of Ayurvedic crude drugs, Part vrn - Shatavari-2, Structure elucidation of bioactive shatavarin-1 and other glycosides. *Indian J Chem* 1988; 27B:12-16.

Kanitkar U K. Identification of *Asparagus racemosus* Willd. *J Univ Poona Sci Tech* 1970; 38: 5-11.

Kishore P, Pandey PN, Pandey SN, Dash S. Treatment of duodenal ulcer with *Asparagus racemosus* Linn. *J Res Ayur Siddha* 1980; 1: 409-416.

Mehra PN. Bhatnagar JK. A comparative study of' satawar' and its supposed botanical source. *Indian J Pharm* 1958b 20: 33-40.

Rao SB. Saponins Sapogenins from Indian medicinal plants. Part I. *Indian J Pharm* 1952; 14:131-132.

Rege NN, Nazareth HM, Isaac A, Karandikar SM, Dahanukar SA. Immunotherapeutic modulation of intraperitoneal adhesion by *Asparagus racemosus*. *J Postgrad Med* 1989; 35: 199-203.

Asparagus sarmentosus L.

Ayurvedic names: *Mahashatavari, urdhvakantha, pivari, dhuivari, vari, abhiru, bahupatra and mahapurusadantika.*

Family: Liliaceae.

Habitat: India.

General Description: Shrub.

Chemical constituents: Saccharine and mucilage.

Specific action as per Ayurveda: Demulcent and spermopiotic.

Therapeutics: Hemorrhoids, malabsorption an eye ailments.

Parts used: Tubers.

80. SPRKKA

Current representative: *Melitotus officinalis*

Possible representative: *Anisomeles malabarica*

Anisomeles malabarica (L.) R.Br.

Common name: Malabar catmint.

Family: Lamiaceae.

Habitat: India.

General Description: A small perennial shrub grows up to 1.5 meters in height. Leaves simple, opposite, very thick, aromatic, oblong-lanceolate, acute, pale above, white below, crenate-serrate, softly woolly: flowers purple, in dense whorls of more or less interrupted spikes; fruits nutlets, bearing ellipsoid and compressed seeds (Choudhary, Bijjem & Kalia, 1972; Brinda, Rukmani & Purushothaman, 1981, Brinda, Rukmani & Purushothaman, 1982; Singh, Uvarani & Raghu, 2003).

Chemical constituents: Diterpenes:anisomelic acid and dovatodiolide.

Action: Stomachic, Diaphoretic and astringent.

Therapeutics: *A. malabarica* pacifies vitiated vata, kapha, epilepsy, hysteria, dementia, anorexia, dyspepsia, colic, flatulence, intestinal worms, fever arising from teething in children, intermittent fevers, arthritis, especially gout, swellings and diarrhea.

Parts used: Whole plant.

Bioactivity: Antibacterial (Babu *et al.,* 2006), anti-Inflammatory, anti-platelet and anti-arthritic Lavanya *et al.,* 2010), and antiepileptic (Singh *et al.,* 2010; Choudhary, Bijjem & Kalia, 2011).

Further reading

Babu G. *et al.* Antibacterial screening of aerial parts of *Anisomeles malabarica* (L) R.Br. *Aryavaidyaidyan* 2006; 20, 1, 53-56.

Brinda P, Rukmani B, Purushothaman KK. Pharmacognostic studies on *Anisomeles malabarica* R.Br. *Bull Medico-Ethno-Biol Res* 1981; 2: 74-84.

Brindha P, Rukmani P, Purushothaman KK. Pharmacognostical studies on *Anisomeles malabarica* R.Br. *Bull Medico-Ethno-Biol Res* 1982; 2:74-84.

Choudhary N, Bijjem KR, Kalia AN. Antiepileptic potential of flavonoids fraction from the leaves of *Anisomeles malabarica. J Ethnopharmacol* 2011; 135:238-242.

Kannabiran B, Krishnamurthy KH. Pharmacognostic study of *Anisomeles malabarica* L. (bhutankusha). *J Res Indian Med* 1972; 7: 43-49.

Lavanya R, Maheshwari U, Harish G, Bharath Raj J, Kamali S, Hemamalani D, Bharath Varma J, Umamaheswara Reddy C. Investigation of *In-vitro* anti-Inflammatory, anti-platelet and anti-arthritic activities in the leaves of *Anisomeles malabarica* Linn. *Res J Pharm Biol Chem Sci* 2010; 1: 745-752.

Singh I, Singh M, Singh T, Bijjem KV, Kalia AN. Anticonvulsant potential of *Anisomeles malabarica* leaves against experimentally induced convulsions in rats. Int *J Green Pharm* 2010; 4:199-205.

Singh S, Uvarani RM, Raghu RS. Pharmacognostical and phytochemical studies on leaves of *Antisomeles malabarica* R.Br. *Ancient Sci Life* 2003; 22: 106-110.

82. SUNISANNAKA

Current representative: *Marselia minuta*

Possible representative: *Celosia argentia*

Marsilea minuta L.

Common name: Water fern.

Family: Marsileaceae.

Habitat: Northeastern Himalayas, Eastern Ghats, Western Ghats, Central Deccan Plateau, Indo-Gangetic Plain, Outlying Islands.

General Description: The plant is highly variable. It appears cushion-like on dry lans, but in water is spreading and creeping. Internodes 0.5-20 cm long. Petioles 2-8 cm long on land and upto

25 cm or more in water. Leaflets a maximum of 2.5 cm. Sporocarp stalks simple, numerous, arising one above the other near base of leaf stalk. These sporocarps are usually crowded and in groups of 2-3.

Chemical constituents: Macrocyclic ketone: marsilin, methylamine, ß-sitosterol and hexahydroxy triterpene: marsileagenin-A (Chatterjee *et al.*, 1963).

Action: Hypno-sedative.

Therapeutics: Insomnia, epilepsy and behavioral disorders.

Parts used: Whole plant.

Bioactivity: Sedative and anticonvulsant (Chatterjee *et al.*, 1963), hypocholesterolemic (Gupta *et al.*, 2000), antifertility (Gupta *et al.*, 2002), antidepressant (Bhattamisra *et al.*, 2008), and anti-aggressive (Tiwari *et al.*, 2010).

Further reading

Bhattamisra SK, Khanna VK, Agrawal AK, Singh PN, Singh SK. Antidepressant activity of standardised extract of *Marsilea minuta* Linn. *J Ethnopharmacol* 2008; 117: 51-57.

Chatterjee A, Datta CP, Chowdhury B, Chatterjee C, Dey PK, Dey CD, Mukherjee SR. The chemistry and pharmacology of marsilin, a sedative and anticonvulsant principle isolated from *Marsilea minuta* L. and *M. rajasthanensis* Gupta. *J Exper Med Sci* 1963; 7:53.

Gupta M, Mazumdar UK, Datta I, Bhattacharya S, Mukherjee S, Manikandan I. Studies of antifertility activity of *Marislea minuta* Linn. *Indian J Pharm Sci* 2002; 64: 176-178.

Gupta RS, Kumar P, Sharma A, Bharadwaj TN, Dixit VP. Hypocholesterolemic activity of *Marsilea minuta* in gerbils. *Fitoterapia* 2000; 71: 113-117.

Tiwari OP, Bhattamisra SK, Tripathi PK, Singh PN. Anti-aggressive activity of a standardized extract of *Marsilea minuta* Linn. in rodent models of aggression. *Biosci Trends* 2010; 4:190-194.

83. SUVARNA JIVANTI*

Ayurvedic name: *Putrabhadra, priyashankari, madhura, jivpristha* and *yashakari.*

Actions and therapeutics: Sweet, cold, aphrodisiac, tonic, antibacterial and useful for regulating *parada* (mercury).

* Botanical identity is to be established.

83. SUVARNA KSHIRI

Current representative: *Argemone mexicana*
Possible representative: *Euphorbia thomsoniana*

Argemone mexicana

See details on page 85-86.

Euphorbia thomsoniana Boiss.

English name: Indian plum and coffee palm.

Family: Euphorbiaceae.

Habitat: E. Asia - Himalayas in Afghanistan, Kashmir, Pakistan and Tibet.

General Description: Herbs, 20-30 cm tall. Root terete, few branches, 30-50 cm × 6-10 mm. Stems many branched basally, slightly purplish, glabrous. Leaves alternate; stipules absent; basal scale-leaves purple or light red, ca. 5 × 3 mm, overlapping; leaf blade elliptic, 2-3 × 1-1.6 cm, glabrous, base subrounded, apex rounded; lateral veins inconspicuous. Primary involucral leaves 5-8, broadly ovate, ca. 2.5 × 1-2 cm, base subrounded, apex slightly narrow, primary rays 5-8, 1-4 cm; cyathophylls 2, triangular-ovate, ca. 1.5 × 1.4 cm, base subtruncate, apex acuminate. Cyathium sessile; involucre campanulate, 4-5 × 4-5 mm, lobes triangular, slightly bent inside, densely white pubescent inside and on margin, sparsely white pilose outside; glands 4, dark brown, reniform, shortly stalked, suberect, sparsely shortly pilose. Male flowers many, little exserted from involucre. Female flower: ovary usually not exserted from cup, sparsely shortly pilose; styles connate below middle, persistent; style arms 2-lobed. Capsule subglobose, 6-7 × ca. 7 mm, sparsely shortly pilose. Seeds ovoid-globose, ca. 6.5 × 5 mm, light gray with brown reticulum, smooth; caruncle present.

Chemical constituents: Isorhamnetin.

Action: Purgative and irritant

Therapeutics: Constipation.

Parts used: Sap.

Note: It is toxic plant.

84. TALISHPATRA

Current representative: *Taxus baccata*

Possible representative: *Flacourtia cataphracta and Rhododendron anthopogon*

Taxus baccata Linn.

English name: Common yew.

Ayurvedic names: *Talisha, dhatripatra and shukodara.*

Family: Taxaceae.

Habitat: Temperate Himalayas.

General Description: It is a small- to medium-sized evergreen tree, growing 10–20 metres (33–66 ft) (exceptionally up to 28 m/92 ft) tall, with a trunk up to 2 metres (6 ft 7 in) (exceptionally 4 m/13 ft) diameter. The bark is thin, scaly brown, coming off in small flakes aligned with the stem. The leaves are lanceolate, flat, dark green, 1–4 centimetres (0.39–1.6 in) long and 2–3 millimetres (0.079–0.12 in) broad, arranged spirally on the stem, but with the leaf bases twisted to align the leaves in two flat rows either side of the stem, except on erect leading shoots where the spiral arrangement is more obvious. The leaves are highly poisonous (Joshi, 1995).

Chemical constituents: Alkaloid (taxine), taxol (Vohora & Kumar, 1971; Parmar *et al.,* 1999) and bisflavonoids (Vohora *et al.,* 1980).

Action: Antispasmodic.

Therapeutics: Chronic bronchitis, asthma, cough and indigestion.

Parts used: Leaves.

Bioactivity: Anti-ovulatory (Chaudhury *et al.,* 1970), bronchodilator (Patel, Patel & Gandhi, 2009), anti-inflammatory (Dutta *et al.,* 2010), and analgesic (Dutta, Yadav & Mariappan, 2010).

Further reading

Chaudhury RR, Saxena SK, Garg SK. Preliminary observations in the rabbit on the anti-ovulatory activity present in *Taxus baccata* Linn. *J Rep Fert* 1970;22, 151-153.

Dutta S, Mariappan G, Sarkar D, Sarkar P. Assessment of anti-inflammatory activity of *Taxus baccata* Linn. bark extract. *Ancient Sci Life* 2010; 29: 19-21.

Dutta S, Yadav A, Mariappan G. Analgesic activity of *Taxus baccata* Linn. bark extract. *Int J Pharm Bio Sci* 2010;1:1-7.

Joshi GC, Tewari VP, Tewari KC. *Taxus baccata* Linn (Himalayan yew) - an emerging anticancer plant drug and its cultivation strategy. *Sach Ayur* 1995;47: 611-612.

Parmar VS, Jha A, Bisht KS, Taneja P, Singh SK. Constituents of yew trees. *Phytochem* 1999; 50: 1267-1304.

Patel PK, Patel KV, Gandhi TR. Evaluation of Effect of *Taxus baccata* Leaves Extract on Bronchoconstriction and Bronchial Hyperactivity in Experimental Animals. *Global J Pharmacol* 2009; 3: 141-148.

Vohora SB, Kumar I, Shah SA, Khan MSY. Effect of bisflavonoids of *Taxus baccata* on the central nervous system. *Indian J Med Res* 1980;71: 815-820.

Vohora SB, Kumar I. Phytochemical and behavioural investigations on *Taxus baccata* Linn. *Indian J Physiol Pharmacol* 1971; 15:37 (Abstract).

Flacourtia cataphracta Roxb ex Willd.

Syn: *Flacourtia jangomas* (Lour.) Rauschel

English name: *Indian plum and coffee palm.*

Family: Flacourtiaceae.

Habitat: Sub himalayan foot hill zone in India extending to South East Asia, China.

General Description: Trees small, dioecious. Trunk spiny. Branches unarmed. Stem white-dotted. Leaves oval, glabrous, acuminate; margins serrate. Flowers small in irregular racemes. Fruit globose, c. 1 cm. long, succulent, dark purple.

Chemical constituents:

Action: Astringent.

Therapeutics: Diarrhoea and fevers.

Parts used: Fruit.

Rhododendron anthopogon D.Don.

Family: Eriaceae.

Habitat: Himalayas.

Habit: Small shrubs, often prostrate or decumbent, (0.3–) 0.5–1 (–1.5) m tall; branchlets loose, intricate, scaly, sparsely setose; bud scales deciduous. Petiole 3–8 mm, scaly; leaf blade aromatic, obovate-elliptic, ovate, rarely orbicular, (1.5–) 2.5–3.5 (–4) × 1–

1.6(–2.5) cm; base rounded; apex obtuse or rounded, mucronate; abaxial surface scales overlapping, dark reddish brown and dark yellow-brown, arranged in 2–3 tiers, often those of the lowest tier darker than rest; adaxial surface dark green, shiny, usually sparsely scaly. Inflorescence 4–6(–9)-flowered. Pedicel 0.2–0.4 cm, scaly; calyx lobes 3–5(–6) mm, elliptic or oblong, persisting to enclose mature capsule, scaly, margin densely ciliate; corolla narrowly tubular-funnelform, pink, or yellowish white, tube 6–12 mm, outer surface not scaly; inner surface densely pilose; lobes spreading, 1.2–2 cm; stamens (5–)6–8(–10), included in corolla tube, filaments glabrous; ovary ca. 1 mm, scaly; style short, thick, straight, as long as ovary, glabrous. Capsule ovoid, 3–5 mm, scaly.

Chemical constituents: Essential oil (Guleria *et al.*, 2011).

Parts used: Stem.

Further reading

Guleria S, Jaitak V, Saini R, Kaul VK, Lal B, Babu GD, Singh B, Singh RD. Comparative studies of volatile oil composition of *Rhododendron anthopogon* by hydrodistillation, supercritical carbon dioxide extraction and head space analysis. *Nat Prod Res* 2011; 25:1271-1277.

85. TINTIDIKA

Current representative: *Rhus parviflora*

Possible representative: *Tamarindus indica*

Rhus parviflora Roxb.

English name: Sumac.

Ayurvedic name: Tintidik

Family: Anacardiaceae.

Habitat: South Himalayas.

General Description: An unarmed softly tomentose shrub; branchlets,leaves beneath petioles and panicle densely clothed with soft rusty tomentum. Leaves palmately 3-foliolate; leaflets 1-3 in., sessile or the terminal contracted into a petiole, elliptic oblong-obovate or orbicular,irregularly crenate above the middle, coriaceous;nerves spreading, parallel. Petals twice as long, oblong.

Chemical constituents: Tannins.

Action: Astringent.

Therapeutics: Diarrhoea and fevers.

Parts used: Seed, stem, leaves and roots.

Tamarindus indica L.

Syn: *T. officinalis Hook.*

English name: Indian dates.

Ayurvedic name: *Amlika.*

Family: Fabaceae.

Habitat: Throughout India, especially in the south India and often cultivated.

General Description: The leaves are paripinnate with sour taste and the flowers are yellow. Endocarp and mesocarp constitutes the edible tamarind of commerce. Endocarp is rough, white and leathery encloses the seed inside and with pulpy mesocarp in the outer. Outer exocarp is, hard but fragile and separated from the mesocarp (Singh, Chaudhari & Sathe, 2005).

Chemical constituents: Alkaloid (hordenine), tartaric acid, citric acid, malic acid and tannins. Chrysanthemin is the red pigment in the pulp. Pectin has been reported from seed-pulp (Ghose & Krishna, 1945; Singh, Chaudhari & Sathe, 2005).)

Action: Antiscorbutic and mild laxative.

Therapeutics: Constipation and scurvy.

Parts used: Fruit and leaves.

Bioactivity: Anti-inflammatory and analgesic (Thomas, Gangadharan & Vijayalakshmi, 1998), antihyperglycemiac and antihyperlipidemic (Maiti, Das & Ghosh, 2005) and anti-snake venom (Ushanandini *et al.,* 2006).

Further reading

Ghose TP, Krishna S. Tamarind seed pectin. Tamarind seed pectin. *Curr Sci* 1945;14, 299-300.

Maiti, RK, Das UK, Ghosh D. Attenuation of hyperglycemia and hyperlipidemia on streptozotocin-induced diabetic rats by aqueous extract of seed of *Tamarindus indica. Biol Pharm Bull* 2005; 28:1172-1176.

Singh SP, Chaudhari BG, Sathe MV. Pharmacoognostic amd phytochemical study of *Tamarindus indica* Linn.-stem bark. *Bull Medico-Ethno-Bot Res* 2005; 26: 34-45.

Thomas A, Gangadharan R, Vijayalakshmi Amma S. Anti-inflammatory and analgesic properties of the leaves of *Tamarindus indicus*. *Ancient Sci Life* 1998;18:120-126.

Ushanandini S, Nagaraju S, Harish Kumar K, Vedavathi M, Machiah DK, Kemparaju K, Vishwanath BS, Gowda TV, Girish KS. The anti-snake venom properties of *Tamarindus indica* (Leguminosae) seed extract. *Phytother Res* 2006;20: 851-858.

86. TIVARIYA

Current representative: *Avicennia officinalis*

Possible representative: *Hydnocarpus wightiana*

Avicennia officinalis L.

Syn: *Avicennia alba*

English name: Indian mangrove.

Family: Avicenniaceae.

Habitat: South Asia.

General Description: Grows to 25m. Bark: Young trees are reddish brown and smooth, older trees are grey-brown. Roots: Pencil-like pneumatophores emerge above ground from long shallow underground roots. Leaves: Thick, leathery, spoon-like (rounded at tip with edges slightly rolled under), shiny green above, underneath slightly hairy. Flowers: Small, yellow, several together, forming a cross-shaped inflorescence. Fruits: Flat capsule containing one seed.

Chemical constituents: Tannins (C.S.I.R., 1948-1976), naphthofuran: (-) avicennol C (Anjaneyulu *et al.,* 2003), triterpenoids and steroids.

Action: Aphrodisiac (Kirtikar & Basu, 1975).

Therapeutics: Seminal debility and thrush.

Parts used: Seed, stem, leaves and roots.

Bioactivity: Anti ulcer (Thirunavukkarasu *et al.* 2010).

Further reading

Anjaneyulu ASR, Murthy, YL, Rao VL, Sreedhar K. Chemical examination of the mangrove plant *Avicennia officinalis*. *Indian J Chem* (*B Org Med*). 2003; 42: 3117-3119.

Kirtikar KR, Basu BB. Indian Medicinal Plants. 4 vols. text, 4 vols. plates. 2nd ed., reprint. Jayyed Press. New Delhi, 1975; 6.

Thirunavukkarasu P, Ramanathan T, Ramkumar L, Shanmugapriya R. Anti ulcer effect of *Avicennia officinalis* leaves in albino rats. *World App Sci J* 2010; 9:55-58.

Hydnocarpus wightiana Blume.

Syn: *Hydnocarpus pentandra* (Buch.-Ham.) Oken

English name: Jungle almond.

Ayurvedic name: *Tuvraka*

Family: Flacourtiaceae.

Habitat: India and Sri Lanka.

General Description: A very large and handsome East Indian tree. The leaves are glossy, entire, and alternate; the flowers yellow and sweet-scented. The fruit is round, ash-colored, and when mature, averages in weight from 10 to 20 pounds.

Chemical constituents: Fixed oil: contains chaulmoogric acid and hydnocarpic acid and flavonolignans: hydnowightin, hydnocarpin and neohydnocarpin (Sharma & Hall, 1991).

Specifc action as per Ayurveda: Antileprotic.

Therapeutics: Psoriasis and leprosy.

Parts used: Seeds and oil.

Bioactivity: Hypolipidemic, anti-inflammatory and antineoplastic: flavonolignans (Sharma & Hall, 1991) and antioxidant: hydnocarpin and isohydnocarpin (Reddy *et al.,* 2005).

Further reading

Reddy SV, Tiwari AK, Kumar US, Rao RJ, Rao JM. Free radical scavenging, enzyme inhibitory constituents from antidiabetic Ayurvedic medicinal plant *Hydnocarpus wightiana* Blume. *Phytother Res* 2005; 19: 277-281.

Sharma DK, Hall IH. Hypolipidemic, anti-inflammatory and antineoplastic activity and cytotoxicity of flavonolignans isolated from *Hydnocarpus wightiana* seeds. *Nat Prod* 1991; 54:1298-302.

Subramanian MPS, Chelladurai V. Herbs useful in the treatment of meganoikal (venereal diseases) - a folklore study. J Joseph Thas (Ed.), 2004; 76 (abstract).

87. TRYANTIKA

Current representative: *Delphinium zalil*

Possible representative: *Gentiana kurroa, Gentiana dahurica* and *Ficus heterophylla*

Delphinium zalil Aitch. & Hemsl.

Syn: *Delphinium semibarbatum* - Bien. ex Boiss.

Ayurvedic name: *Sprkka.*

Family: Ranunculaceae.

Habitat: West Asia.

General Description: A perennial herb with bright yellow flowers and having somewhat honey-like smell; the flowers are pubescent; many of them tolerably perfect, resembling in size and shape to those of the common larkspur; the fruit consists of three follicles, which are arranged like those of the aconite and dehisce on the inside, they are marked with prominent longitudinal ribs, have pointed apices and are supported upon a stout curved peduncle; the seeds are numerous, angular and of light brown colour. The seeds are thickly set with white feather-shaped hairs, arranged in rows.

Chemical constituents: Alkaloid: delphinine.

Action: Bitter.

Therapeutics: Spleen and liver diseases.

Parts used: Seed, stem, leaves and roots.

Gentiana kurroa Royle

English name: *Gentian.*

Ayurvedic name: *Trayamana, Trayanthi, Girija* and *Anuja.*

Family: Gentianaceae.

Habitat: Kashmir, Himalayan alpines

General Description: A small perennial herb with stout rhizome bearing decumbent flowering stems; leaves narrowly oblong; radical rosulate, 7.5-12.5 cm by 6-13 mm; stem leaves 2.5 cm; narrower. Flowers blue, spotted with white, 4.5-5 cm long, 2 cm diameter; solitary or racemose. Calyx about half as long as the corolla; lobes 5 linear. Corolla 5-lobed. Capsule oblong.

Chemical constituents: Enzymes, gentiamarin, gentianose, gentisin, glycosides, oil, pectin, and phenolic acid.

Action: Stomachic.

Therapeutics: Fever, wounds, skin disease, diarrhoea, cough, hepes, psuedo abdominal tumours and sprue.

Parts used: Rhizome.

***Gentiana dahurica* Fisch.**

Syn: *G. olivieri* Griseb

English name: Common gentian.

Family: Gentianaceae.

Habitat: E. Siberia, Mongolia, C. Asia, NW. China.

General Description: A small perennial herb Long, narrow, lanceolate leaves grow on branching stems. Deep blue tubular flowers with whitish throats open from leaf axils or in terminal clusters.

Parts used: Rhizome.

88. TVAK

Current representative: *Cinnamomum tamala*

Possible representative: *Cinnamomum cassia and Cinnamomum zeylanicum*

***Cinnamomum tamala* (Buch.-Ham.) Nees & Eberm**

Synonym: *Laurus tamala* Buch.-Ham.

English name: Indian bark.

Ayurvedic name: *Tejapatra*.

Family: Lauraceae.

Habitat: It occurs in tropical, sub- tropical Himalayas between 900-2300 m.

General Description: *C. tamala* is a moderate sized evergreen tree attaining a height of 8 m, and a girth of 150 cm. Its bark produces mucilage. Leaves lanceolate, glabrous; alternately placed, opposite and short stalked. 3-nerved from the base.

Detailed Description of the Leaves

(a) *Macroscopic*

Leaves-12.5-20 cm long, 5-7.5 cm wide at the centre, 3 converging nerves from base to apex young leaves pink, petiole 7.5-13 mm long, margin entire, apex acute or accuminate, both surfaces smooth, stomata paracytic odour, aromatic, taste, slightly sweet, mucilaginous and aromatic.

(b) *Microscopic*

Petiole and midrib-transverse section of petiole and midrib shows epidermis externally covered with cuticle, uniseriate, multicellular (1 to 3 cells), trichomes present, oil cells single or in group, isolated large stone cells, much lignified showing striations found scattered, most of the parenchymatous cells of cortex with reddish-brown contents, pericycle represented by a few layers of sclerenchymatous cells, stele more or less planoconvex as in the midrib of leaf, xylem on upper and phloem on lower side consisting of usual elements, present.

Lamina-transverse section of lamina shows dorsiventral structure, represented by palisade tissue on upper and spongy parenchyma on lower side, epidermis same as in midrib, externally covered with cuticle, below upper epidermis single row of closely packed palisade layer followed by multilayered, irregular, thin-walled cells of spongy parenchyma without intercellular spaces, idioblasts containing oil globules present in mesophyll and also in palisade, lower epidermis covered externally with cuticle, lamina intervened by several small veinlets: vascular bundles covered with thick-walled fibres on both side.

Chemical constituents: Essential oils (d-s phellandrene and eugenol).

Specific action as per Ayurveda: Appetiser and stimulant.

Therapeutics: Hemorrhoids.

Bioactivity: Hypoglycaemic and hypolipidemic (Sharma & Dwivedi et al., 1996), anti-inflammatory (Devi, Kannappan & Anuradha, 2007; Chaurasia & Tripathi, 2011), antidiarrhoeal (Rao *et al.,* 2008), gastroprotective (Eswaran *et al.,* 2010) and immunomodulator (Chaurasia, Mishra & Tripathi, 2010).

Further reading

Chaurasia JK, Mishra A, Tripathi YB. Immunomodulation property of hexane fraction of leaves of *Cinnamomum tamala* Linn. in rats. *Cell Biochem Funct* 2010; 28:454-460.

Chaurasia JK, Tripathi YB. Chemical characterization of various fractions of leaves of *Cinnamomum tamala* Linn. toward their anti-oxidant, hypoglycemic, and anti-inflammatory property. *Immunopharmacol Immunotoxicol* 2011; 33:466-472.

Devi SL, Kannappan S, Anuradha CV. Evaluation of *in vitro* antioxidant activity of Indian bay leaf, Cinnamomum tamala (Buch.-Ham.) T. Nees & Eberm using rat brain synaptosomes as model system. *Indian J Exp Biol* 2007; 45:778-784.

Eswaran MB, Surendran S, Vijayakumar M, Ojha SK, Rawat AK, Rao ChV. Gastroprotective activity of *Cinnamomum tamala* leaves on experimental gastric ulcers in rats. *J Ethnopharmacol* 2010; 128:537-540.

Rao CV, Vijayakumar M, Sairam K, Kumar V. Antidiarrhoeal activity of the standardised extract of *Cinnamomum tamala* in experimental rats. *J Nat Med* 2008; 62:396-402.

Sharma S R, Dwivedi SK *et al.,* Hypoglycaemic and hypolipidemic effects of *Cinnamomum tamala* Nees leaves. *Indian J Exp Biol* 1996; 34: 372-374.

Cinnamomum cassia Blume.

Ayurvedic name: *Tvaka*.

Family: Lauraceae.

Habitat: It is reported to grow wild in isolated places in Mizoram and also cultivated to a limited extent in Tamil Nadu. It is extensively cultivated in China. The major source of Cinnamon in world trade is China.

Description: It is an evergreen aromatic tree. The bark is gray, smooth and thick up to 1.5 cm. The leaves are oblong-elliptic, dark shiny green and with a three prominent nerves from the base. The flowers are small in lax, spreading, terminal, with axillary panicles. The fruits are black, pulpy, aromatic, elliptic drupes with a single seed borne in the cup of the calyx lobe.

Chemical constituents: Cassia bark yields from 1 to 2 per cent of volatile oil, (cinnamyl acetate, cinnamic acid, phenylpropyl acetate and orthocumaric aldehyde),

Specific action as per Ayurveda: Appetiser and stimulant.

Therapeutics: Hemmorhoids

Parts used: Leaves and bark.

Bioactivity: Antioxidant (Dhan et al., 2007), antibacterial (Sharma *et al.,* 2009), and anti-mutagenic (Sharma *et al.,* 2011).

Clincial studies: Non insulin dependent diabetes mellitus (Soni & Bhatnagar, 2009)

Further reading

Dhan P, Suri S, Upadhyay G, Singh BN. Total phenol, antioxidant and free radical scavenging activities of some medicinal plants. *Int J Food Sci Nut* 2007; 58: 18-28.

Sharma A, Chandraker S, Patel VK, Ramteke P. Antibacterial activity of medicinal plants against pathogens causing complicated urinary tract. *Indian J Pharm Sci* 2009; 71: 136-139.

Sharma N, Trikha P, Athar M, Raisuddin S. Inhibition of benzo pyrene- and cyclophosphamide-induced mutagenicity by *Cinnamomum cassia. Mut Res* 2011; 480-481:179-188.

Soni R, Bhatnagar V. Effect of cinnamon (*Cinnamomum cassia*) intervention on blood glucose of middle aged adult male with non insulin dependent diabetes mellitus (NIDDM). *Studies Ethno-Med* 2009; 3:141-144.

Cinnamomum zeylanicum Breyn

Ayurvedic name: *Tejapatra.*

Family: Lauraceae.

Habitat: *C. zeylanicum* is native to China and Sri Lanka but now is cultivated throughout Indonesia, India and Vietnam.

General Description: *C. zeylanicum* is a small to medium evergreen tree from the family Lauraceae. Ranging in height between 6-12m, the heavily foliated tree is known for its rolled quills of aromatic bark.

Detailed Description of the Inner Bark

(a) *Macroscopic*

Bark pieces about 0.5 mm thick, brittle, occurs as single or double, closely packed compound quills, upto a metre or more in length and upto about 1 cm in diameter, outer surface, dull yellowish-brown, marked with pale wavy longitudinal lines with occasional small scars or holes, inner surface darker in colour, striated with longitudinally elongated reticulation, fracture, splintery, free from all but traces of cork, odour, fragrant, taste, sweet, aromatic with sensation of warmth.

(b) *Microscopic*

Transverse section of bark (devoid of cork and c.ortex) shows except at certain places pericyclic sclerenchyma, 3 or 4 rows of isodiametric cells, sometimes tangentially elongated, inner and radial walls often being thicker than the outer, some containing starch grains, small groups of pericylic fibres embedded at intervals in the sclerenchyma, phloem of tangential bands of sieve tissue alternating with parenchyma, and containing axially elongated secreting cells containing volatile oil or mucilage, phloem fibres with very thick walls, upto 30 μ in diameter, isolated or in short tangential rows, sieve tubes narrow with transverse sieve plates, collapsed in outer periphery, medullary rays of isodiametric cells, mostly 2 cells wide, cortical parenchyma and medullary rays containing small starch grains mostly below 10 μ in diameter, minute acicular crystals of calcium oxalate present.

Chemical constituents: The essential oil of cinnamon bark (max. 4%) is dominated by the two phenylpropanoids cinnamaldehyde and eugenol. The main chemical components of the essential oil, obtained from the leaves, are eugenol, eugenol acetate, cinnamic aldehyde and benzyl benzoate.

Specific action as per Ayurveda: Appetiser and stimulant.

Therapeutics: Tvak is pungent, bitter and sweet in taste, pungent in the post digestive effect and has hot potency. It alleviates kapha and vata dosas, but aggravates the pitta dosha. It possesses light, dry and sharp attributes. Tvak is an appetizer, anti-colic, aromatic, deflatulent, analgesic and anti-emetic in properties. It is used in the diseases like pruritus, distaste, diseases due to ama, heart diseases, piles, worms, sinusitis, cough, asthma, diseases of urinary bladder etc.

Parts used: Hemorrhoids.

Bioactivity: Hypoglycaemic, fungitoxic (Tiwari, Dixit & Dixit, 1994), pro-healing (Kamath, Rana & Chowdhury, 2003), anti-inflammatory and analgesic (Ganapaty & Beknal, 2005) and cytotoxic (Singh *et al.*, 2009).

Further reading

Ganapaty S, Beknal AK. Anti-inflammatory and analgesic activity of *Cinnamomum zeylanicum* leaf oil. *Indian Drugs* 2005. 32: 373.

Jayaprakasha GK, Jaganmohan Rao L, and Sakariah. Chemical composition of the volatile oil from the fruits of *Cinnamomum zeylanicum* Blume. *Flav Fragr J* 12, 331-333.

Kamath JV, Rana AC, Chowdhury AR. Pro-healing effect of *Cinnamomum zeylanicum* bark. *Fitoterapia* 2003;73: 111-115.

Singh R, Koppikar SJ, Paul P, Gilda S, Paradkar AR, Ghanekar R. Comparative analysis of cytotoxic effect of aqueous cinnamon extract from *Cinnamomum zeylanicum* bark with commercial cinnamaldehyde on various cell lines. *Pharm Biol* 2009;47: 1174-1179.

Tiwari R, Dixit R, Dixit SN Studies on fungitoxic properties of essential oil of *Cinnamomum zeylanicum* Breyn. *Indian Perf* 1994;38: 98-104.

89. UDDALAKA

Current representative: *Paspalum scrobiculatum*

Possible representative: *Bauhinia variegata* and *Cordia myxa*

Paspalum scrobiculatum L.

Syn: *Paspalum bispicatum* Hack.

English name: Ditch millet.

Ayurvedic name: *Kodrava.*

Family: Fabaceae.

Habitat: South America.

General Description: Perennial. Culms tufted, 30–50 cm tall. Leaf blades 5–15 × 0.2–0.6 cm, pilose, rarely glabrous. Spikelets single, usually suborbicular, 2–2.3 mm; upper glume and lower lemma 5–7-veined, veins often concolorous with back; upper lemma dark brown at maturity (Ayyar & Narayanaswamy, 1948).

Chemical constituents: α-amyrin and β-sitosterol. N-alkanes and alpha-palmitin from seeds (Misra & Siddiqi, 2000). A toxic principle has been reported (Bhide & Aimen, 1959).

Specific action as per Ayurveda: *Medhya* (nootropic).

Therapeutics: Indigestion, loss of memory, obstinate constipation and general debility.

Parts used: Whole plant.

Bioactivity: Tranquillizer (Bhide & Aimen, 1959; Deo & Bhinde, 1961), wound-healing (Hegde, Anitha & Chandra, 2005) and antidiabetic (Jain *et al.*, 2010).

Toxicity: Chronic toxicity has been reported (Antony, Shukla & Janardhanan, 2003).

Further reading

Antony M, Shukla Y, Janardhanan KK. Potential risk of acute hepatotoxicity of kodo poisoning due to exposure to cyclopiazonic acid. *J Ethnopharmacol* 2003; 87:211-214.

Ayyar KV, Narayanaswamy K. "Varagu", *Paspalum scrobiculatum* (Linn.) *Curr Sci* 1948; 17: 367.

Bhinde NK, Aimen RA. Pharmacology of a tranquilizing principle in *Paspalum scrobiculatum grain*. *Nature* 1959; 183:1735-1736.

Deo VR, Bhinde NK. Effect of *Paspalum scrobiculatum* extract on acutely disturbed schizophrenic patients. A prelimiinary report. *Psychopharmacologia* 1961; 2:295-6.

Hegde PS, Anitha B, Chandra TS. *In vivo* effect of whole grain flour of finger millet (*Eleusine coracana*) and kodo millet (*Paspalum scrobiculatum*) on rat dermal wound healing. *Indian J Exp Biol* 2005; 43:254-258.

Jain S, Bhatia G, Barik R, Kumar P, Jain A, Dixit VK. Antidiabetic activity of *Paspalum scrobiculatum* Linn. in alloxan induced diabetic rats. *J Ethnopharmacol* 2010; 127:325-328.

Misra L, Siddiqi SA. n-alkanes and alpha-palmitin from *Paspalum scrobiculatum* seeds. *Zeitschrift fuer Naturforschung* Section C *J Biosci* 2000;55: 500-502.

Sethi N. Chronic toxicity study of *Paspalum scrobiculatum* extract in rats. *Indian J Physiol Pharmacol* 1969; 13:185-188.

Shah LP, Doongaji DR, Bagadia, Vahia NS, Sheth UK. Clinical observations on *Paspalum scrobiculatum* (Harik). *Indian J Psych* 1964; 6: 131-135.

Shah LP, Vasavda MP, Bagadia VH, Sheth UK. Further experience with a stable fraction of harik - *Paspalum scrobiculatum*. *Indian J Psych* 1966; 8: 260-264.

90. VACHA

Current representative: *Acorus calamus*

Acorus calamus L.

English name: Sweet flag.

Family: Araceae.

Habitat: Europe and India.

General Description: *A. calamus* is a semi-aquatic, perennial, aromatic herb with creeping rhizomes. The rhizome is horizontal, jointed, somewhat vertically compressed and spongy within. It is pale to dark brown in colour; the leaves are grass-like or sword shaped, long and slesnder; flowers are small, yellow-green and in a spadix.

Detailed Description of the Rhizome

(a) *Macroscopic*

Drug occurs in simple or rarely with thumb-like branches at nodes; sub cylindrical to slightly flattened, somewhat tortuous or rarely straight, cut pieces of 1-5 cm long, and 0.5-1.5 cm thick; upper side marked with alternately arranged, large, broadly, triangular, transverse leaf scars which almost encircle the rhizome; at nodes leaf sheath mostly having an appearence present; lower side shows elevated tubercular spots of root scars; light-brown with reddish-tinge to pinkish externally, buff coloured intemally; fracture, short; odour, aromatic; taste, pungent and bitter.

(b) *Microscopic*

Rhizome - Shows single layered epidermis; cortex composed of spherical to oblong, thin-walled cells of various sizes, cells towards periphery, smaller, somewhat collenchymatous, more or less closely arranged cells towards inner side, rounded and form a network of chains of single row of cells, enclosing large air spaces, fibro-vascular bundles and secretory cells having light yellowish-brown contents, present in this region; endodermis distinct; stele composed of round, parenchymatous cells enclosing large air spaces similar to those of cortex and several concentric vascular bundles arranged in a ring towards endodermis, a few vascular bundles scattered in ground tissues; starch grains simple, spherical, measuring 3-6 μ in dia., present in cortex andg round tissue.

Powder - Buff coloured; shows fibres, reticulate, annular vessels and simple spherical starch grains, measuring 3-6 μ in diameter.

Chemical constituents: Bitter principle: acorin, resin: acoretin, alkaloid: calamine and volatile oil containing β-asarone (Srivastava & Syamasunder, 2003).

Specific action as per Ayurveda: *Medhya* (nootropic).

Therapeutics: In the Ayurvedic system of medicine, the rhizomes are used for the treatment of a host of diseases such as epilepsy and other mental ailments, chronic diarrhoea, dysentery, bronchial catarrh, intermittent fevers, glandular and abdominal tumours. They are also used to treat kidney and liver troubles, rheumatism and eczema.

Parts used: Roots.

Bioactivity: Neuroprotective (Shukla *et al.,* 2002; Shukla *et al.,* 2005), antimicrobial (Sabitha *et al.,* 2003; Deepak, 2009), antioxidant (Manikandan *et al.,* 2005) and antifungal (Modhumita, 2006).

Further reading

Deepak G. Antimicrobial activity of *Acorus calamus* (L.) rhizome and leaf extract. *Acta Biol* 2009;53: 45-49.

Manikandan S, Srikumar R, Jeya PN, Sheela DR Protective effect of Acorus calamus Linn. on free radicle scavengers and lipid peroxidation is discrete regions of brain against noise stress exposed rat. *Biol Pharm Bull* 2005; 28: 2327-2330.

Modhumita G. Antifungal Properties of Haem Peroxidase from *Acorus calamus*. *Ann Bot* 2006; 98:1145-1153.

Sabitha RA, Satyakala M, Sandya DV, Suryanarayana MU. Evaluation of antibacterial activity from rhizome extract of *Acorus calamus* Lin. *J Sci Ind Res* 2003; 62: 529-550.

Shukla PK, Khanna VK, Ali MM, Maurya R, Khan MY, Srimal RC. Neuroprotective effect of *Acorus calamus* against middle cerebral artery occlusion-induced ischaemia in rat. *Hum Exp Toxicol* 2006; 25:187-194.

Shukla PK, Khanna VK, Ali MM, Maurya RR, Handa SS, Srimal RC. Protective effect of *Acorus calamus* against acrylamide induced neurotoxicity. *Phytother Res* 2002; 16: 256-260.

Srivastava SK, Syamasunder KV. Essential oil composition of *Acorus calamus* L. from the lower region of the Himalayas. *Flav Frag J* 2003; 18:18–20.

91. VAJRAVALLI

Current representative: *Euphorbia nivulia*

Possible representatives: *Celosia argentea* and *Cissus quandrangularis.*

Celosia argentea L. var. *cristata* Kuntze

Syn: *Celosia cristata* L.

English name: *Common Cockscomb.*

Family: Amaranthaceae.

Habitat: Native to Asia.

General Description: Leaves - Alternate, glabrous, petiolate, lanceolate, linear-lanceolate or lanceolate-ovate, to +8cm long, + 4 cm broad, entire. Inflorescence - Terminal cristate cluster to +12 cm broad, domed, chaffy. Flowers - Perianth variously colored, typically red in this variety, 5-merous, corolla absent. Stamens 5.

Chemical constituents: Phytosterols.

Specific action as per Ayurveda: Astringent.

Therapeutics: Bloody diarrhoea and mennorhagia.

Parts used: Seeds and flowers.

Cissus quadrangularis L.

English name: Edible stemmed vine.

Family: Vitiaceae.

Habitat: *C. quadrangularis* is probably native to India or Sri Lanka, but is also found in Africa, Arabia, and Southeast Asia.

General Description: *C. quadrangularis* reaches a height of 1.5 m and has quadrangular-sectioned branches with internodes 8 to 10 cm long and 1.2 to 1.5 cm wide. Along each angle is a leathery edge. Toothed trilobe leaves 2 to 5 cm wide appear at the nodes. Each has a tendril emerging from the opposite side of the node. Racemes of small white, yellowish, or greenish flowers; globular berries are red when ripe.

Detailed Description of the Dried Stem

(a) *Macroscopic*:

Drug occurs as pieces of stem of varying lengths; stern quadrangular, 4-winged, internodes constricted at nodes; a tendril occasionally present at nodes; internodes 4-15 cm long and 1-2 cm thick; surface smooth, glabrous, buff coloured with greenish tinge, angular portion reddish-brown; no taste and odour.

(b) *Microscopic*:

Mature stern shows squarish outline with prominent projection at each anular point; epidermis single layered, covered externally with thick cuticle; epidermal cells thin-walled, rectangular and tangentially elongated, followed by 2-3 layers of cork and single layered cork cambium; cortex composed of 8-16 layers of thin-walled, circular to oval parenchymatous cells; four patches of collenchymatous cells present in all the four angular points embedded in cortical region like an umbrella arching over large vascular bundles; in the projected portion of angular region cortical cells filled with brown-red contents present; endodermis not distinct; stele consists of a large number of vascular bundles varying in size arranged in the form of a ring separated by rays of parenchyma; 3 -4 vascular bundles larger in size, in each angular region, below collenchymatous patch, while rest of bundles smaller in size; vascular bundles collateral and open type, capped by sclerenchymatous sheath which is well developed in larger bundles; cambium and interfascicular cambium quite distinct; central region occupied by a wide pith composed of thin-walled, circular to oval parenchymatous cells; idioblasts containing raphides and isolated acicular crystals of calcium oxalate present in the outer region of cortex and also in a number of cells throughout the region; rosette crystals of calcium oxalate also found in most of the cells in cortical region; starch grains present throughout the cortical and the pith regions.

Powder - Brown; shows fragments of vessels, fibres, parenchymatous cells and a few rosette crystals of calcium oxalate, starch grains and idioblast containing raphides and isolated acicular crystals of calcium oxalate.

Chemical constituents: Minerals and vitamins. Ketosteroid (Sen, 1964), two unsymmetric tetracyclic triterpenoids (Bhutani *et al.,* 1984), water soluble glucoside, onocer-7-ene-3 alpha, 21 beta-diol, delta-amyrin, delta-amyrone and 3,3',4,4'-tetrahydroxybiphenyl (Mehta, Kaur & Bhutani, 2001) and friedelin (Aswar *et al.,* 2010).

Specific action as per Ayurveda: Astringent.

Therapeutics: Diarrhoea and dysentery.

Parts used: Bark.

Bioactivity: Promotes the healing process of experimentally induced fractures (Singh & Udupa, 1962; Prasad & Udupa, 1963; Deka *et al.,* 1994; Sanyal, Ahmad & Sastry, 2005; Parisuthiman *et al.,* 2009), cholinergic (Singh *et al.,* 1984), analgesic (Singh *et al.,* 1994), atiosteoporotic (Shirwaikar, Khan & Malini, 2003), antibacterial and antioxidant (Chidambra Murthy *et al.,* 2003) and gastroprotective (Jainu & Devi, 2006).

Further reading

Aswar UM, Bhaskaran S, Mohan V, Bodhankar SL. Estrogenic activity of friedelin rich fraction (IND-HE) separated from *Cissus quadrangularis* and its effect on female sexual function. *Pharmacog Res* 2010; 2: 138-145.

Chidambara Murthy KN, Vanitha A, Mahadeva Swamy M, Ravishankar GA. Antioxidant and antimicrobial activity of *Cissus quadrangularis* L.. *J Med Food.* 2003; 6:99-105.

Das PK, Sanyal AK. Studies on *Cissus quadrangularis* Linn. i. acetylcholine like action of the total extract. *Ind J Med Res* 1964; 52:63-67.

Deka DK. *et al.*, Effect of *Cissus quadrangularis* in accelerating healing process of experimentally fractured radius-ulna of dog. *Indian J Pharmacol* 1994; 2: 44-45.

Jainu M, Devi CS. Gastroprotective action of *Cissus quadrangularis* extract against NSAID induced gastric ulcer: role of proinflammatory cytokines and oxidative damage. *Chem Biol Interact* 2006;161:262-270.

Mehta M, Kaur N, Bhutani KK. Determination of marker constituents from *Cissus quadrangularis* Linn. and their quantitation by HPTLC and HPLC. *Phytochem Anal* 2001;12:91-95.

Parisuthiman D, Singhatanadgit W, Dechatiwongse T, Koontongkaew S. Cissus quadrangularis extract enhances biomineralization through up-regulation of MAPK-dependent alkaline phosphatase activity in osteoblasts. In Vitro Cell Dev Biol Anim. 2009; 45:194-200. *In Vitro Cell Dev Biol Anim* 2009;45:194-200.

Prasad GC, Udupa KN. Effect of *Cissus quadrangularis* on the healing of cortisone treated fractures. *Indian J Med Res* 1963;51:667-76.

Sanyal A, Ahmad A, Sastry M. Calcite growth in *Cissus quadrangularis* plant extract, a traditional Indian bone healing aid. *Curr Sci* 2005; 89:1742-5.

Sen SP. Study of the active constituents (ketosteroids) of *Cissus quadrangularis* Wall. *Ind J Pharm* 1964; 26: 247-248.

Shirwaiker A, Khan S, Malini S. Antiosteoporotic activity of Ethanol extract of *Cissus quadrangularis* Linn on ovariectomized rat. *J Ethnopharmocol* 2003;89:245-50.

Singh LM, Udupa KN. Studies on "*Cissus Quadrangularis*" in fracture by using phosphorus 32. III. *Ind J Med Sci* 1962;16:926-31.

Singh SP, Mishra N, Dixit KS, Singh N, Kohli RP. An experimental study of Analgesic Activity of *Cissus quadrangularis*. *Indian J Pharmacol* 1994;16:162-163.

92. VARAHI

Current representative: *Dioscorea bulbifera*

Possible representative: *Tacca aspera*

Dioscorea bulbifera L.

English name: Air yam.

Ayurvedic names: *Varahi kanda.*

Family: Dioscoreaceae.

Habitat: Himalayas.

General Description: *D. bulbifera* is a perennial vine with broad leaves and two types of storage organs. The plant forms bulbils in the leaf axils of the twining stems, and tubers beneath the ground.

These tubers are like small, oblong potatoes. The stems are round to slightly angled in cross section and they twine coutner-clockwise. Conspicuous aerial tubers (called bulbils) are pale, round to globose in shape, up to 13 cm wide and are formed in leaf axils (Datta, Datta *et al.,* 1984).

Chemical constituents: Starch (Mulla & Kulkarni, 2001), steroid saponins: diosbulbin B, diosbulbin D and disogenin (Kate, Deshmukh & Saoji, 1978; Shriram *et al.,* 2008). P-Hydroxyacet-ophenone derivatives have been reported (Gupta & Singh, 1989).

Specific action as per Ayurveda: Tonic.

Therapeutics: Hemorrhoids, skin diseases and general debility

Parts used: Tubers.

Bioactivity: anorexient (Jindal, Kelkar & Doctor, 1969) and antimicrobial (Seetharam *et al.,* 2003)

Further reading

Datta K, Datta SK *et al.* Pharmacognostic evaluation of potential yams (Dioscorea). *J Eco Taxon Bot* 1984;5:181-196.

Gupta D, Singh J. p-Hydroxy acetophenone derivatives from *Dioscorea bulbifera Phytochemistry* 1989; 28:947-949.

Jindal MN, Kelkar VV, Doctor RB. The anorexient activity of Kalio-kund (*Dioscorea bulbifera* Linn.), methylphenidate and cocaine in rats: a preliminary study. *Indian J Med Res* 1969;57:1075-80.

Kate DR, Deshmukh VK, Saoji AN. Studies on diosgenin content of Balanites, fenugreek and bulbs of *Dioscorea bulbifera* Linn. *Indian J Pharm Sci* 1978;40: 228

Mulla RM, Kulkarni AR. Starch from bulbils of *Dioscorea bulbifera* Linn. *J Eco Taxon Bot* 2001;18: 517-519.

Seetharam YN, Jyothishwaran G, Sujeeth H, Barad A, Sharanabasappa G, Shivkumar D. Antimicrobial activity of *Dioscorea bulbifera* bulbs. *Indian J Pharm Sci* 2003; 65: 195-196.

Shriram V, Jahagirdar S, Latha C, Kumar V, Puranik V, Rojatkar S, Dhakephalkar PK, Shitole MG. A potential plasmid-curing agent, 8-epidiosbulbin E acetate, from *Dioscorea bulbifera* L. against multidrug-resistant bacteria. *Int J Antimicrob Agents* 2008;32:405-10.

Tacca aspera Roxb.

Ayurvedic names: *Surana.*

English name: Fiji Arrowroot, Tahiti Arrowroot.

Family: Cyperaceae.

Habitat: Tropical India.

General Description: These are interesting, evergreen, perennial and herbaceous plants that grow wild in many tropical climates. They can be grown in humid, tropical greenhouses. *T. pinnatifida* (leontopetaloides) contains a lot of starch in its rhizomes, which, when extracted and prepared for food, is called East Indian Arrowroot or South Sea Arrowroot. The kinds in cultivation have creeping rhizomes or rootstocks and large, oval, crinkly foliage. The purplish-brown flowers are produced in bunches atop long, thick stems, 18 to 24 inches high. Long whiskers are suspended from the lovely flowers.

Chemical constituents: The tuber contains gamma-amino-butyric acid, glycine, leucine, valine, quercetin-3-arabinoside, D (-) - ribose, n-triacontanol, betulinic acid, castano-genin and taccalin.

Action: Tonic.

Therapeutics: General debility.

Parts used: Tubers.

Further reading

Tripathi RP, Tiwari KP. Phytochemical Investigation of the roots of *Tacca aspera*. *Planta Med* 1981; 41:414-415.

93. VASUKA

Current representative: *Calotropis gigantia*

94. VRDDHA DARU

Current representative: *Argyeria speciosa*

Possible representative: *Salvia plebeia, Marsdenia tenassisimus* and *Thespesia populnea*.

Argyreia speciosa Sweet.

Syn: *Argyreia nervosa* (Burm.f.) Bojer

Common name: Elephant creeper; Wooly morning glory.

Other Ayurvedic names: *Vriddhadarā*.

Family: Convolvulaceae.

Habitat: India.

General Description: A very large climber; stem stout, white-tomentose. Leaves are 7.5 - 30.0 cm in diameter, acute, ovate, glabrousabove and persistently white-tomentose beneath, base cordate, petioles 5 - 15 cm long and white-tomentose. Flowers in subcapitate cymes; penduncles 7.5 – 15 cm long, stout, white-tomentose; bracts large, ovatelanceolate with a long acumen, thin, veined, pubescent outside, glabrous inside, deciduous the outer sometimes 5 cm long; pedicels very short often almost 0, whitetomentose, calyx white-tomentose outside; corolla 5 - 6.3 cm long, tubular-infundibuliform, the bands silky pubescent outside, tube somewhat inflated, white pubescent outside, rose purple and glabrous inside. Ovary glabrous, fruit glabrous, 2.0 cm in diameter, apiculate.

Detailed Description of the Leaf (Singh, 1965; Singh, 1972; Agarwal & Rastogi; 1974a; Prasad & Chauhan, 1975; Sasikala *et al.,* 1991).

(a) *Macroscopic*

The lower surface of the leaf is entirely covered with hair, which gives the leaf a silvery soft wooly appearance; the upper surface is green, glabrous and shows the markings of nerves by slight depressions. The mature leaf is dorsiventral, unicostate with a strong midnerve and several faint lateral nerves, alternate, petiole, acute at the apex and cordate at the base. The margin is entire but slightly wavy near the base. Lateral nerves 14 - 20 pairs arise alternatively on the midrib; the single nerves bifurcate before reaching the edge, the-anterior branch unites with the posterior one of the neighbouring nerve; an arched nervule connects the two branches slightly above the point of bifurcation. Neither the main secondary nerves nor their branches reach the margin. Petiole stout and cylindrical, a little shorter than the length of the blade is completely covered with wooly tomentum.

(b) *Microscopic*

The transverse section of the leaf near the apex shows a prominent ridged midrib on the lower surface and a small groove on the upper surface, while a section through the basal region presents a small ridge on the upper side as well. The ventral cuticle is stratified while the dorsal is thin and simple. The epidermal cells of the upper side have synclinous walls with rubiaceous type of sunken stomata. The openings of the latex canals are bound by 5 - 6 cells, the epidermal cells or the under side differ from those of the upper in possessing smaller cells and about twice the number of stomata and openings of latex canals. The cells of the epidermis along the veins on both sides of the leaf are roughly rectangular straight walled and completely devoid of appendages. The spongy tissue is composed of rounded cells enclosing air spaces and a few latex canals. The palisade cells are nearly rectangular, roughly four times longer than broad and are seen in the section usually in a single row only and rarely in two rows. A few latex canals are sometimes present in this zone as well. The vascular bundles are hexagonal in transverse section and occur in characteristic, continuous single row chains.

Chemical constituents: Alkaloid: ergometrine (Agrawal & Rastogi, 1974b; Nair, Daniel & Sabnis, 1987), resin, tannin, fixed from seeds (Biswas, Tiwari & Dutt, 1947; Khan, Kamil & Ilyas, 1992), aryl esters and a coumarin (Shrivastava A, Shukla, 1998), hexadecanyl p-hydroxycinnamate and scopoletin (Shukla *et al.*, 1999).

Specific action as per *Ayurveda*: Tonic.

Therapeutics: Bronchitis, anuria and constipation.

Parts used: Leaves and roots.

Bioactivity: Antibacterial (Mishra & Chaturvedi,1978; Batra & Mehta, 1985; Shukla *et al.*, 1999), antifungal (Mishra & Chaturvedi,1978), antiviral, antifilarial (Parveen, Khan & Singhal, 1990), anti-inflammatory and antiarthritic (Srivastava, Kant & Tewari, 1972; Gokhale *et al.*, 2002), immunomodulator: (Gokhale, Damre & Saraf, 2003; Hanumanthachar, Navneet & Jyotibala, 2007), nootropic (Hanumanthachar, Navneet & Jyotibala, 2007); aphordiasic (Subramonium *et al.*, 2007), hypoglycemic and antihyperglycemic (Hemet *et al.*, 2008) and anticonvulsant (Vyavhare & Bodhankar, 2009).

Further reading

Agarwal SR, Rastogi RP. Pharmacognostical and Preliminary Phytochemical Studies of *Argyreia nervosa* Burm. *Indian J Pharmacol* 1974a; 35: 118-119.

Agrawal SK, Rastogi RP. Ergometrine and other constituents of *Argyreia speciosa* sweet. Indian J Pharmacol 1974b; 36: 118-119.

Gokhale AB, Damre AS, Saraf MN. Investigations into the immunomodulatory activity of *Argyreia speciosa*. *J Ethnopharmacol* 2003;84:109-114.

Batra A, Mehta BK . Chomatographic analysis and antibacterial activity of the seed oil of *Argyreia speciosa*. *Fitoterapia* 1985;56: 357-359.

Biswas B, Tiwari LD, Dutt S. Chemical composition of the fixed oil from the seeds of *Argyreia speciosa*. *Ind. Soap J* 1947; 13:51-54.

Gokhale AB, Damre AS, Kulkarni KR, Saraf MN. Preliminary evaluation of anti-inflammatory and antiarthritic activity of *S. lappa, A. speciosa* and *A. aspera*. *Phtyomedicine* 2002;9: 433-437.

Gokhale AB, Damre AS, Saraf MN. Investigation into the lmmuno-Modulatory activity of *Argyreia speciosa*. *J. Ethanopharmacol* 2003; 84: 109-114.

Hanumanthachar J, Navneet K, Jyotibala C. Evaluation of nootropic effect of *Argyreia speciosa* in mice, *J Health Sci* 2007; 53: 382-388.

Hemet LE, Satyanarayana T, Ramesh A, Durga Prasad, Routhu Y, Srinivas KV. Hypoglycemic and antihyperglycemic effect of *Argyreia speciosa* Sweet. In normal and in alloxan induced diabetic rats. *J Nat Rem* 2008;8: 203-208.

Khan MS, Kamil SM, Ilyas M Phytochemical investigation on the leaves of *Argyreia speciosa*. *J Indian Chem Soc* 1992; 69:110-113.

Mishra SH, Chaturvedi SC. Antibacterial and antifungal activity of the oil and unsaponifiable matter of *Argyreia speciosa* Sweet. *Indian Drugs* 1978;13: 29-31.

Nair GG, Daniel M, Sabnis SD. Ergolines in the seeds of some Indian convolvulaceae. *Indian J Pharm Sci* 1987;49: 100-102.

Parveen N, Khan NU, Singhal KC. Antifilarial activity of *Argyreia speciosa* against *Setaria cervi in vitro*. *Phytother Res* 1990; 4:162-164.

Prasad S, Chauhan RBPS. Pharmacognostical studies on vidhara. Part III. *Argyreia speciosa* Sweet. *J Res. Indian Med* 1975;10:23-31.

Sasikala F, Brindha P, Ali SU, Kundu AB. Contribution to the Pharmacognostic anatomy of *Argyreia speciosa* Sweet-Leaf. *Indian Drugs* 1991;28: 403-407.

Shrivastava A, Shukla YN. Aryl esters and a coumarin from *Argyreia speciosa*. *Indian J Chem* 1998;37B:192-194.

Shukla YN, Srivstava A, Sunil Kumar, Sushi Kumar. Phytotoxic and antimicrobial constituents of *Argyreia speciosa* and *Oenotherabiennis*. *J. Ethnopharmacol* 1999; 67:241-245.

Singh P. Pharmacognosy of leaf of *Argyreia speciosa* Sweet. *J Sci Ind Res* 1957;16: 204-206.

Singh P Pharmacognostic study of root and stem of *Argyreia speciosa*. *Acta Phytother* 1972;19:46-52.

Singh P . Pharmacognostic study of root and stem of *Argyreia speciosa* Sweet. *Quart J Crude Drug Res* 1965;5: 774-783.

Srivastava MC, Kant V, Tewari JP. Anti-inflammatory activity roots of *Argyreia speciosa* (Samundrashokha). *Mediscope* 1972;15:219-222.

Srivatasav A, Shukla Tau SP, Kumar S. Aryl esters and a coumarin from Argyreia speciosa. *J Arom Med Plants* 1998;20: 774-778.

Subramonium A, Madhavachandran V, Ravi K, Anuja VS. Aphordiasic property of the elephant creeper *Argyreia nervosa*. *J Endocrinol Rep* 2007; 11: 82-85.

Vyavhare NS, Bodhankar SL. Anticonvulsant activity of *Argyreia speciosa* in mice. *Indian J Pharm Sci* 2009;71: 131-134.

Salvia plebeia R.Br.

Syn: *Salvia brachiata* Roxb.; *Salvia parviflora* Roxb.

Family: Lamiaceae.

Habitat: *S. plebeia* is found in Cagayan Province in Luzon. It probably occurs also as a weed in and about towns in various other provinces at low altitudes. It also occurs in India to China, and in Malaya and Australia.

General Description:: *S. plebeia* is an annual, hairy herb. The stems are stout, erect, hoary, and 15 to 45 centimeters in height. The leaves are oblong or ovate, 2.5 to 7.5 centimeters long, and narrowed and pointed at both ends. The spikes are panicled, often fastigiated. The flowers are hardly 6 millimeters long, lilac or nearly white, occurring in small, very numerous whorls in numerous, slender, panicled, glandular racemes. The calyx is stalked, bell-shaped, and 10 to 12 millimeters long; the upper calyx-lip is entire, and the lower one obtusely 2-toothed. The corolla-tube is very short, and the included upper lip is short, nearly straight, slightly flattened, and concave. The nutlets are very minute and ellipsoid.

Therapeutics: The seeds of the plant are used in the treatment of diarrhoea, gonorrhoea, menorrhagia and haemorrhoids.

Further reading

Chopra RN, Nayar SL, Chopra IC. *Glossary of Indian Medicinal Plants (Including the Supplement).* Council of Scientific and Industrial Research, New Delhi, 1986.

Thespesia populnea Soland. ex. Corr.

Common name: Portia tree.

Ayurvedic names: *Vriddhadara*.

Family: Malvaceae.

Habitat: India.

General Description: This evergreen tree is bushy when young but thins out with age. It grows to 40 ft (12 m) or more with a spread of 10–20 ft (3–6 m). It has heart-shaped leaves and cup-shaped yellow flowers that are produced intermittently throughout the year in warm climates. Each flower has a maroon eye that ages to purple.

Chemical constituents: Gossypol (Akhila & Rani 1993), tannin, red coloring matter and oil. Flavonoids have been reported from flowers (Shirwaikar & Srinivasan, 1996). Alkaloids, carbohydrates, protein, tannins, phenols, flavonoids, gums, mucilage, saponins, terpenes have been reported from the ethanolic extract of the bark (Vasudevan, Gunnam & Parle, 2006).

Specific action as per Ayurveda: Astringent, acrid, cooling, depurative, vulnerary, and alternative.

Therapeutics: According to Ayurveda, it is useful in skin related troubles, leprosy, diseases of blood and urinary system, diarrhea, dysentery, Cholera, diabetes, ascites and skin diseases.

Parts used: Bark.

Bioactivity: Antihepatotoxic (Shirwaikar & Kumar *et al.,* 1995), antioxidant (Illavarasan *et al.,* 2003), anti-inflammatory and analgesic: ethanolic extract of the bark (Vasudevan, Gunnam & Parle, 2006) and anti-dementia (Vasudevan, Gunnam & Parle, 2006).

Further reading

Akhila A, Rani K. Biosynthesis of gossypol in *Thespesia populnea. Phytochemistry* 1993;33: 335-340.

Ilavarasan R, Vasudevan M, Anbazhagan S, Venkataraman S. Antioxidant activity of *Thespesia populnea* bark extracts against carbon tetrachloride-induced liver injury in rats. *J Ethnopharmacol* 2003; 87:227-230.

Kavimani, S., R. Ilango, *et al.* (1999). Antisteroidogenic activity of floral extract of *Thespesia populnea* Corr. in mouse ovary. *Indian J Exper Biol* 37: 1241-1242.

Lakshmanan KK, Naryanan ASS. Antifertility herbals used by the tribals in Anaikkatty Hills, Coimbatore District, Tamil Nadu (India). *J Econ Taxon Bot* 1990; 14: 171-173.

Shirwaikar A, Kumar AV, Krishnanand BR, Sreenivasan. KK. Chemical investigation and antihepatotoxic activity of *Thespesia populnea. Int J Pharmacog* 1995; 33: 305-310.

Shirwaikar A, Srinivasan KK. New flavonoids from the flowers of *Thespesia populnea. J Med Arom Plant Sci* 1996;18: 266-269.

Vasudevan M, Parle M. Pharmacological actions of *Thespesia populnea* relevant to Alzheimer's disease. *Phytomedicine* 2006;13:677-87.

Vasudevan, M.; Gunnam, K.K. & Parle, M. (2006). Antinociceptive and anti-inflammatory effects of *Thespesia populnea* bark extract. *J Ethnopharmacol* 2007; 109:264-70.

Note: See notes on *Marsdenia tenassisimus.*

95. VRSCHIKALI

Current representative: *Pergularia extensa*

96. YAVASA-DHANA YAVASA

Current representative: *Alhagi camelorum*

Possible representative: *Fagonia arabica*

Alhagi camelorum Fisch.

Syn: *Alhagi pseudalhagi* (M. Bieb.) Desv. ex B. Keller. & Shap.; *Alhagi persarum* Boiss. & Buhse.; *Hedysarum pseudalhagi* M. Bieb.

Common name: Camel thorn.

Ayurvedic names: *Duralabha.*

Family: Fabaceae.

Habitat: This shrub is native to the region extending from the Mediterranean to Russia, but has been introduced to many other areas of the world, including Australia, southern Africa, and the western United States.

General Description: The perennial plant grows from a massive rhizome system which may extend over six feet into the ground. New shoots can appear over 20 feet from the parent plant. Above the ground, the plant rarely reaches four feet in height. It is a heavily-branched, gray-green thicket with long spines along the branches. It bears small, bright pink to maroon pea flowers and small legume pods, which are brown or reddish and constricted between the seeds. The seeds are mottled brown beans.

Chemical constituents: Sucrose.

Specific action as per Ayurveda: Diuretic, expectorant and laxative.

Therapeutics: Bronchitis, anuria and constipation.

Parts used: Whole plant and extract (manna).

Fagonia arabica L.

Syn: *Fagonia cretica* L.

Common name: Khorasan thorn.

Ayurvedic names: *Dhanayasaka.*

Family: Zygophyllaceae.

Habitat: Afghanistan and India.

General Description: Perennial or annual, pubescent to glabrous shrubs and herbs. Leaves compound tri-or unifoliolate; stipules conspicuous, frequently transformed into spines. Flowers pink to rose, 5-fid, stamens 10 without appendages. Fruit a pyramidal, 5-carpelled capsule. Seeds flat, oblong with foveolate-punctate testa. Lower leaves tri-, upper unifoliolate, spines longer than leaves. Leaflets flat. Capsule over 4 mm. across, calyx deciduous. Flowers up to 15 mm. across, plant dark green. Fruit-peduncle reflexed, shorter than capsule. Capsule 3-5 mm. long, 5-7 mm. broad, calyx deciduous.

Chemical constituents: Steroid saponins.

Specific action as per Ayurveda: Febrifuge.

Therapeutics: Fever.

Parts used: Whole plant.

Bioactivity: Antibacterial (Godbole & Pendse, 1960), cardio protective (Prasad *et al.*, 2007; Das & Kaushik, 2010)

Further reading

Das R, Kaushik A. Synergistic activity of *Fagonia arabica* and *Heteropneustes fossilis* extracts against myocardial, cerebral infarction, and embolism disorder in mice. *J Pharm Bioallied Sci* 2010; 2: 100–104.

Godbole SH, Pendse GS. Antibacterial property of some plants. *Indian J Pharm* 1960; 22: 39.

Prasad S, Kashyap RA, Taori JM, Daginawala HF. Effect of *Fagonia arabica* (dhamasa) on *in vitro* thrombolysis. *BMC Comp Alter Med* 2007;7:36.

Sharma VN, Singh KP. Pharmacological actions of *Fagonia arabica* - an Indian indigenous drug. *Indian Pract* 1960;13: 265.

97. YAVATIKTA

Ayurvedic names: *Mahatikta, dridpada, visarpini, nakuli, netramila, shankhini, patratanduli, tanduli, akshapida, sulshampatri, yashasvini, maheshvari, tiktayava, yavi* and *tikta.*

Action: Bitter and sour in taste, appetizer, laxative.

Therapeutics: Worm infestation, skin diseases, poisoning and indigestion.

Note: Some of the authors have quoted *Andrographis paniculata* as *Yavatikta*.

INDEX OF BOTANICAL NAMES